Computer-Supported Collaboration

Computer-Supported Collaboration

Theory and Practice

Weidong Huang
University of Technology Sydney
Sydney, Australia

Mark Billinghurst
University of Auckland
Auckland, New Zealand

and

University of South Australia
Adelaide, Australia

Leila Alem
University of Technology Sydney
Sydney, Australia

Chun Xiao
University of Technology Sydney
Sydney, Australia

Troels Rasmussen
Aarhus University
Aarhus, Denmark

WILEY

Published by John Wiley & Sons, Inc., Hoboken, New Jersey.
Published simultaneously in Canada.

For general information on our other products and services or for technical support, please contact our Customer Care Department within the United States at (800) 762-2974, outside the United States at (317) 572-3993 or fax (317) 572-4002.

Wiley also publishes its books in a variety of electronic formats. Some content that appears in print may not be available in electronic formats. For more information about Wiley products, visit our web site at www.wiley.com.

Library of Congress Cataloging-in-Publication Data applied for:

Hardback: 9781119719762

Cover Design: Wiley
Cover Image: © XH4D/Getty Images

Set in 9.5/12.5pt STIXTwoText by Straive, Chennai, India
Printed and bound by CPI Group (UK) Ltd, Croydon, CR0 4YY

C9781119719762_230524

Contents

About the Authors

Weidong Huang received his PhD in computer science from the University of Sydney. He is currently Associate Professor at the University of Technology Sydney. He also has formal training in experimental psychology and professional experience in psychometrics. He is the author of over 150 publications. His research interests are in human–computer interaction, visualization, and data science. He is a founding chair of the Technical Committee on Visual Analytics and Communication for IEEE SMC Society and a guest editor of a number of SCI indexed journals. He has served as a conference chair, a program committee chair, or an organization chair for a number of international conferences and workshops.

Mark Billinghurst is currently Director of the Empathic Computing Laboratory and Professor at the University of South Australia in Adelaide, Australia, and also at the University of Auckland in Auckland, New Zealand. He earned a PhD in 2002 from the University of Washington and conducts research on how virtual and real worlds can be merged, publishing over 650 papers on augmented reality, virtual reality, remote collaboration, empathic computing, and related topics. In 2013, he was elected as a Fellow of the Royal Society of New Zealand, and in 2019 he was given the ISMAR Career Impact Award in recognition for lifetime contribution to AR research and commercialization. In 2022, he was selected for the ACM SIGCHI Academy, for leading human–computer interaction researchers, and also selected for the IEEE VGTC VR Academy for leading VR researchers.

Leila Alem is an adjunct professor at the UTS faculty of engineering and information technology. She was a principal research scientist at the CSIRO Digital Productivity and Services based in Sydney. Her formal training is in artificial intelligence and cognitive psychology. Over the past 23+ years of her career at CSIRO, she has designed, evaluated, and delivered ICT solutions and services for industries such as aviation, mining, manufacturing, and health. Leila has an established international profile in human–computer interaction with a special interest in enhancing user experiences through novel ways of working together using mobile and wearable technologies. She is the editor of two books with Springers on mobile collaborative augmented reality systems. Her main focus of research has been in the area of human factors in computer mediated collaboration settings. Drawing on cognitive psychology, social science and human–computer interface research she has investigated the media factors, the cognitive factors, and the social factors at play in telepresence systems and environments.

Chun Xiao received her PhD in computer science from Otto-von-Guericke University Magdeburg. She was a postdoctoral researcher at the University of Technology Sydney, conducting research on remote collaboration. She has published papers in human–computer interaction and visualization.

Troels Rasmussen received his PhD in computer science from Aarhus University Denmark. His research interests are in augmented reality and human–computer interaction. He was a postdoctoral researcher at the Aarhus University.

Acknowledgments

Our research on remote collaboration on physical tasks started from an R&D project with an industry partner in the mining industry more than 15 years ago. The project was to support the changing role of mining operators in the face of mining automation. The digitally enabled collaborations that take place in these remote operations involved an operator collocated with the object of the collaboration, which is a physical object (a piece of mining equipment) and a remote expert. Our initial research has drawn on previous work on remote collaboration on physical tasks by Susan Fussel and Mark Billinghurst. Since then, we have collaborated with researchers, students, and a number of organizations including one start-up and a commercial partner. They have contributed to the research results that are presented in this book directly or indirectly in various ways, and we express our sincere gratitude to them. Without their contribution, participation, and support, this book would not have come to fruition. In particular, we thank our coauthors: Franco Tecchia, Seungwon Kim, Mathew Wakefield, and Henry Been-Lirn Duh, for giving us permissions to include the coauthored published works in this book. For the same, we thank the publishers as well: Springer, J.UCS Consortium, Bentham Open, and Elsevier. We also thank Tiare Feuchtner and Kaj Grønbæk for their contribution to the work discussed in Chapters 11 and 12. Finally, we thank Boeing R&D for the opportunity to conduct a trial within their operation in Seatle, which has led to the successful commercialisation of our research effort.

We extend our sincere thanks and appreciation to the editors Indirakumari Siva, Ranjith Kumar Thanigasalam, Teresa Netzler, and Victoria Bradshaw and the production team from John Wiley & Sons for their professional support throughout this project.

1

Remote Collaboration on Physical Tasks

1.1 Introduction

Remote collaboration on physical tasks (or remote guidance/remote assistance) typically involves one or more remote helpers guiding one or more local workers to work collaboratively on the manipulation of physical objects [1, 2]. In this type of remote collaboration, both the workers and helpers are physically distributed. On the one hand, the workers have direct access to the physical objects to be worked on but do not have full skills or knowledge on how to operate or manipulate them; thus, they need to receive help from the remote helpers. On the other hand, the remote helpers know how, but do not have physical access to the objects [3]. Technologies that support remote guidance can greatly improve the productivity and safety of tasks by allowing experts to provide timing guidance and training to individuals remotely without having to travel on-site. It has a wide range of applications in industrial domains (e.g. [4]) and has the potential to revolutionize those industries in terms of how the business operates and how service can be provided to their customers, from manufacturing and construction to healthcare and education, to name a few.

With recent advances in networking, augmented reality (AR), virtual reality (VR), mobile and wearable technologies, it has become increasingly possible in practice to enable helpers to remotely guide individuals in performing complex physical tasks with precision and efficiency [5]. Given the increasing demand for remote guidance technologies from industries and increasing interest and effort in research from academics, this research book explores the latest and typical developments in remote guidance technologies and provides comprehensive reviews of the current state-of-the-art research in this field, including our own research findings and developments in the past 15 years.

Computer-Supported Collaboration: Theory and Practice, First Edition.
Weidong Huang, Mark Billinghurst, Leila Alem, Chun Xiao, and Troels Rasmussen.
© 2024 The Institute of Electrical and Electronics Engineers, Inc. Published 2024 by John Wiley & Sons, Inc.

1.2 Remote Collaboration in Perspective

The rest of the book has 12 chapters, each focusing on a specific aspect of remote collaboration research. Both technology and communication are essential elements of remote collaboration, and understanding whether and how technology impacts communication behaviors is important for the design of remote collaboration systems. However, this is an area that has not been well-researched. The technology impact can be predicted using communication models. Thus, we dedicate the next chapter of our book to the discussion of how existing communication models can be used to predict the impact of different AR technologies in remote collaboration and if a new communication model needs to be developed. More specifically, we provide a review of various existing communication models and show how they can be used to analyze communication in both AR and non-AR interfaces for remote guidance on physical tasks. We also discuss the limitations of current models, identify research gaps, and explore possible further developments.

The third chapter provides a review of communication cues in remote collaboration. It starts with an overview of the research landscape over the past three decades and then investigates the communication context based on which a remote collaboration is conducted. We categorize communication cues in remote collaboration systems as verbal, visual, haptic, and empathic communication cues and review the systems and experiments that studied each of them to identify advantages and limitations under different situations. Finally, we summarize and address the challenges of multimodality communication modeling and system design for high usability and suggest potential future research directions for augmented remote collaboration system design aiming at effectiveness, reliability, and ease of use.

For remote guidance on physical tasks, in addition to verbal communications, how to convey other communication cues effectively has been researched extensively. Given the importance and variety of possible communication cues as outlined in the third chapter, we presented a review in the fourth chapter to summarize the communication cues being used, approaches that implement the cues, and their effects on remote guidance on physical tasks [6]. In this chapter, we categorize the communication cues into explicit and implicit ones and report our findings. Our review indicates that a number of communication cues have been shown to be effective in improving system usability and helping collaborators to achieve optimal user experience and task performance. More specifically, there is a growing interest in providing a combination of multiple explicit communication cues to cater for the needs of different task purposes and in providing combination of explicit and implicit communication cues.

Although technology for remote collaboration is becoming increasingly more essential and affordable, and eye gaze is an important cue for human–human communication, there is much that remains to be done to explore the use of gaze in remote collaboration, especially for collaboration on physical tasks. Recent advancement in eye tracking technologies enables gaze input to be added to collaborative systems, especially for remote guidance and is expected to bring more promising opportunities to reduce misunderstanding and improve effectiveness. The fifth chapter surveys publications with respect to eye tracking-supported collaborative physical work under remote guidance. We categorize the prototypes and systems presented according to four metrics ranging from eye-tracked subjects to gaze visualization. Then, we summarize the experimental and investigation findings to have an overview of the eye tracking mechanism in remote physical collaboration systems, as well as the roles that eye gaze and its visualization play in common understanding, referential, and social copresence practices.

The sixth chapter provides a summary of how to conduct evaluation studies of AR-based remote guidance systems. As previously discussed in this book, communication is an essential part of remote collaboration, and many technologies have been developed to enable people to better connect and communicate with one another. However, the impact of these technologies can only be measured through conducting evaluation studies and measuring how the technologies change communication behavior between real people. Therefore, the purpose of this chapter is to help the readers become more proficient in their own evaluation studies and create research outputs that will inspire others in the field. More specifically, in this chapter, we present evaluation case studies, derive a number of design guidelines, and discuss methods that can be used to create robust evaluation studies. Finally, this chapter concludes with a list of possible research directions.

From the seventh chapter, we introduce a range of typical remote guidance systems. These systems were developed with different configurations to meet different collaboration requirements and to serve as platforms for us to investigate specific research questions. First, in this chapter, we present a remote guidance system called HandsOnVideo [7], a system that uses a near-eye display to support mobility and unmediated representations of hands to support remote gestures, enabling a remote helper guiding a mobile worker working in nontraditional-desktop environments. The system was designed and developed using a participatory design approach, which allowed us to test and trial a number of design ideas. It also enabled us to understand from a user's perspective some of the design tradeoffs. The usability study with end users indicated that the system is useful and effective. The users were also positive about using the near-eye display for mobility and instructions and using unmediated representations of hands for remote gestures.

The eighth chapter introduces HandsInAir [8], a wearable system for remote guidance. This system is designed to support the mobility of the collaborators and provide easy access to remote expertise. HandsInAir draws on the richness of hand gestures for remote guiding and implements a novel approach that supports unmediated remote gestures and allows the helper to perform natural gestures by hands without the need for physical support. A usability study was also conducted demonstrating the usefulness and usability of HandsInAir. More specifically, the participants were positive about the mobility support provided by the system to the collaborators. According to their feedback, the mobility support allows workers to access a remote helper more easily. Also, helpers are enabled to continuously engage with the system and their partner when they move around during the guiding process. Participants who played the role of helper also considered gesturing in the air as being intuitive and effective.

The ninth chapter introduces HandsInTouch [9], which supports a unique remote collaboration gesture interface by including both raw hand gestures and sketch cues on a live video or still images. We also conducted a user study comparing remote collaboration with the interface that combines hand gestures and sketching (the HandsInTouch interface) to one that only used hand gestures when solving two tasks: Lego assembly and repairing a laptop. It was found from the study that adding sketch cues improved the task completion time, only with the repairing task, as this had complex object manipulation, and that using gesture and sketching together created a higher task load for the user. The implications of our findings for system design and application are also discussed in the chapter.

The tenth chapter describes Handsin3D [10], a system that uses three-dimensional (3D) real-time capturing and rendering of both the remote workspace and the helper's hands and creates a 3D shared visual space as a result of colocating the remote workspace with the helper's hands. The 3D shared space is displayed on a head-tracked stereoscopic hand-mounted display (HMD) that allows the helper to perceive the remote space in 3D as well as guide in 3D. A user study conducted with the system reveals that the unique feature of HansIn3D is the integration of the projection of the helper's hands into the 3D workspace of the worker. Not only does this integration gives users flexibility in performing more natural hand gestures and ability in perceiving spatial relationship of objects more accurately but also offers greater sense of copresence and interaction.

The eleventh chapter introduces a component-based tailorable remote assistance system called RAK. The design and development of RAK were informed by the results and findings of an interview study with employees of a manufacturing industry. Then, an experimental simulation with RAK that was conducted at a technical college for plastic manufacturing was briefly described. A large part of the chapter was devoted to our discussion and reflection on the results and observations of the user studies. It is encouraging that we are able to derive some meaningful and unexpected new insights, which could guide the directions of future work. These include the tailoring behaviors of both workers and helpers, sharing machine sound from the workspace to the helper, and supporting workspace awareness with multi-camera setups.

The twelfth chapter introduces two multi-camera AR research prototypes, SceneCam and CueCam. These two systems are developed to help collaborators maintain awareness of each other in large workspaces. Multi-camera remote assistance has some benefits over using one camera from the point of view of the worker, most notably the view independence of the helper. However, in this chapter, we point out the challenges that stand in the way of obtaining good workspace awareness when using multiple cameras and demonstrate with the two systems how AR visualization and tracking can be used to address these awareness challenges in various ways.

The final chapter introduces some industrial systems that support remote guidance on physical tasks. Each of these industrial systems was designed to meet specific design and/or business purposes. Current challenges and possible future directions are also discussed. These include ergonomically tested devices and privacy and ethical aspects of remote guidance, network connection, and information delay, reproducing the environment of face-to-face collaboration for remote collaboration, and replacing a communication cue with another cue of a different modality. Apart from these topics, the chapter concludes the book with other possible directions being mentioned, including artificial intelligence and cloud-based remote guidance support, embedment, and integration of cognitive, physiological, empathic, and multimodal communication cues, investigation of possible effects of human factors, language, social and cultural factors, and more rigors and empirically validated evaluation frameworks, design principles, metrics, and methodologies for remote collaboration on physical tasks.

1.3 Book Audience

This book is for researchers, engineers, scientists, and practitioners who are interested in the research of remote collaboration and its potential applications in various industrial domains. Academics and postgraduate students in science and engineering will also find this book useful as a comprehensive reference book. It provides a comprehensive overview of and detailed insights into the current state-of-the-art research and the potential future directions for the topic. We hope that this book will inspire new research and innovation, and ultimately lead to new theories and development of more effective and efficient remote collaboration systems and tools to meet real-world needs.

References

1 Kraut, R.E., Fussell, S.R., and Siegel, J. (2003). Visual information as a conversational resource in collaborative physical tasks. *Human–Computer Interaction* 18 (1-2): 13–49. https://doi.org/10.1207/S15327051HCI1812_2.

2 Gergle, D., Kraut, R.E., and Fussell, S.R. (2013). Using visual information for grounding and awareness in collaborative tasks. *Human–Computer Interaction* 28 (1): 1–39. https://doi.org/10.1080/07370024.2012.678246.

3 Kim, S., Billinghurst, M., and Kim, K. (2020). Multimodal interfaces and communication cues for remote collaboration. *Journal on Multimodal User Interfaces* 14: 313–319. https://doi.org/10.1007/s12193-020-00346-8.

4 Kiber. Kiber 3. https://kiber.tech/company/ (accessed 3 July 2022).

5 Wang, P., Bai, X., Billinghurst, M. et al. (2021). AR/MR remote collaboration on physical tasks: a review. *Robotics and Computer-Integrated Manufacturing* 72 (C): https://doi.org/10.1016/j.rcim.2020.102071.

6 Huang, W., Wakefield, M., Rasmussen, T.A. et al. (2022). A review on communication cues for augmented reality based remote guidance. *Journal on Multimodal User Interfaces* 16: 239–256. https://doi.org/10.1007/s12193-022-00387-1.

7 Alem, L., Huang, W., and Tecchia, F. (2011). Supporting the changing roles of maintenance operators in mining: a human factors perspective. *The Ergonomics Open Journal.* 4: 81–92. https://doi.org/10.2174/1875934301104010081.

8 Huang, W. and Alem, L. (2013). Gesturing in the air: supporting full mobility in remote collaboration on physical tasks. *Journal of Universal Computer Science.* 19: 1158–1174. https://doi.org/10.3217/jucs-019-08-1158.

9 Huang, W., Kim, S., Billinghurst, M., Alem, L. (2019) Sharing hand gesture and sketch cues in remote collaboration. *Journal of Visual Communication and Image Representation.* 58, 428-438. https://doi.org/https://doi.org/10.1016/j.jvcir.2018.12.010.

10 Huang, W., Alem, L., Tecchia, F., and Duh, H. (2018). Augmented 3D hands: a gesture-based mixed reality system for distributed collaboration. *Journal on Multimodal User Interfaces* 12: 77–89. https://doi.org/10.1007/s12193-017-0250-2.

2

Communication Models for Remote Guidance

2.1 Introduction

Communication is an essential part of remote collaboration, so understanding the impact of technology on communication is important in the system design process. For example, understanding how communication will change if one person cannot see what their remote collaborator is doing or if they have the ability to point or draw in their field of view. One way to do this is by using communication models; these are theoretical frameworks that can be used to predict communication behaviors when people use different collaboration technologies.

One important element of remote collaboration is to understand the impact of technology on communication behaviors. For example, Whittaker reviews using audio only to audio and video conferencing in a collaboration task and finds that people performed equally well but had very different communication patterns [1]. Previous researchers have developed a range of different communication models to explain how people communicate with one another and to predict the impact of technology on remote collaboration. For example, Clark and Brennon's grounding model of communication [2] has been used to predict communication behavior in video conferencing, especially when compared to audio-only conferencing [3].

In this book, our main focus is on Augmented Reality (AR), a collection of display, input, and tracking technologies that can be used to seamlessly overlay video imagery in the real world [4]. The ability to provide virtual visual and spatial cues makes AR an ideal technology for enhancing face-to-face and remote collaboration [5]. Previous collaborative AR systems have overlaid virtual video of remote collaborators in a user's real space [6], used shared virtual content to enhance face-to-face collaboration [7], and enabled a remote user to place virtual cues in a local person's workspace [8]. Studies with these systems have found that remote people feel a higher degree of social presence when using AR than when using video conferencing [6], they collaborate more naturally [9], and use behaviors similar to face-to-face collaboration [7].

Computer-Supported Collaboration: Theory and Practice, First Edition.
Weidong Huang, Mark Billinghurst, Leila Alem, Chun Xiao, and Troels Rasmussen.
© 2024 The Institute of Electrical and Electronics Engineers, Inc. Published 2024 by John Wiley & Sons, Inc.

In this chapter, we discuss how existing communication models can be used to predict the impact of different AR cues in remote collaboration and if a new communication model needs to be developed. Unfortunately, this is an area that has not been well-researched. Despite the potential of AR for remote collaboration, there are relatively few formal user studies conducted with collaborative AR systems. For example, In a survey of all AR user studies conducted between 2005 and 2014, Dey et al. [10] found less than 5% of studies involved collaborative systems, and very few of those collected communication measures. Marques et al. [11] suggested that there is "… minimal support of existing frameworks and a lack of theories and guidelines to guide the characterization of the collaborative process using AR." So, there has been relatively little previous work done on exploring communication models in AR for remote collaboration, and there is a need for more research on this topic.

There are many different types of collaborative AR systems, but the focus of this chapter is specifically on head-worn AR systems for remote collaboration on physical tasks. A typical example is a system that uses a see-through head-mounted display (HMD) with a camera mounted on it that allows a local worker to stream a view of their workspace to a remote helper. The remote helper in turn can add virtual content to the local worker's view to help assist them with the physical task that they are doing (see Figure 2.1). Figure 2.1a shows a typical version of such a system with a depth-sensing camera added to an Epson AR display. Figure 2.1b shows the view through the AR HMD and the remote expert view, where the expert is drawing on the live video feed to provide AR visual cues back into the local workers' view. This type of system could be used in many applications, such as a remote expert helping a mechanic fix a car or an expert surgeon remotely assisting a novice doctor.

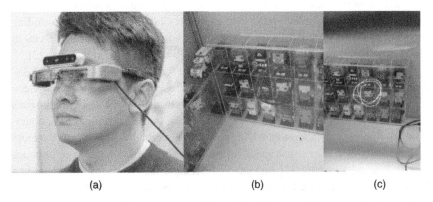

(a)   (b)   (c)

Figure 2.1 A simple example of an HMD-based collaborative AR system. (a) The HMD with depth-sensing camera attached, (b) remote expert view with live annotation, (c) AR view.

There are many examples of research that have a similar setup, such as [6, 12–16]. This type of configuration is also becoming increasingly common in industrial applications. For example, Microsoft's Remote Assist application uses the Hololens2 AR HMD to allow a local worker to collaborate with remote helpers [17]. Remote Assist streams the Hololens2 camera view to one or more remote users viewing the content on the web, who are then able to talk to the local worker, see what they are seeing, and place virtual arrows or other cues in the field of view.

Although not widely used, examples of systems like this are not new. Research on AR systems for remote collaboration dates back to the 1990s with the Shared-View work [18], and British Telecom's CamNet system [19]. Since then, dozens of research papers have been published, but there have been relatively few studies of these systems from a communications perspective. Being able to evaluate this research from a communications perspective will help identify the research areas that should be further investigated, provide guidelines for improving the user experience, and establish the limitations of the current communication models. Just as using communication models improved video conferencing, the same type of approach could be used to improve AR systems for remote collaboration.

In this chapter, we review various communication models and show how they can be used to analyze communication in different AR interfaces for remote guidance on physical tasks. In the remainder of this chapter we first provide a historical review of communication models, especially focusing on remote communication (Section 2.2). Next, we show how communication models have been applied to analyze non-AR remote collaborative systems (Section 2.3) and research on the application of communication models to collaborative AR (Section 2.4). In Section 2.5, we discuss the limitations of current communication models and explore how they could be extended to accommodate all of the communication affordances of AR systems for remote assistance. Finally, we identify the research gaps that should be explored in the next generation of collaborative AR systems (Section 2.6).

The goal of this chapter is to provide the reader with enough understanding of communication models that they can use to predict the impact of various technology elements on AR systems for remote collaboration. This should enable them to develop better systems and to improve their own research in this area.

2.2 Overview of Communication Models

Communication theories attempt to describe and explain how people share knowledge and information with each other. Communication models are formalized concepts of the information-sharing process. They can be simple or complex and there have been a wide variety of models developed. In this section, we provide a quick overview of some of the most important historical communication models.

Formal models of communication date back thousands of years to Aristotle and his work on rhetoric [20]. In this classic work, he proposed a simple communication model with three parts: a speaker, a message, and a listener. Each of these parts is essential. For example, a speaker and their message do not communicate if there is no listener. These three elements of speaker, message, and listener have been used in many subsequent models, with Kumar noting that "Western theories and models of communication have their origin in Aristotle's Rhetoric" ([21], p. 16).

In a similar way, Green et al. [22] present a simple human-to-human communication model that has three key components:

- The communication channels available.
- The communication cues provided by each of these channels.
- The affordances of the technology that affect the transmission of these cues.

They say that there are three main types of communication channels available: audio, visual, and environmental, where visual and audio cues are those that can be seen and heard, and environmental channels support interactions with the surrounding world. Depending on the communication medium, different communication cues may be able to be transmitted between collaborators. For example, using text chat will enable text messages to be sent between people but will prevent the communication of audio or environmental cues.

In face-to-face communication, a wide variety of communication cues are used when people collaborate together. These can be classified into Visual, Audio, and Environmental cues (see Figure 2.2). Audio cues include speech, paralinguistic, para-verbals, prosodics, intonation, and other types of audio. Visual cues are those generated by the user and include gaze, gesture, facial expression, and body position, among others. Finally, environmental cues include actions of the user in the environment to support communication, such as object manipulation, writing or

Figure 2.2 Face-to-face communication cues.

(a)   (b)

Figure 2.3   Introducing a separation between task space and communication space.
(a) Face-to-face collaboration with task space a subset of communication space.
(b) Remote collaboration with task space separate from the communication space.

drawing, object presence, and the spatial relationships between objects, among others. One of the goals of a communication model is to understand how variation in these elements can affect communication.

In addition to using a range of different communication cues, when people are collaborating on a task, there is a distinction between the task space and communication space (Figure 2.3a). The task space is the physical workspace that people are focusing on to complete a particular task, while the communication space is the space where people are able to see each other and share communication cues.

When people are collaborating face to face they can easily see each other and the range of different communication cues being used, so the task space is a subset of the communication space (Figure 2.3a). Ishii describes this as seamless communication because there is no functional separation between the task and communication space [23]. However, when people are working remotely, the task space is often separated from the communication space (see Figure 2.3b). For example, when video conferencing, people may have the face of their collaborator on one screen, while looking at a shared document on another. In this case, it is impossible to see many of the remote collaborator's communication cues at the same time as looking at the task space. So, there is an artificial seam between the communication space and task space. This is the type of impact of remote communication technology that needs to be predicted through communication models.

Green et al. [22] point out that the benefit of communication models is that they can be used to predict collaborative behavior and the impact of technology on collaboration. For example, if two people are talking over the phone, they are likely to use more verbal cues than if they were using a video conferencing link capable of sharing audio and visual cues. In the case of text-only communication, communication is reduced to one content-heavy visual channel, with a number of possible effects, such as less verbose communication, use of longer phrases, increased time to reach grounding, slower communication, and fewer interruptions.

2.2.1 Linear Communication Models

Most communication models trace their roots back to Shannon and Weaver's 1949 linear communication model [24]. Their model has a source, a transmitter, a signal, a receiver, and a destination (see Figure 2.4). Following Aristotle, the source is the equivalent of the speaker, and the destination is the same as the listener. Aristotle's message gets converted into a signal at the transmitter. This signal is called a sent signal, and while it is being transmitted, noise is added to it, resulting in the received signal that reaches the receiver. For example, applying this model to a telephone call, the speaker's voice is converted to an electrical signal conveyed over telephone lines, but the signal is degraded by additional noise in the telephone line that can make it difficult for the listener to hear.

The Shannon and Weaver communication model is unique as it was initially developed to describe communication over technology, namely telephones and radios. Shannon was focusing on the noise caused by the technology and correctly decoding the sender's message. Although this model was designed for telecommunication, it has been widely used in other areas.

Around the same time, Berlo [25] developed a model that he described as "a model of the ingredients in communication." It had four main parts: a source, a message, a channel, and a receiver (see Figure 2.5). The source and receiver were identical, with both having the same five characteristics: communication skills, attitudes, knowledge, social system, and culture. The message was composed of five elements: structure, content, treatment, and code, while the channel had the five senses: seeing, hearing, touching, smelling, and tasting.

Berlo believed that for effective communication to take place, the source and receiver had to be at the same level, such as having the same communication skills or similar knowledge. However, there are some limitations to this model, including not considering noise, having a lack of feedback, and it is a linear model. Most significantly, it assumes that people need to have the same knowledge or skill level for effective communication, which very rarely happens in everyday life.

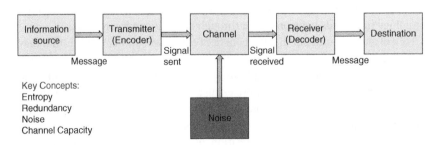

Figure 2.4 The Shannon and Weaver communication model.

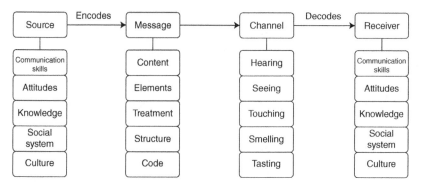

Figure 2.5 Berlo's source message channel receiver (SMCR) model.

2.2.2 Nonlinear Communication Models

Until the 1970s, many of the prevailing communication models were based on linear information transmission, like Shannon and Weaver's [24] or Berlo's [25]. Further contributions came from Schramm [26], who considered communication from a human behavior perspective. He introduced concepts such as shared knowledge and a feedback loop as part of the communication. Schramm argued that the received message could be different to the intended message. It is possible for two different interpretations to be gained from the same message. These interpretations are formed by the different experiences and cultures of the sender and receiver. This led to Schramm's use of a Venn diagram to show the message that the sender intended and the message the receiver interpreted, with the overlap between the two being the signal part of the message that was correctly interrupted (see Figure 2.6). Schramm discusses these two parts as frames of reference and the overlap as the shared knowledge that is being communicated.

Schramm also discussed the importance of feedback, with a model showing each communicator with messages going between them. In this case, feedback is the information that flows back from the receiver to the sender and provides

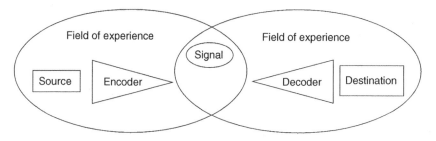

Figure 2.6 The Schramm experience model.

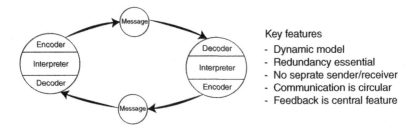

Key features
- Dynamic model
- Redundancy essential
- No seprate sender/receiver
- Communication is circular
- Feedback is central feature

Figure 2.7 The Osgood and Schramm circular model.

an important cue as to how well they are doing and how much of the message is understood. The sender and receiver are identical in that both have three parts: an encoder, an interpreter, and a decoder (see Figure 2.7). This is referred to as the Schramm–Osgood circular model.

The Osgood and Schramm circular model introduced feedback [26], and Kincaid extended this into the convergence theory of communication [27] (see Figure 2.8). In this model communication is considered as a process and sharing of knowledge, not a one-way linear information flow. The participants continue to

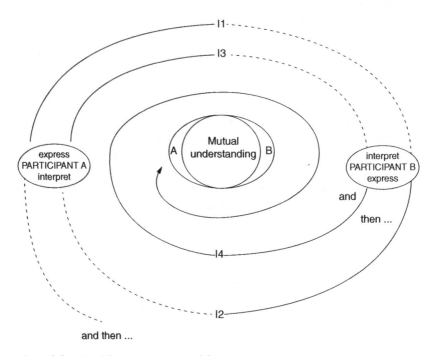

Figure 2.8 Kincaid's convergence model.

share information with each other in a self-correcting process that converges on a common understanding. Kincaid was critical of the view that communication was linear and so proposed a nonlinear cyclical model where the communicators worked to reach mutual understanding.

A similar model that focuses on the simultaneous sending and receiving of messages between people is Barnlund's Transactional Model of Communication [28]. In Barnlund's model, communication is seen as a reciprocal system where people communicating interact with and influence one another. There are a set of public, private, and nonverbal cues shared between communicators, some of which will be perceived at any particular time. These cues help establish the communication meaning over time, in a cumulative manner as more and more cues are shared. Figure 2.9 shows the information flow; here circles show the participants encoding and decoding messages, arrows show the messages being sent, and jagged lines the cues available to be perceived. As Figure 2.9 shows, in Barnlund's model, communication is the evolution of meaning, and it is dynamic, circular, continuous,

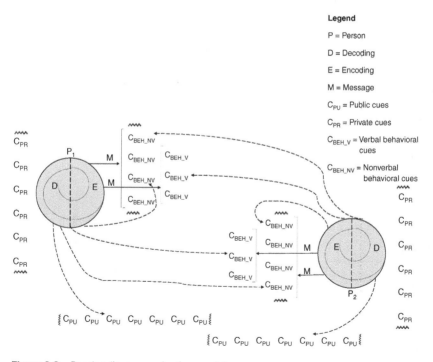

Figure 2.9 Barnlund's communication model.

complex, unrepeatable, and irreversible. Barnlund's model also includes noise and filters and represents social, cultural, and relational factors.

Clark and Brennan [2] built on these feedback models to create the concept of Grounding, a process for establishing common ground in the form of mutual knowledge. They also described the feedback loop as contributions to conversations and acknowledgments. Grounding is the action where collaborators engage in shared communication and express confirmation of comprehension through words or bodily movements until they arrive at a shared understanding. So in a conversation people may keep on offering clarifying statements until they arrive at the same shared understanding (see Figure 2.10).

They outline eight grounding constraints, or factors that contribute to the grounding, stating that *"when a medium lacks one of the constraints, it forces people to use alternative techniques"* [2] (see Figure 2.11). For example, telephone calls have the constraint of Audibility, while video conferencing provides both Audibility and Visibility. So on a telephone call the participants cannot share nonverbal cues and might need to compensate through additional spoken language compared to a video conference. Several studies have shown that achieving common ground is more difficult when technologies have been used to communicate at a distance compared to face-to-face collaboration.

In addition to the communication models described, there are other theories around the impact of technology on communication, such as Social Presence

Alan: Now, um, do you and your husband have a j- car?	B: How would you describe the color of this flower?
Barbara: - have a car?	S: You mean this one [pointing]?
Alan: Yeah.	B: Yes.
Barbara: No.	S. It's off-yellow.
(a)	(b)

Figure 2.10 Examples of grounding in conversation. (a) Asking a clarifying question. (b) Pointing to clarify an object.

Medium	Copresence	Visibility	Audibility	Cotemporality	Simultaneity	Sequentiality	Reviewability	Revisability
Face to face	•	•	•	•	•	•		
Telephone			•	•	•	•		
Video conference		•	•	•	•	•		
Two-way chat				•	•	•	•	•
Answering machine			•			•		
E-mail							•	•
Letter							•	•

Figure 2.11 Clark and Brennan's grounding constraints across different media.

[29, 30], Media Richness Theory [31, 32], and the computer-mediated communication interactivity model [33]. Social Presence [30] is defined as the degree to which a person is perceived as a " in mediated communication [29] and is a quality of the medium itself, depending on the ability of the medium to transmit communication cues (facial expression, gaze, nonverbal cues, etc.). Media Richness Theory is based on the idea that communication media vary in their ability to enable people to communicate with one another. For example, face-to-face communication is much richer than a phone call, which is richer than a letter. Overall, richer, personal communication media is more effective for communicating, and so media can be arranged from leaner to richer mediums in terms of how effective they are at supporting communication (see Figure 2.12).

One important concept is that in mediated communication, it is important to fit the richness of the communication medium to the task complexity [34] (see Figure 2.13). For example, people do not need a face-to-face meeting to complete a simple communication task, such as finding out the weather. Similarly, a person is unlikely to propose marriage or conclude an important business deal using a mobile phone text message. Daft and Lengel [34] propose that there is a band of effective media use where the media richness matches the task complexity requirements. If the media richness is much higher than what is required by the task, then there will be no performance improvements, and the additional information conveyed may even negatively impact communication. Similarly, if the media richness is much less than what is required by the task, then there can be a communication breakdown.

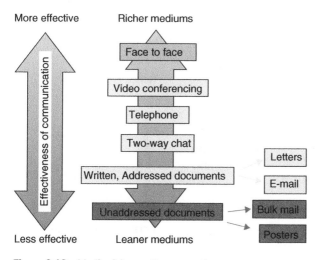

Figure 2.12 Media richness theory continuum.

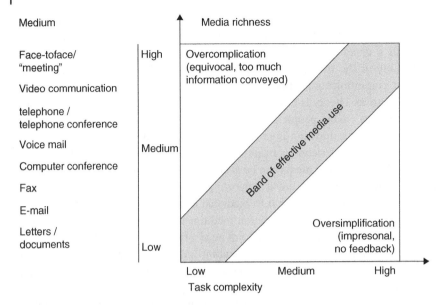

Figure 2.13 The band of effective media use.

2.2.3 Summary

In summary, from this research there are a number of key concepts that can be learned. First, communication is a complex, continuous, and dynamic process. Participants communicating with one another work together to cocreate meaning, using the least effort possible. In this process there are multiple types of communication cues that can be used, including verbal, nonverbal, and environmental. Figure 2.14 shows an abstract representation of current communication

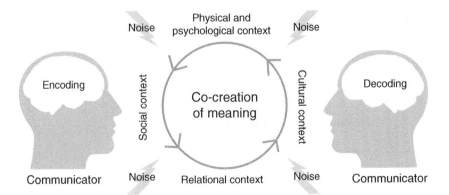

Figure 2.14 A typical communication model.

models. In this case, two communicators work together using a variety of cues to cocreate meaning. There are different contexts which contribute to helping to understand what the communicator is meaning, and there is noise added to any communication taking place.

Second, technology can be used to enable people to communicate at a distance, and enable mediated communication. In this case, there are often multiple channels that can be used to convey the communication cues. Depending on their properties, different channels provide different amounts of Media Richness, and Social Presence. So, the type of communication channel being used can affect the grounding process, and also needs to fit to the task requirements.

The value of communication models is using them to understand communication in practice, especially technology mediated communication. In Section 2.3, we discuss how to apply these communication models to real remote collaborative systems.

2.3 Applying Communication Models

One of the main values of communication models is that they can be used to predict the impact of technology on communication behavior. Whittaker provides a review of communication theories applied to remote collaboration technology such as video conferencing [35]. For example, Fussell has shown how conversational grounding can be applied in wearable teleconferencing [12] to predict how shared views of the workspace can maintain situational awareness, and promote a sense of copresence. In her study, a local worker was assembling a toy robot with help from a remote expert. Fussell found that workers used significantly more words to complete the task in an audio-only condition compared to other conditions that shared visual cues of the workspace. This is expected based on the communication models discussed above. The addition of a visual channel showing the task space should reduce the need for the local workers to describe what they are doing to achieve common ground. The face-to-face condition will enable the sharing of a rich array of nonverbal communication cues and so should reduce the need for verbal communication even further to achieve common ground.

According to Whittaker [35], mediated communication models try to explain communication, usually by (a) characterizing how technologies differ in terms of their communication affordances, (b) describing how the affordances of a given technology differ from those of face-to-face communication, and (c) explaining how these different affordances produce differences between mediated and face-to-face communication in process, content, or outcome of communication. In this case, affordances refer to the innate ability of the communication media to convey communication cues. For example, an audio-only phone call affords

Table 2.1 Technologies and their communication affordances.

Affordance		Interactivity	
		Interactive	Noninteractive
Mode	Linguistic	Phone, Audioconference, Chat Instant Messaging	Email, answerphone, voicemail, FAX, letter, usenet
	Linguistic and visual	Videoconference, VideoPhone, Shared Workspace	VideoMail

sharing of audio and linguistic cues, while video conferencing technology affords sharing of visual cues in addition. Whittaker proposes that the technology affordances of modality and interactivity are sufficient to classify most communication technologies (see Table 2.1). The different modes that communication technologies support are either linguistic, or visual and linguistic, while the technology can be either interactive or noninteractive. Traditional video conferencing provides poor support for spatial cues [36]. However, AR technologies allow spatial cues to be added to this table because they enable people to perceive the spatial relationships between objects, the environment, and each other.

Considering the affordances of different communication technologies, communication theories can also predict how these affordances will affect communication behavior. For example, face-to-face communication uses behavior such as head nods, gaze, and gesture to mediate turn taking [35]. However, with audio-only communication these cues are no longer visible, so the technology does not afford these behaviors, and turn-taking could be affected. User studies have shown that turn taking is indeed affected by these systems [37, 38]. Similarly, changing the video-frame rate can also affect communication behavior, as expected [39].

Researchers have applied communication models to audio and video conferencing with the goal of validating the predictions from the communication models. For example, Doherty-Sneddon et al. [40] compared communication patterns in face-to-face and video-mediated communication, finding that when participants in a collaborative problem-solving task could both see and hear each other, the structure of their dialogues differed compared to when they only heard each other. O'Conaill et al. evaluated speech patterns over two different quality video links, finding that as latency decreased, the communication became closer to face-to-face communication patterns in terms of spoken turn length and distribution, back channels, and interruptions [38]. These behaviors are predicted by the communication models described in Section 2.2.

These studies focused on seeing the face of a remote collaborator, but other studies have explored communication behaviors when seeing the task space of a remote collaboration. In this case tele conferencing cameras point at the user's workspace, not their face. Kraut et al. [41] explored the use of shared video on a collaborative problem-solving task, finding that how people communicated varied depending if audio only or audio plus video conferencing technology was used. They found that *"... help was more proactive and coordination was less explicit when the pairs had video connections."* They repeated the experiment with a face-to-face condition, finding a significant improvement in performance and demonstrating that value of copresence for grounding [42]. Similarly, Fussell et al. [12] used a conversation grounding model to explore communication with head-mounted and scene cameras, finding that the number of words spoken differed significantly between face-to-face conditions, audio only and remote camera conditions, as predicted by the model.

From these studies there are a number of predictions that can be made about systems for remote collaboration on physical tasks:

1. *Text Only:* Using a text only communication channel should lead to use of longer written phrases compared to other communication channels, require an increased time to reach grounding, and produce slower communication and fewer interruptions.
2. *Audio:* Using an audio channel with no visual cues could require more speaking to achieve common ground, and conversational turn taking may be affected. In collaboration on a physical task, audio-only communication will require extensive verbal description of the task space.
3. *Visual View of the Workspace:* In a remote collaboration setting, providing a visual view of the workspace will significantly reduce the verbal description needed. It will support more implicit help where the remote helper can automatically see what the local worker is trying to do without them explaining it.
4. *Remote Pointing:* Providing remote pointing ability in a collaborative interface will lead to increased use of deictic phrases and reduce the need for the collaborators to explicitly refer to the objects that they are about to collaborate with.
5. *View of Collaborator:* Providing a view of the remote collaborator enables the use of nonverbal cues for turn taking, and showing comprehension. This will reduce the need for confirmation audio and support richer nonverbal communication cues, which will lead to faster grounding.
6. *Three-Dimensional View of Task Space:* Providing a three-dimensional view of the task space will increase the environment spatial cues, and reduce the amount of visual descriptors needed to communicate about physical tasks.

7. *Three-Dimensional View of Collaborator:* Providing a three-dimensional representation of a remote collaborator will increase support for spatial cues such as pointing at objects in the scene. This will also create a high sense of Social Presence, which will lead to faster grounding.

As can be seen, as remote collaboration becomes more similar to face-to-face collaboration, the communication behavior should also become more similar to face-to-face communication behavior. In Section 2.4, we review selected papers in collaborative AR, and see if the same predictions hold true for these systems.

2.4 Communication Behaviors in AR Conferencing

So far we have discussed the evolution of communication models, and predictions that can be made from these models. In many cases, user studies are informed by communication models and the researchers collect conversational behaviors such as turn taking, number of back channels, or interruptions to validate these models. Although, this type of research has been commonly conducted with audio and video conferencing, it is rare in collaborative AR experiences. In this section, we review examples of collaborative AR systems, focusing on those that explore conversational behaviors to see if the observed behaviors match the predictions.

Most of the early research on AR conferencing was focused on creating working prototypes and evaluating the impact on task performance or social presence rather than communication. For example, one of the first AR conferencing systems was developed by Billinghurst and Kato [6], who used an HMD and AR tracking to superimpose live video avatars of remote collaborators over the real world. This allowed a user to see a life sized image of a remote collaborator in their real space. In a user evaluation of the system, people reported that they felt that AR conferencing was more similar to face-to-face communication than video conferencing or audio-only conferencing [43]. The 3DLive system improved upon this by using multiple cameras to create the illusion of a 3D virtual video avatar appearing in the real world [44]. Most recently, the Microsoft HoloPortation project [45] uses real-time volumetric capture to superimpose 3D captured virtual avatars over the real world. However, with these projects, no evaluation was conducted of the communication behaviors demonstrated with the systems.

Other researchers have explored how to add AR cues to desktop video conferencing. For example, de Souza Almeida et al. [46] developed HANDY, a desktop AR conferencing system using a separate camera to capture the user's hand and enabling people to introduce gesture cues into a remote user's video image. In a user study compared to normal video conferencing user's felt that the HANDY system provided a higher degree of social presence, it was easier to communicate, and participants felt closer to each other. Similarly, Barakonyi et al. [47] presented a desktop AR conferencing system that allowed users to attach AR content to markers that they both see on their video feeds. This combined the advantages of face-to-face video conferencing with tangible AR interaction methods using physical objects. However, these papers did not present any results analyzing communication behavior.

Researchers have also focused on using AR to enhance task space collaboration. Kurata et al. [48] developed a wearable camera/laser projector system that enabled a remote user to project AR imagery into a local user's workspace, and assist them with real world tasks. Similarly, Tait et al. [49] developed a portable projected AR system that captured a 3D model of the local user's workspace and projected onto it. Other wearable AR systems have been developed for comparison between handheld and wearable [50] or to improve the mobility of the users [23].

A number of papers have focused on communication behaviors in face-to-face AR collaborative systems. Kiyokawa et al. [7] conducted two experiments comparing the communication behaviors of face-to-face users of collaborative AR systems. They found that differences in real-world visibility with different display technology severely affect communication behaviors, and that the spatial relationship between the task and communication spaces also severely affected communication behaviors. These experiments used a simple pointing task and focused on collecting communication process measures such as counting deictic phrases, positional phrases, pointing gestures, and numbers of words spoken. Similarly, Billinghurst et al. [51] conducted an experiment comparing collaboration in a face-to-face AR system with a projected desktop interface, finding that speech and gesture communication with the AR system was closer to natural face-to-face communication than with the desktop interface. Prytz et al. [52] explored the impact of sharing gaze cues in face-to-face collaboration AR systems and found that the amount of eye contact was increased in a face-to-face collaboration where people were not wearing AR displays but that the decrease in eye contact with HMDs did not seem to have a direct effect on the collaboration.

However, there have been far fewer examples of remote collaborative AR systems exploring communication behavior. We conducted a search on Google Scholar with the terms "Remote Collaboration" AND "Augmented Reality" AND "Conversational analysis" and found the papers shown in Table 2.2. The term conversational analysis was used because communication behavior cannot be measured in detail without conversational analysis. For each of the listed papers, we provide a short description of the collaborative task, the conditions compared in the user study, and the results found from the conversational analysis. In addition to the conversational behaviors, most of the user studies also collected additional measures such as task performance, and subjective feedback.

Table 2.2 Remote collaborative AR studies with conversational analysis.

	Technology and task	Conditions	Results found
[53]	Desktop monitors Annotation tool **Task** Robot construction 28 pairs Remote expert	Video only, Video + manual erase. Video + auto erase **Measures** Number of words Types of Deixis phrases Deixis per minute	Expert used over three times words of worker Experts used fewer words when the annotation tool was available. Workers used mostly local deixis. Expert changed use of deixis as a function of media Pointing most common type of annotation.
[54]	Desktop monitor Projected AR **Task** 12 pairs 2D block layout Remote expert	Different pointer types; – None – Mouse – Gaze **Measures** Number of phrases Conversational coding	Expert used over four times words of worker Worker mainly confirming expert instructions Expert used fewer phrases in mouse/gaze condition that none, less in mouse than gaze Significantly fewer acknowledgments and procedural messages in mouse condition.
[8]	AR HMD Desktop **Task** Object Arranging 18 pairs Remote expert	Video only Fixed view Freeze view Independent view **Measures** Speaking time Types of phrases Number of instructions Time moving objects	Less speaking in the independent view condition than in the freeze or fixed view Less adjustment instructions in the independent view than the fixed view Fewer queries in the independent view than the fixed view and the freeze views More view change instructions in fixed and freeze view than in the independent view.

Ref	Setup / Task	Measures	Findings
[55]	AR HMD Desktop **Task** Replacing controller in cabinet 25 pairs Remote expert	see-through HMD, spatial projection, video-mixing tablet **Measures** Phrase types (8 types), Phrases categorized into "reactive" and "proactive"	Worker and expert needed more statements using the HMD than in the Tracking or Screenshot condition. The best result was achieved by the Projector Application Significant difference in "worker questions" between Tracking and HMD conditions.
[56]	AR HMD Desktop collaboration **Task** Object searching 20 pairs	Map vs. list interface MR system guidance vs. spatial information **Measures** Number of words spoken References to objects or locations	MR guidance significantly reduced the words used Experts made significantly more references to objects in the worker's environment when they used the map than list interface Experts made fewer references with system guidance.
[13]	AR HMD Desktop **Task** Lego Assembly 12 pairs Remote expert	Evaluated three cues: – None – Pointer – Eye tracking – Pointer + eye tracker **Measures** Number of words Types of phrases used	Use of a pointer reduced the amount of communication needed. Remote helpers using different words for identifying objects or specifying directions. Without the pointing cue, objects described in terms of color, shape, or size. With the pointing cue, participants used the phrase "this one" while pointing at the object. Significant effect of pointer on number of phrases said by local and remote users.

(Continued)

Table 2.2 (Continued)

	Technology and task	Conditions	Results found
[57]	Remote Collaboration VR/Leds Wiring equipment **Task** Wiring switches 18 pairs	voice guidance with no visual LED feedback. **Measures** Number of questions Number of deictic phrases	Difference in errors, but no difference in communication patterns, questions, and deictic as well as explanatory expressions.
[58]	AR HMD VR HMD **Task** Finding Lego blocks 12 pairs Remote expert	A1. Nonspatialized voice: A2. Spatialized voice: A3. Nonspatialized voice + spatialized auditory beacon: A4. Spatialized voice + spatialized auditory beacon: **Measures** 4 Type of utterance	With visual cues, the conversation became more intuitive. When seeing the head frustum and the hand gestures, some participants said "I can see you moving," When visual cues were invisible in their current FoV, some participants intuitively asked "where is your hand/head?" When the expert gave explicit instructions in A1 and A2, most utterances were complete references When auditory beacons were included (A3 and A4), the complete references significantly decreased, and most utterances were backchannel and reference pronoun For the local user in A1 and A2, a large portion of their utterances were backchannel/acknowledgment

| [59] | AR HMD

Task

collaboratively building molecule models

Eight pairs

Expert guide | Six different collocation conditions and model conditions:

face to face vs. remote, real vs. virtual vs. augmented

Measures

Deictic phrases | The guide did the majority of talking. Both guide and builder made heavy use of deictic speech (e.g. "this," "that" and "here")

The builder's responses were most often used short to acknowledge a reference or to clarify a reference by asking a question

The builder would often clarify a reference by selecting the sphere the guide was pointing to, if the sphere had not already been selected

While working in the remote/virtual and remote augmented scenarios, the virtual pointer was heavily used

In the colocated/augmented scenario, a majority of the participants preferred to make references using their hands

All guides and a majority of the builders felt that sharing video was helpful when making references, especially in the remote augmented scenario |

Some studies just used simple view sharing and pointing on a desktop interface to enable a remote expert to help a local worker complete a task. For example, [53] describes an experiment where a remote export is helping a local worker assemble a toy robot. Both the expert and the local worker communicate over desktop monitors, showing a live video view of the workspace. The remote expert is able to draw on the video, and an experiment was conducted to explore the impact of having annotation cues that automatically disappeared or manually had to be removed by the user. Pairs completed the task faster, and the experts used significantly fewer words when the drawing tool was used. Pointing was the most common type of annotation done by the expert (over 75% of the annotations were for pointing), and there was a significant difference in the type of deixis language used by the expert depending on the type of annotation removal used.

In [54], the impact of the type of pointers on collaboration languages was explored. In this case a worker on a table was assembling 2D puzzle blocks with a pointer appearing projected in the blocks. A remote expert was shown a live camera view of the workspace on their desktop monitor and could use either their mouse, or gaze to drive the remote pointer. So there were three pointing conditions: none, mouse, and gaze. Video was recorded of the users completing the task and a transcription made. From this was recorded the task completion time, and the number of phrases needed to complete the task. The phrases were also classified into one of five categories: procedural, task statue, referential, internal state, and acknowledgment.

There was a significant impact of the type of pointer used on the task performance and the user language. Not surprisingly, using a mouse or gaze-driven pointer allowed users to complete the task faster than with no pointer at all, but the mouse pointer was also faster than using the gaze pointer. Using a pointer also reduced the ambiguity of the language, and so with the mouse pointer, there were significantly fewer phrases needed to complete the task than with the gaze-driven point or no pointer. The gaze-driven pointer also required less phrases than using no pointer. In terms of low-level conversational analysis, people used significantly fewer acknowledgment and procedural messages when using a mouse pointer than gaze pointer or no pointer, and similarly, fewer acknowledgment and procedural messages when using a gaze pointer than no pointer. Overall, this showed that using a pointer enabled people to communicate more clearly and achieve common ground more quickly, leading to a reduction in task performance time and amount of communication needed, as well as changing the type of communication needed.

Other systems used more complicated AR cues. For example, Tait and Billinghurst [8] developed a collaborative system that enabled a desktop expert to share AR cues with a local worker to help them complete an object arrangement task. The local worker wore an AR display and was guided by a remote helper to

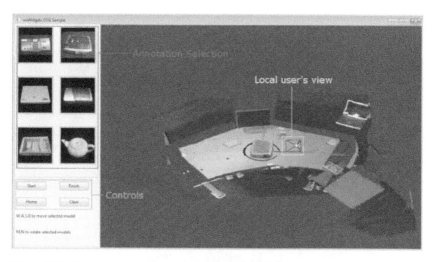

Figure 2.15   Remote user's desktop interface for 3D object placement.

place real objects at certain locations in the real world. The remote helper could guide the local worker by placing AR objects in the real world at the location where the real objects should be placed. The remote helper used a desktop 3D interface that showed a 3D model of the local worker's task space and allowed the remote helper to place the virtual objects in the 3D scene model, as well as see a representation of the local worker and where they were looking (see Figure 2.15). As the remote user places virtual objects, the local worker could see them appearing in the real world as AR objects (see Figure 2.16).

The use of a desktop interface for the remote helper meant that they could have an independent view into the shared task space. So a user study was conducted to explore the impact of view independence on remote collaboration in AR interface. In particular the authors compared four conditions:

1. *Video Only:* where the remote helper could see the task space from a video camera mounted on the worker's HMD, and they can speak to the worker but not place AR models. In this case, there is no independent view.
2. *Fixed:* The remote helper can see a live video view from the local worker, and 3D scene model, but they cannot change their view. They can talk to the local worker and also show AR models to help them complete the task. There is no independent view.
3. *Freeze:* The remote user can see the 3D scene from the perspective of the worker and can freeze the scene when needed to place AR objects. When the scene is frozen, the view will snap back to the worker's current view, so it is semi-independent.

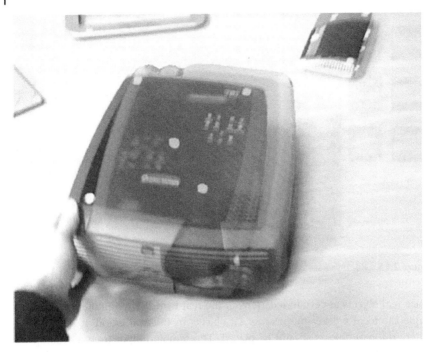

Figure 2.16 Local user's view of the AR object being placed in the real world.

4. *Independent:* In this case, the remote helper has full independent control over the view of the 3D scene and sees the local worker's position represented in the scene. This is the only condition where the remote helper has full control over their viewpoint.

Based on the predictions from Section 2.3, it would be expected that the Video-only condition would require the most verbal communication, while the independent view condition would require the least verbal communication. It would also be expected that the remote helper would make a lot of requests for the worker to change their viewpoint in the fixed view condition. To test this, conversational measurements were made of the amount of time spent speaking in each condition, and the types of phrases spoken, as well as performance time and some subjective survey measures.

As predicted, the authors found that there was significantly less time spent speaking in the independent view, than the fixed or freeze view, while there was no difference between fixed and freeze view conditions. Similarly, there were significantly more spoken instructions needed to complete the task in the fixed view than the independent view. There were also significantly more instructions for adjusting the viewpoint in the fixed and freeze view than the independent view, and the local user asked the remote helper significantly more questions in the fixed and freeze conditions than the independent view condition. Overall people completed the task significantly faster in the independent view condition, than either the fixed or freeze conditions.

These results showed that the communication behavior was as expected, following the predictions from Section 2.3. Similarly, looking at the results from the table we also get agreement with many of the predictions of the impact of technology. For example, [13] explored the impact of using a remote pointer on collaboration between an AR user in a HMD and remote expert on a desktop completing a Lego block assembly task. They found that using the remote pointer reduced the amount of spoken communication needed, and changed how the remote expert referred to words; with the pointer they would say "this one" while pointing at an object, but without the pointer they would refer to objects by color, shape, or size. This is what would be expected from a Grounding Model, where additional spoken communication would be needed if there was no pointing reference to clearly indicate the object of interest. Similarly, in [58] when audio beacons were used to indicate object location, the amount of spoken references to object location significantly decreased.

There can also be an impact of different AR display types on communication. For example, [55] compared using a see-through AR HMD, to a tablet, to a projected AR interface for providing remote assistance in a cabinet wiring task. They measured conversational grounding by classifying spoken phrases into three categories of "question," "description," and "acknowledgment," and also into "reactive" or "proactive" types, depending if the speaker initiated the conversation or not. They found that the worker and expert took significantly more time with the HMD condition compared to the AR projector condition and also spoke 50% more statements on average in the HMD condition than a projector. This is expected because in the HMD condition, the remote expert could only see the portion of the workspace that the local user was looking at, rather than the whole workspace as viewed in the projector condition. This meant that the expert had to talk more to get the worker to look at the points of interest. Also, the AR HMD had a smaller field of view than the AR projected display, meaning that the worker could not always see the AR annotations that the expert was providing, requiring more discussion and time. The authors also had an audio-only phone condition, and as expected, this performed significantly worse than in the visual conditions.

The results in Table 2.2 show that the interface elements used in the AR interfaces for remote collaboration can have a significant impact on communication. This is shown by the work in [56] where the author compared using a map-based spatial interface with a nonspatial list interface for performing a spatial task, and using AR guidance or no AR guidance. In the AR guidance condition the remote helper was able to put AR cues on physical icons in the real world (seen in a

Hololens HMD), in the no AR guidance condition, the remote user could only tell the user where to look to find the target icons. They found that providing AR guidance significantly reduced (by 50%) the number of words that needed to be communicated to complete the task. Remote helpers also made significantly more references to objects in the worker's space with the map interface than with the list interface. Using the Grounding model, this is to be expected. Providing a visual AR cue of the target object should significantly reduce the spoken language needed. Similarly, the map made it easier to reference real-world objects. The research provides a good example of how AR can be used to offload the referential process, and so make communication more efficient between a local worker and remote helper. Similar results were found in [59] where using a remote pointer for referencing was found to be key in establishing common ground between a local worker and remote expert in a molecular assembly task.

It should be noted that all of the papers mentioned in Table 2.2 involved communication between a helper who had expert knowledge and a worker with less knowledge. The remote expert was guiding the local worker to complete a task in their physical environment. This is common for systems using AR to improve collaboration on a physical task. However, it would be interesting to see if there were differences in communication behavior for tasks where the collaborators had the same role/knowledge level. There are many examples of AR remote collaboration when the participants have the same role (e.g. [60–62], among others), but unfortunately communication behaviors in such systems have not been studied.

2.5 A Communication Model for AR

As has been shown earlier in this chapter, communication models have been successfully used to predict the impact of technology on remote communication (telephone and video conferencing). Fussell et al. [12] experimented with communication models and video conferencing. Hauber et al. [63] utilized communication models for VR. In Section 2.4, we showed how communication behaviors in collaborative AR systems could also be predicted using Grounding and other communication models.

However, these existing communication models may not be entirely suitable for collaborative AR interfaces because they do not capture all of the communication affordances. AR has some unique affordances that potentially can improve communication, especially for remote collaboration on physical tasks. This includes the ability to provide virtual spatial cues directly in the real environment, showing virtual communication cues that are normally unseen, and removing the artificial seam between the communication space and task space. Consequently, there is a need for a communication model that captures these affordances.

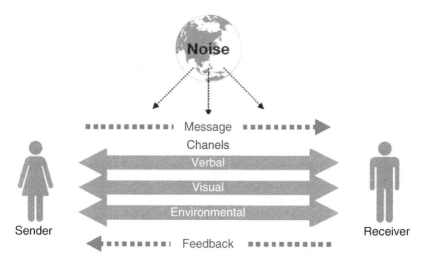

Figure 2.17   AR communication model.

Figure 2.17 shows a simple communication model that could be applied to AR, based on a variation of the diagram showing face-to-face communication at the beginning of this chapter (see Figure 2.2). In this case, communication channels are classified into three types: (1) Verbal, (2) Visual, and (3) Environmental. The verbal channel encompasses all of the audio cues that people make when communicating with one another, including speech, paralinguistic, para-verbals, prosodics, intonation, and other types of audio. Visual cues are the nonverbal cues generated by the person communicating, such as gaze, gesture, facial expression, and body position, among others. Environmental cues include elements of the person's surroundings that can help with communication such as objects, writing or drawings, or other elements.

A person communicating sends messages using these channels, while at the same time getting feedback from the listener indicating if the messages are being received and understood. Noise is also introduced into the model based on a number of personal, technological, environmental, and contextual factors.

Considering this model, when developing an AR system for remote collaboration on physical tasks we need to decide on the visual, verbal, and environmental factors that we want to use to facilitate the collaboration. For example, in terms of verbal factors, is the system going to support audio communication, spatial audio or nonspatial, and what quality of audio? In terms of visual factors, what type of virtual cues are going to be provided, are there going to be virtual communication cues to show nonverbal communication cues (e.g. using AR cues to show gaze lines), and how are the users going to be represented? Finally, in terms of environmental factors, which parts of the workspace need to be shared, how is this to be done (such as using 2D video, 360 degree views, or a 3D model), and does the remote collaborator have an independent or shared view of the workspace.

Reducing any of these communication channels will have an impact on the collaboration. For example, many collaborative AR systems provide the remote collaborator with a live camera view of the local user's workspace from a head worn camera (e.g. in [13] or [18]). However, this only provides a limited view of the local workers environment, and does not support view independence reducing the situational awareness, and environmental cues that can be used for communication. For some tasks, this may take longer to complete the task as the remote collaborator tries to understand what is happening, or needs to spend time telling the local worker to look toward objects of interest.

It should be noted that Figure 2.17 just provides a high-level communication model, and more research is needed to provide the detail to enable this to be used as a predictive tool. The relative importance of the visual, verbal, and environmental channels needs to be explored further, as do the perceptual factors of AR displays that can introduce noise and uncertainty into the communication process. However, as researchers begin to introduce conversational analysis into their collaborative research this will enable more detailed communication models to be developed.

2.6 Conclusions

In this chapter we have reviewed the development of communication models from Aristotle, until the more recent work of Clark and Brennan [2] (Grounding), Gunawardena and Zittle [29]) (Social Presence), Daft and Lengel [34] (Media Richness), and others. These models have evolved from simple linear message passing, to multi-channel feedback loops where people work together to cocreate meaning.

Communication models can be used to predict the impact of different technology on communication behaviors. In Section 2.3, we review how existing communication models can be applied to audio and video conferencing systems. A list of seven predictions are made that can be used to predict the impact of different communication channels in systems for remote collaboration on physical tasks.

Section 2.4 then returns to AR collaborative systems, and reviews the few research papers that have performed a conversational analysis in AR applications for remote collaboration on physical tasks. Results from these studies show that the conversational behaviors can be predicted by using the existing models, such as the Grounding model of Clark and Brennan. For example, [54] showed that having a remote pointer significantly reduced the conversation needed to achieve common ground and clearly identify the objects of interest.

However, there are some improvements that could be made to existing communication models to take into account the unique affordances of AR systems. This was discussed in Section 2.5, where the importance of considering the visual, verbal, and environmental communication channels was reviewed. For collaboration on physical tasks, being able to share and experience a local workers physical workspace is particularly important, as is being able to enable the remote helper to place virtual cues into that environment.

There is still considerable research that could be done on the development and application of communication models for collaborative AR systems. As mentioned, many collaborative systems being developed do not have formal user studies or only capture very superficial usability measures, such as task performance. Apart from the few papers mentioned in Section 2.5, almost no work has been done on conversational analysis that can be used to more deeply understand the impact of the different types of AR technology on communication.

There are also new developments in AR/VR display and sensing technologies that could provide even more insights into collaboration models. For example, the latest AR HMDs have eye-tracking and VR displays are coming that have a wide range of physiological sensors (EEG, EDA, EMG, EOG, etc.) integrated into the display itself [64]. These could be used to capture and share different physiological cues that are not normally apparent in face-to-face communication, but which could provide deeper understanding about people's emotional and cognitive state. One particularly exciting area is new research being conducted on how brain activity can be synchronized in collaborative VR experiences [65], and provide great insight into social neuroscience aspects of collaboration. It is clear that there are exciting opportunities for future research in this space.

References

1 Whittaker, S. (1995). Video as a technology for interpersonal communications: a new perspective. *Multimedia Computing and Networking* 2417: 294–304.
2 Clark, H.H. and Brennan, S.E. (1991). Grounding in communication. *Perspectives on Socially Shared Cognition* 13 (1991): 127–149.
3 Kirk, D., Rodden, T., and Fraser, D.S. (2007). Turn it this way: grounding collaborative action with remote gestures. *Proceedings of the SIGCHI conference on Human Factors in Computing Systems*San Jose, California, USA (28 April–03 May 2007). pp. 1039–1048.
4 Azuma, R.T. (1997). A survey of augmented reality. *Presence: Teleoperators & Virtual Environments* 6 (4): 355–385.
5 Billinghurst, M. and Kato, H. (2002). Collaborative augmented reality. *Communications of the ACM* 45 (7): 64–70.

6 Billinghurst, M. and Kato, H. (1999). Real world teleconferencing. In: *CHI'99 Extended Abstracts on Human Factors in Computing Systems*, 194–195. New York, NY, United States: Association for Computing Machinery.

7 Kiyokawa, K., Billinghurst, M., Hayes, S.E., et al. (2002). Communication behaviors of co-located users in collaborative AR interfaces. *Proceedings of the International Symposium on Mixed and Augmented Reality*, Darmstadt, Germany (30 September–01 October 2002). IEEE, pp. 139–148.

8 Tait, M. and Billinghurst, M. (2015). The effect of view independence in a collaborative AR system. *Computer Supported Cooperative Work (CSCW)* 24: 563–589.

9 Gauglitz, S., Nuernberger, B., Turk, M., and Höllerer, T. (2014). World-stabilized annotations and virtual scene navigation for remote collaboration. *Proceedings of the 27th Annual ACM Symposium on User Interface Software and Technology*, Honolulu Hawaii USA (58 October 2014). pp. 449–459.

10 Dey, A., Billinghurst, M., Lindeman, R.W., and Swan, J.E. (2018). A systematic review of 10 years of augmented reality usability studies: 2005 to 2014. *Frontiers in Robotics and AI* 5: 37.

11 Marques, B., Teixeira, A., Silva, S. et al. (2022). A critical analysis on remote collaboration mediated by augmented reality: Making a case for improved characterization and evaluation of the collaborative process. *Computers & Graphics* 102: 619–633.

12 Fussell, S.R., Setlock, L.D., and Kraut, R.E. (2003). Effects of head-mounted and scene-oriented video systems on remote collaboration on physical tasks. *Proceedings of the SIGCHI Conference on Human Factors in Computing Systems*, Ft. Lauderdale, Florida, USA (5–10 April 2003). pp. 513–520.

13 Gupta, K., Lee, G.A., and Billinghurst, M. (2016). Do you see what I see? The effect of gaze tracking on task space remote collaboration. *IEEE Transactions on Visualization and Computer Graphics* 22 (11): 2413–2422.

14 Huang, W. and Alem, L. (2013). Gesturing in the air: supporting full mobility in remote collaboration on physical tasks. *Journal of Universal Computer Science* 19 (8): 1158–1174.

15 Kim, S., Lee, G.A., and Sakata, N. (2013). Comparing pointing and drawing for remote collaboration. *2013 IEEE International Symposium on Mixed and Augmented Reality (ISMAR)*, Adelaide, Australia (01–04 October 2013). IEEE, pp. 1–6.

16 Madeira, T., Marques, B., Alves, J., et al. (2021). Exploring annotations and hand tracking in augmented reality for remote collaboration. *Human Systems Engineering and Design III: Proceedings of the 3rd International Conference on Human Systems Engineering and Design (IHSED2020): Future Trends and Applications, September 22-24, 2020, Juraj Dobrila University of Pula, Croatia 3*. Springer International Publishing, pp. 83–89.

17 Microsoft (2024) Microsoft Remote Assist. Accessed February 18, 2024. Website: https://dynamics.microsoft.com/en-au/mixed-reality/remote-assist/

18 Kuzuoka, H. (1992). Spatial workspace collaboration: a SharedView video support system for remote collaboration capability. *Proceedings of the SIGCHI Conference on Human Factors in Computing Systems,* Monterey California USA (03–07 May 1992), pp. 533–540.

19 Matthews, M.R., Cameron, K.H., Heatley, D., and Garner, P. (1993). Telepresence and the CamNet remote expert system. *Proceedings of Primary Health Care Specialist Group.* British Computer Society, pp. 12–15.

20 Aristotle (1952). Rhetoric. In: *The Works of Aristotle,* vol. II (ed. R.M. Hutchins), 593–675. Chicago: Encyclopedia Britannica Inc.

21 Kumar, K.J. (2005). *Mass Communication in India,* 3e. Bombay: Jaico.

22 Green, S.A., Billinghurst, M., Chen, X., and Chase, J.G. (2008). Human-robot collaboration: a literature review and augmented reality approach in design. *International Journal of Advanced Robotic Systems* 5 (1): 1.

23 Ishii, H., Kobayashi, M., and Arita, K. (1994). Iterative design of seamless collaboration media. *Communications of the ACM* 37 (8): 83–97.

24 Shannon, C.E. and Weaver, W. (1949). *The Mathematical Theory of Communication.* Urbana, Illinois: University of Illinois Press.

25 Berlo, D.K. (1960). *The Process of Communication* (ed. Bernieri). New York: Holt, Rinehart & Winston.

26 Schramm, W. (1954). *The Process and Effects of Mass Communication.* University of Illinois Press.

27 Kincaid, D.L. (1980). The Convergence Model of Communication. Papers of the East-West Communication Institute, No. 18.

28 Barnlund, D.C. (1970). A transactional model of communication. In: *Language Behavior: A Book of Readings in Communication* (ed. J. Akin, A. Goldberg, G. Myers, and J. Stewart), 43–61. The Hague: Mouton & Co.

29 Gunawardena, C.N. and Zittle, F.J. (1997). Social presence as a predictor of satisfaction within a computer-mediated conferencing environment. *American Journal of Distance Education* 11 (3): 8–26.

30 Short, J., Williams, E., and Christie, B. (1976). *The Social Psychology of Telecommunications.* Toronto; London; New York: Wiley.

31 Daft, R.L. and Lengel, R.H. (1986). Organizational information requirements, media richness and structural design. *Management Science* 32 (5): 554–571.

32 Suh, K.S. (1999). Impact of communication medium on task performance and satisfaction: an examination of media-richness theory. *Information & Management* 35 (5): 295–312.

33 Lowry, P.B., Romano, N.C., Jenkins, J.L., and Guthrie, R.W. (2009). The CMC interactivity model: how interactivity enhances communication quality and process satisfaction in lean-media groups. *Journal of Management Information Systems* 26 (1): 155–196.

34 Daft, R.L. and Lengel, R.H. (1983). *Information Richness. A New Approach to Managerial Behavior and Organization Design*. Texas A and M Univ College Station Coll of Business Administration.

35 Whittaker, S. (2003). Theories and methods in mediated communication: Steve Whittaker. In: *Handbook of Discourse Processes*, 246–289. Routledge.

36 Buxton, W.A., Sellen, A.J., and Sheasby, M.C. (1997). Interfaces for multiparty videoconferences. *Video-Mediated Communication* 385–400.

37 Daly-Jones, O., Monk, A., and Watts, L. (1998). Some advantages of video conferencing over high-quality audio conferencing: fluency and awareness of attentional focus. *International Journal of Human-Computer Studies* 49 (1): 21–58.

38 O'Conaill, B., Whittaker, S., and Wilbur, S. (1993). Conversations over video conferences: an evaluation of the spoken aspects of video-mediated communication. *Human-Computer Interaction* 8 (4): 389–428.

39 Jackson, M., Anderson, A.H., McEwan, R., and Mullin, J. (2000). Impact of video frame rate on communicative behaviour in two and four party groups. *Proceedings of the 2000 ACM Conference on Computer Supported Cooperative Work*, Philadelphia, PA, USA (2–6 December 2000). pp. 11–20.

40 Doherty-Sneddon, G., Anderson, A., O'malley, C. et al. (1997). Face-to-face and video-mediated communication: a comparison of dialogue structure and task performance. *Journal of Experimental Psychology: Applied* 3 (2): 105.

41 Kraut, R.E., Miller, M.D., and Siegel, J. (1996). Collaboration in performance of physical tasks: effects on outcomes and communication. *Proceedings of the 1996 ACM Conference on Computer Supported Cooperative Work*, Boston Massachusetts USA (16–20 November 1996). pp. 57–66.

42 Fussell, S.R., Kraut, R.E., and Siegel, J. (2000). Coordination of communication: effects of shared visual context on collaborative work. *Proceedings of the 2000 ACM Conference on Computer Supported Cooperative Work*, Philadelphia, PA, USA (2–6 December 2000). pp. 21–30.

43 Billinghurst, M. and Kato, H. (2000). Out and about—real world teleconferencing. *BT Technology Journal* 18 (1): 80–82.

44 Prince, S., Cheok, A.D., Farbiz, F., et al. (2002). 3D live: Real time captured content for mixed reality. *Proceedings. International Symposium on Mixed and Augmented Reality*, Darmstadt, Germany (30 September–01 October 2002). IEEE, pp. 7–317.

45 Orts-Escolano, S., Rhemann, C., Fanello, S., et al. (2016). Holoportation: Virtual 3d teleportation in real-time. *Proceedings of the 29th Annual Symposium on User Interface Software and Technology*, Tokyo, Japan (16–19 October), pp. 741–754.

46 de Souza Almeida, I., Oikawa, M.A., Carres, J.P., et al. (2012). AR-based video-mediated communication: a social presence enhancing experience. *2012 14th Symposium on Virtual and Augmented Reality*, Rio de Janiero, Brazil (28–31 May 2012). IEEE, pp. 125–130.

47 Barakonyi, I., Frieb, W., and Schmalstieg, D. (2003). Augmented reality videoconferencing for collaborative work. *Proceedings of the 2nd Hungarian Conference on Computer Graphics and Geometry*.

48 Kurata, T., Sakata, N., Kourogi, M., et al. (2004). Remote collaboration using a shoulder-worn active camera/laser. *Eighth International Symposium on Wearable Computers* (Vol. 1. IEEE, pp. 62–69).

49 Tait, M., Tsai, T., Sakata, N., et al. (2013). A projected augmented reality system for remote collaboration. *2013 IEEE International Symposium on Mixed and Augmented Reality (ISMAR)*, Adelaide, Australia (01–04 October 2013). IEEE, pp. 1–6.

50 Johnson, S., Gibson, M., and Mutlu, B. (2015). Handheld or handsfree? Remote collaboration via lightweight head-mounted displays and handheld devices. *Proceedings of the 18th ACM Conference on Computer Supported Cooperative Work & Social Computing*, Vancouver, Canada (14–18 March 2015). pp. 1825–1836.

51 Billinghurst, M., Kato, H., Kiyokawa, K. et al. (2002). Experiments with face-to-face collaborative AR interfaces. *Virtual Reality* 6: 107–121.

52 Prytz, E., Nilsson, S., and Jönsson, A. (2010). The importance of eye-contact for collaboration in AR systems. *2010 IEEE International Symposium on Mixed and Augmented Reality*, Seoul South Korea (13–16 October 2010). IEEE, pp. 119–126).

53 Fussell, S.R., Setlock, L.D., Yang, J. et al. (2004). Gestures over video streams to support remote collaboration on physical tasks. *Human-Computer Interaction* 19 (3): 273–309.

54 Akkil, D. and Isokoski, P. (2018). Comparison of gaze and mouse pointers for video-based collaborative physical task. *Interacting with Computers* 30 (6): 524–542.

55 Aschenbrenner, D., Rojkov, M., Leutert, F., et al. (2018). Comparing different augmented reality support applications for cooperative repair of an industrial robot. *2018 IEEE International Symposium on Mixed and Augmented Reality Adjunct (ISMAR-Adjunct)*. IEEE, Munich, Germany (16–20 October 2018), pp. 69–74.

56 Johnson, J.G., Gasques, D., Sharkey, T., et al. (2021). Do you really need to know where "that" is? enhancing support for referencing in collaborative mixed reality environments. *Proceedings of the 2021 CHI Conference on Human Factors in Computing Systems*, Yokohama Japan (08–13 May 2021). pp. 1–14.

57 Ladwig, P., Dewitz, B., Preu, H., and Säger, M. (2019). Remote guidance for machine maintenance supported by physical leds and virtual reality. *Proceedings of Mensch und Computer 2019*, Hamburg Germany (08–11 September 2019), pp. 255–262).

58 Yang, J., Sasikumar, P., Bai, H. et al. (2020). The effects of spatial auditory and visual cues on mixed reality remote collaboration. *Journal on Multimodal User Interfaces* 14: 337–352.

59 Chastine, J.W., Nagel, K., Zhu, Y., and Yearsovich, L. (2007). Understanding the design space of referencing in collaborative augmented reality environments. *Proceedings of Graphics Interface*, Montréal, Québec, Canada (28–30 May 2007). pp. 207–214.

60 Kim, S., Lee, G., Sakata, N., and Billinghurst, M. (2014). Improving co-presence with augmented visual communication cues for sharing experience through video conference. *2014 IEEE International Symposium on Mixed and Augmented Reality (ISMAR)*, Munich, Germany (10–14 September 2014). IEEE (pp. 83–92).

61 Piumsomboon, T., Dey, A., Ens, B. et al. (2019). The effects of sharing awareness cues in collaborative mixed reality. *Frontiers in Robotics and AI* 6: 5.

62 Yoon, B., Kim, H.I., Lee, G.A., et al. (2019). The effect of avatar appearance on social presence in an augmented reality remote collaboration. *2019 IEEE Conference on Virtual Reality and 3D User Interfaces (VR)*, Osaka, Japan (23–27 March 2019). IEEE, pp. 547–556.

63 Hauber, J., Regenbrecht, H., Billinghurst, M., and Cockburn, A. (2006). Spatiality in videoconferencing: trade-offs between efficiency and social presence. *Proceedings of the 2006 20th Anniversary Conference on Computer Supported Cooperative Work*, Banff Alberta Canada (4–8 November 2006). pp. 413–422.

64 Bernal, G., Hidalgo, N., Russomanno, C. and Maes, P. (2022). Galea: a physiological sensing system for behavioral research in virtual environments. *2022 IEEE Conference on Virtual Reality and 3D User Interfaces (VR)*. IEEE, Christchurch, New Zealand (12–16 March 2022), pp. 66–76.

65 Gumilar, I., Sareen, E., Bell, R. et al. (2021). A comparative study on inter-brain synchrony in real and virtual environments using hyperscanning. *Computers & Graphics* 94: 62–75.

3

Communication Cues in Augmented Remote Collaboration

3.1 Introduction

The means of communication have shifted to remote mode worldwide since the outbreak of the pandemic. As shown in Google Trends of the topic *remote* and search term *work from home*, both reached their peak in interest in March 2020, which is in line with the global lockdown time point and remains at a higher level of interest than ever before (Figure 3.1). From a technical perspective, *"the current pandemic situation is going to push users and developers of remote collaboration technologies to find out what works and what does not, and what solutions really need to be delivered"* [1]. The shift to remote collaboration comes with significant benefits and big challenges in how people communicate remotely.

It is well known that emails, phone calls, and videos are proved to be reliable telecommuting methods for remote collaboration among distributed teams and participants, where people communicate by verbal and visual information in real-time synchronous or asynchronous ways. Recent augmented/virtual/mixed reality (AR/VR/MR) technologies have led us to an exciting immersive remote collaborative world. Virtual meetings enable the sharing of voice, images, screen, and streamed videos with augmented markup [2]. In this chapter, we use the term *augmented remote collaboration* to describe remote collaborations in which AR/VR/MR technologies are applied. Typical application scenarios of augmented remote collaboration include, but are not limited to, remote maintenance of large machines or control rooms, crime scene forensics, emergency response, remote acting performance [3–5], remote medical, healthcare and diagnostic

Computer-Supported Collaboration: Theory and Practice, First Edition.
Weidong Huang, Mark Billinghurst, Leila Alem, Chun Xiao, and Troels Rasmussen.
© 2024 The Institute of Electrical and Electronics Engineers, Inc. Published 2024 by John Wiley & Sons, Inc.

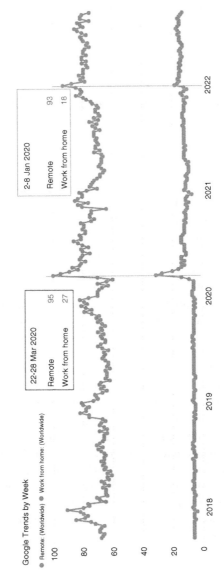

Figure 3.1 Google trends worldwide of the topic "remote" and search term "work from home" during the past five years.

services [6–9], training courses including construction equipment hazard detection and operator training [10–12], virtual patient-based medical and surgery training [13], and critical risk management and training in oil and gas industry [14]. In these scenarios, participants communicate not only using fundamental audio and text messages, but also with augmented visual and haptic information. It is generally agreed that more and more tasks in our work and life can be and will be conducted through augmented remote collaboration. Meanwhile, the remote communication mode has been compared with in-person communication more often due to people's increased participation of remote collaborations. In order to complete the task successfully, verbal and nonverbal communication cues have been employed in remote collaboration to establish common ground, exchange ideas, and share experiences.

Before moving forward, we would like to clarify that we will mainly discuss human–human remote collaboration in this chapter, although human-robot collaboration could be mentioned as references. In a remote collaboration scenario, how participants share what they see, what they do, what they want to do, and how they feel will determine different collaboration system setups, thus determine how people communicate [15]. This motivates us to survey the communication cues that can be and how to be employed in remote collaboration.

In this survey, we start with an overview of the research landscape over the past three decades, then investigate the communication context based on which a remote collaboration is conducted. We categorize communication cues in remote collaboration systems as verbal, visual, haptic, and empathic communication cues, and review the systems and experiments that studied these cues to identify advantages and limitations under different situations. Finally we discuss the challenges in multimodality communication modeling and system design for high usability, and suggest potential future research directions for augmented remote collaboration system design aiming at effectiveness, reliability, and ease of use.

3.2 The Research Landscape – Trends Over Time

We collected publications since the year of 1990 by searching the key term "*remote collaboration*" in the Scopus database. A normalization process has been applied to the key terms to solve nonalphabetic, typical initials, as well as singular and plural forms. Then, we analyzed the corpus of the collected 993 publications to review the research topics, the major contributed countries, as well as research organizations and researchers over time.

The network visualization of Figure 3.2 shows the main co-occurred key terms related to communication cues in the corpus, where the node size represents the occurrence of the term, and the node darkness represents the average year of the publications for the term. More specifically, the light gray nodes represent much recent research topics, while those slightly darker but larger nodes represent research topics being studied more intensively in recent years, but over a relatively longer time period, such as remote collaboration. We list some interesting findings as follows:

- The term "remote collaboration" is closely related to "augmented reality," "virtual reality" and "mixed reality" technologies, where they co-occur with some most recent topics like "hand gesture," "eye gaze," "facial expression," and "shared experience."
- The bar chart indicates that the number of publications has significantly increased in the recent two decades, given the fact that hardware and technologies have advanced greatly during the time.
- Key terms related to augmented communication cues can be found as "non-verbal communication," "computer vision," "hand gesture," "gesture communication," "pointing," "video," "knowledge sharing," "shared experience," "avatars," "annotation," "whiteboard," and "eye gaze." To some extent, Figure 3.2 can be regarded as a semantic web of concepts in the research field. We can categorize the communication method being investigated in the research community as face-to-face and computer-mediated communication. Terms regarding communication cues can be clustered as verbal and visual cues, and visual cues consist of annotation, gesture (especially hand gesture), sketch (which is semantically related to whiteboard), eye gaze, and physiological cues for immersive remote collaboration (which are semantically related to shared experience). The terminologies shown in Figure 3.2 outline issues and concepts which will be discussed in this chapter.

Figure 3.3 shows the countries based on the weighted author affiliation counts. The United States published the most publications, followed by Japan, Australia, and Germany. Figure 3.4 presents the organizations as well as the author contribution. Universities, research institutes, and Microsoft are the major research organizations that contribute the most to the research community.

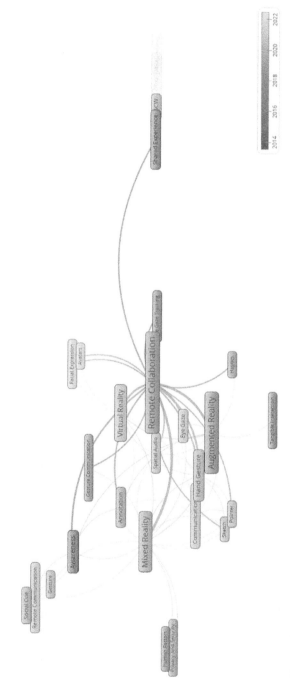

Figure 3.2 Key terms related to communication cues in remote collaboration.

Figure 3.3 Weighted contribution by countries, each publication is weighted 1.

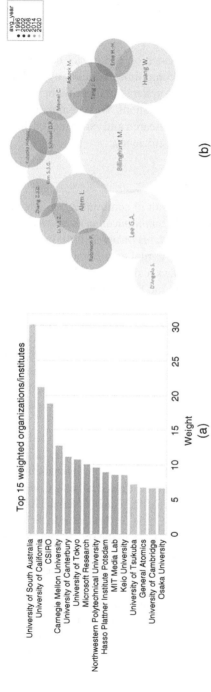

Figure 3.4 Top 15 weighted contribution by organizations and authors. Each publication is weighted 1, which is shared equally by all authors of the publication. (a) Sum of weights by organizations/institutes. (b) Contribution by authors, bubble color represents average publication year of an author, bubble size represents the sum of adjusted weights.

3.3 Communication in Augmented Remote Collaboration

In this section, we will survey research on context for communication and verbal, visual, haptic, and empathic communication cues applied and investigated in augmented remote collaboration systems to identify their usability, effectiveness, and limitation under different situations. This examination will shed light on the challenges for potential further research and experimental directions aimed at improving system performance and efficiency.

3.3.1 Context for Communication

A successful collaboration is the result of communication interactions based on common ground. "*The context of the communication interaction involves the setting, scene, and expectations of the individuals involved*" [16]. In remote collaboration systems with a local workspace or virtual environment, a shared visual view of the workspace or environment plays a significant role in establishing common ground in addition to common ground built up via verbal conversations. In particular, participants in a remote collaboration system can share the scene of the workspace and environment, and further share their specific view of interest as visual awareness to conduct the collaboration, usually with conversations and a visual awareness indicator such as a pointer or annotation as a companion. The workspace scene can be shared in an independent or dependent view from the perspective of a local user, either on a 2D screen or 3D mixed reality environment [17]. Different perspectives of the shared view lead to different levels of communication effort to share the awareness.

The remote user of the mobile collaboration system reported in [18] can explore the local scene independently from the local user's current camera position, and can set spatial annotations in the scene, which are immediately shared with the local user in augmented reality to guide a physical task. Compared with collaborations under the condition that the live video is streamed by the local user's camera and shared with the remote user, or the live video with additional non-sticky markers created by the remote user on the screen and shared back with the local user, the collaboration mode with independent view from the remote user demonstrates overwhelming user preference, short task completion time and a high level of usability.

The CoVAR remote collaborative system developed by Piumsomboon et al. [19] enabled different dependent view modes. In the Miniature mode, the VR user is scaled down to a fraction of the size of the AR user, while in the God mode, the VR user is exaggerated bigger than the AR user, which leads to various dependent view awareness in different modes.

Some walk-through remote collaboration systems such as ShowMe [20] and Cai and Tanaka's MR collaboration system [21] share the dependent view from the local user wearing a head-mounted display with the remote user, natural gestures can be rendered in the systems to greatly improve the immersion and real-time communication during remote collaboration tasks like navigation or virtual shopping. This also applies to the physical work system with a restricted workspace such as HandsIn3D [22].

Tait and Billinghurst [23] studied the remote view independence in augmented collaborations. The study compared the remote collaboration by asking participants to perform an object placement task in conditions of video only, fixed view, freeze view, and independent view. Verbal communication is available in all conditions. Under the video-only condition, the remote user sees the scene of workspace from the local user's perspective as a live video. In the case of fixed view, the remote user sees a virtual 3D model of the scene and live video from the perspective of the local user, where the view is fixed to the position of the head-mounted headset worn by the local user, augmented annotations are available in addition to verbal communication. Both video only and fixed view are dependent view conditions. Under the semi-independent view condition of freeze view, the remote user can freeze or unfreeze their perspective, allowing them to continue observing an area of interest even if the local user looks somewhere else, based on the fixed view condition. In the case of full independent view, the remote user sees the 3D model of the scene from an independent camera he/she can control, a virtual view frustum representing the view of the local user is seen within this independent scene. Results suggested that the increased view of independence in a relatively big work environment leads to faster task completion, more user confidence, and less time spent on verbal communication during the task.

Investigations also indicate that with an independent view of workspace, additional visible communication cues can help share the specific awareness more effectively by reducing ambiguity. In a user study investigating whether additional eye gaze pointers could help in indicating the visual awareness in a remote cowriting task, Kütt et al. [24] found that"*shared gaze improved task correctness and led collaborators to look at and talk more about shared content*," and that the shared view on the workspace can be regarded as independent to each other. In a physical collaboration scenario, the visible hand pointer is necessary in an independent shared view, but it does not help much in a dependent shared view where the hand gesture can be fully understood [25].

The studies and investigations suggest that both the shared scene of the workspace or environment and the view dependence are important factors that construct the common ground for a remote collaboration. The view dependence over a shared scene determines how much context information has been shared, thus leading to different communication effort.

3.3.2 Verbal Communication Cues

Most remote collaboration systems and interfaces support verbal communication. Conversations can be usually regarded as the fundamental communication cues in remote collaboration if applicable, where participants communicate with each other by live talking or recorded audio.

It is commonly believed that the remote collaboration benefits from a video conferencing interface where participants can see each other [26]; however, in the study of Reinhardt et al. [27], who investigated how a visual or invisible embodied agent for a smart speech assistant helps, it is suggested that a realistic embodied agent is preferred as it provides additional communication cues such as eye contact and gaze during conversation. But if visual attention other than the speaker is required, e.g. when being mobile or in a multitask situation, visual embodied agent is not necessary.

In remote collaboration with both verbal and visual cues, some research found that participants failed to respond to auditory stimuli more often than they failed to respond to the visual stimuli, which is known as the Colavita visual dominance effect. Further, the study of Fort et al. [28] revealed that humans react more rapidly to the combination of auditory or visual information compared with either one alone. An early study by Bauer et al. [29] compared verbal-only communication with verbal plus a telepointer indicated that a visual telepointer improves the remote collaboration based on verbal-only communication. The investigation of Kütt et al. [24] recently proved that again. They compared voice-based communication with or without gaze visualization in remote collaboration, where the participants shared the same content on their individual screen. Voice-based communication without gaze visualization resulted in more task completion time and cognitive workload. However, it is argued that the collaboration performance may be determined not only by the communication cues available but by other factors such as the expertise of participants and whether the participants trust their collaborators [30], or behaviors of participants. Participants constantly describing their thoughts or actions, and/or consistently initiating conversations even in few phrases to keep the feeling of copresence alive and to exchange information, can collaborate much more effectively with their remote collaborator, as observed in the study of Piumsomboon et al. [31].

In case verbal communication is unavailable, or not effective enough to explain spatial information and delicate processing, especially when context information such as a map or common understanding is missing or incomplete, other communication cues instead or as supplements are needed. For example, Waldow and Fuhrmann [32] developed MASR-engine (mixed reality audio speech recognition) for people with hearing disabilities, where speech is recorded via the integrated microphone of the head-mounted display (HMD) and further processed to text, that is sent over network to be visualized on a remote client end. This enables deaf or hard-of-hearing people to participate in augmented remote collaborations. Audio information also helps spatial referencing in a large workspace. Yang et al. [33] applied spatial auditory cues for large workspace operations in remote-guided mode, which is reported very effective to help participants in real world to locate the site in a large space.

3.3.3 Visual Communication Cues

In this section, we will review typical visual communication cues that have been employed for remote collaboration in previous investigations and systems. These cues are augmented annotations, pointers, body gestures including hand gestures and head gestures, eye gaze, as well as virtual replica of objects.

3.3.3.1 Annotations

Annotations with text, symbol, sketch (drawing) are visual indicators to present information explicitly, which additionally come with an advantage of cross-lingual representation if well-designed. In particular, it benefits users with poor audio conditions or deficiency. Augmented annotations are often used to share position and/or professional knowledge such as instructions of maintenance operations in remote collaborations or in remote education and training, where texts, symbols, and sketches are often used as different annotation forms. Annotations may apply to image, freeze image of video, live video, and virtual environment.

Kim et al. [34] compared the use of pointers and drawing annotations on still images and live videos. Their study suggested that drawing annotations require fewer inputs by creating and require less cognitive load for the recipient. Moreover, annotation-based remote collaboration is reported to be faster and with less mistakes than pointer-based remote collaboration [35].

In the *StickyLight* system developed by Adcock and Gunn [36], a remote user could point and draw annotations, and project the annotations on the real objects by the projector installed on a helmet worn by a local user or a tripod to guide and support physical tasks. User study showed that *sticky* annotations were preferred much more than *nonsticky* annotations, as annotations are dependent on the specific target and situation.

The remote user in the mobile remote collaboration prototype system developed by Gauglitz et al. [18] could set live spatial annotations in an independent camera scene, which were immediately shared with the local user in augmented reality to support collaboration. Their results suggested that the live video annotation-based remote collaboration mode was preferred over the other two compared modes of video only and video with static markers.

Augmented annotations drawn by mobile users of SoAR system can be shared with the other participants in field works, with the same point of view [11], which was reported as very helpful for a supervisor to monitor the status and give further instructions.

In the live 360 panorama-based MR remote collaboration prototype system developed by Teo et al. [37], users were able to draw augmented 3D annotations, and the hand-pointing gesture could be visualized as ray cast. Their results showed that participants acting as local hosts were able to perform tasks faster and precisely with the help of visual annotations than the visual hand pointer only.

In the remote collaboration system studied in the work of Madeira et al. [38], a remote expert could create instructive annotations based on the on-site content shared by the on-site professional and share back. Annotations based on real-time video stream from a remote expert were also available. Annotations could be adjusted, aligned, and re-scaled by tracked hand gestures to improve visualizations and enrich the on-site user experience.

The tangible chess game system CheckMate [39] annotated relevant tiles on an augmented chessboard as current place, possible next move, and possible attacks by different colors and symbols. It is a good practice in augmented gaming and training.

Augmented 3D annotations can be drawn with HoloLens for many application scenarios including environmental science exploring [40] or construction site hazard detection [12], where an on-site user wearing HoloLens shares the field of view with the remote user, and customized contextual information as well as drag-and-drop holographic annotations controlled by hand gestures enable effective communications between both end users.

The experiments and investigations above show that augmented annotations as important visual communication cues improve remote collaborations. However, specific annotations such as text or symbols need to be well designed to fit specific application scenarios, in many cases, annotations need to be explained or addressed verbally or by gestures.

The visibility of annotations is dependent on the specific context, therefore, they need to be adjusted or removed with a change in the context. For example, annotations can be manipulated by hand tracking, enabling the adjustment of their position and scale in the real-world according to the context, thus enriching the on-site professional experience and improving visualization of information [38]. The study of Kim et al. [34] indicated that drawing annotations need to be removed when a specific step of task is done.

3.3.3.2 Pointers

A pointer is a specific annotation that functions as indicator of visual awareness focus by highlighting the physical location of interest. It is usually accompanied with more verbal explanations or other communication cues than the text, symbol or drawing annotations [29, 34]. A pointer can be produced directly by an input device such as a mouse, or by tracking body gestures such as hand, head or eye gaze, depending on different system setups, and visualized as symbols in augmented remote collaboration systems.

The study of Bauer et al. [29] compared the performance of remote collaboration under two conditions: verbal communication only and verbal communication in addition to a telepointer. It revealed that a telepointer significantly improves the remote collaboration based on verbal communication by showing the position of interest in a shared view streamed by a head-amounted camera, which leads to less extra referential speech effort.

Fussell et al. [41] compared the effectiveness of a cursor pointer in video, video-only, and side-by-side conditions for a physical task. Results showed that the cursor pointer was of value for referring to objects and locations in the work environment during remote collaboration, but that the pointer did not improve performance time over video-only condition. It suggests that cursor pointing is valuable for remote collaboration on physical tasks; however, additional gestural support will be required to make performance using video systems as good as performance working side-by-side.

Sakata et al. [42] developed a wearable active camera with laser pointer (WACL) as a human interface device for telecommunications to support remote instructions. The laser pointer on the wearable equipment could be controlled by a remote instructor to indicate objects and directions in real-world environment. In the follow-up study, the WACL was compared with a HMD and a head-mounted camera-based headset interface; results showed that the WACL is more user-friendly without a significant difference in task completion time compared to the head-mounted headset [43]. Augmented laser pointers were also applied successfully in automobile industries for spot welding inspection [44].

Yamamoto et al. [45] investigated the difference between visual hand gesture pointer with and without full-body avatar for an object placement remote collaboration task. The results suggested that a visual embodiment of the remote user can improve the performance of remote collaboration compared to the case of visible hand gesture pointers only, as the remote user's nonverbal cues like body motion, orientation, and location have a positive impact on the task completion.

Pejoska-Laajola et al. [11] presented the application of their social augmented reality app (SoAR) based on augmented video stream with pointing and drawing. Tested on selected field sites, including a facility maintenance service site, a construction site and a construction training center, the results suggested that augmented video calls not only improve remote collaboration by allowing participants to point at task objects and locations, but also potentially improve informal workplace learning within the group.

Pointers as instant visual awareness indicators play a very important role both in face-to-face and remote communications. As in face-to-face communication, a visible representation of pointer cannot apply, additional communication cues like conversations or body gestures are always necessary to share more details about the focus of attention. While in augmented remote collaboration, we can take the advantage of AR/VR/MR technologies to monitor and track a visible augmented pointer, which makes the pointer very informative and helpful. However, the visibility of a pointer may need to be designed well according to different shared view and other communication cues available. For example, a mouse pointer works with less ambiguity than an eye gaze pointer for pointing purpose if available [46–49], due to the versatile functions of eye gaze that may cause confusion. Another example is, that a visible hand gesture pointer destination is not necessary in a dependent view, as discussed in Section 3.3.1. Overall, augmented pointers greatly improve the remote collaboration performance and the user experience.

3.3.3.3 Body Gestures

Body gestures can be interpreted as a kind of sign language that co-occur very often with spoken language, these nonverbal signals make up a huge part of our daily communication. The term *"gestures"* is often used as a representation of body language, if described without prejudice or evaluative implications, can refer to visible body actions [50]. While some of the interaction behaviors like shaking hands can we not use in remote mode the same way as in face-to-face communication, we still use a lot of body language such as gestures of hand, head, arm, and leg, and open or close postures in remote collaboration for purposes like pointing, object operation, or emotion expression and social interaction. Pavlocvic et al. [51] categorized gestures into intentional and unintentional movements from the behavioral and cognitive psychological perspective, where intentional movements stand for body movements conducting physical operations or conveying meaningful information for communication purpose, whereas unintentional movements do not. We will focus on the studies of intentional hand and head gestures in this section.

Hand Gesture The functionality of hand gestures for communication purpose has been studied for a long time. In early video-mediated remote collaborations, Fussell et al. [52] categorized hand gesture functions for physical tasks as (1) iconic gestures describing object features, (2) kinetic/motion for object manipulation such as grabbing and pinching, (3) deictic pointing gestures, and (4) spatial/distance gestures. Kirk et al. [53] defined a corpus of hand gestures to promote awareness during collaboration, where flashing hands stand for establishing reciprocity of perspectives, mimicking hands indicating how physical pieces should be joined together, and so on. Hand gestures can be recognized and interpreted in a much explicit way in augmented remote collaboration systems and interfaces with the advantage that the mid-air gestures can be visualized and embodied, the target or direction can be rendered in the shared view, and users can point to as well as manipulate the virtual objects for demonstration purpose via gestures, all bring benefits to improve the communication efficiency and user experience.

Research on video-mediated gestures has been a main topic over the decades. Recently Li et al. [54] reviewed the technologies of hand gesture interfaces in VR. It can be seen that great progress has been made for hand gesture tracking and recognition in terms of hardware devices and software during past years. In this section, we will focus on the evolution of hand gesture embodiment in remote collaboration systems, as embodiment is the visual representative during communication. In addition, the manipulative hand gestures, as well as the combination of eye gaze and hand gestures that are used to support communication, will be also discussed.

Hand gestures have been embodied in a variety of forms such as shadows on whiteboard, robot with laser pointer, digital arm shadows, Kinect arms, 2D representation overlaid in the shared 2D view, or 3D model in virtual reality to support remote collaboration. Table 3.1 is a summary of typical forms of hand gesture embodiment in remote collaboration systems.

In early 1990s, Tang and Minneman developed a video-based VideoDraw prototype to support remote collaborative drawing activity, which conveyed concurrent hand gestures by full-colored video in remote mode [55]. Later, their VideoWhiteboard prototype conveyed hand gestures by shadows on the whiteboard [56].

Kuzuoka et al. developed video-based remote collaboration robots GestureCam and GestureMan that enabled a remote instructor to guide and monitor a local operator conducting the task by controlling the robot [57, 58]. The robot functions as a physical embodiment of the remote user who can use non-verbal expressions, see and show gestures, arrange its body, tools, and gestural expression sequentially and interactively. Despite the constrains of hardware and networking at the time, their studies revealed significant factors essential for a successful remote collaboration, which are still applicable nowadays. The essentials include a shared independent view of workspace as well as a shared dependent view, in addition to the visual awareness addressed by a laser pointer (could be augmented pointer or annotations nowadays), and the sharing of sympathy between/among participants.

Tang et al. [59] studied mixed presence groupware (MPG) remote collaboration systems in which both colocated and distributed participants work over a shared visual workspace. To solve the weak presence problem of distributed participants in a conventional groupware remote collaboration system, they developed MPGSketch system by determining virtual seating positions of participants and replacing conventional telepointers with digital arm shadows that extend from a

Table 3.1 Embodiment of hand gestures in remote collaboration systems.

System	Reference	Embodiment	Application scenario	Year
VideoWhiteboar	[56]	Shadow on whiteboard	Paired remote guidance	1991
GestureCam/	[57, 58]	Robot with camera and laser pointer	Paired remote guidance	1994, 2000
MPGSketch	[59]	Digit arm shadow in teleconferencing	Groupware teleconferencing	2005
Remote lag	[60]	Kinetic arm traces	Groupware teleconferencing	2011
KineticArm	[61]	Kinetic Arms	Groupware teleconferencing	2013
HandsInAir	[62]	2D hand gestures	Paired remote guidance	2013
ShowMe	[20]	3D hand gestures in VR/MR	Paired remote guidance	2015
HandIn3D	[22]	3D mid-air gestures in MR	Paired remote guidance	2017
SharedSphere	[17, 63]	3D mid-air gestures in VR/MR	Paired remote guidance	2017
Go together	[21]	3D mid-air gestures in VR/MR	Paired teleconferencing	2019

participant's side of the table to their pointed location. The digital arm shadows functioning as participant embodiments indicate virtual seating positions, convey person-specific orientations, and increase awareness of actions, which benefit the collaboration.

Yamashita et al. [60] investigated three problems in a group collaboration concerning the invisibility of remote gestures, which are occlusion, diverted attention, and the supposition of visibility. They proposed a technique called "remote lag" that visualizes the motion history of an arm embodiment in the shared workspace, which enables participants with instant playback of remote gestures to recover from the missed context of coordination. Their experimental results showed that remote lags effectively alleviated the invisibility problems and decreased the overall workload of workers. Genest et al. [61] developed KinectArms toolkit that helped users build arm embodiments representing rich and subtle gestures in distributed tabletop groupware. KinectArms consists of a capture module and a display module. The capture module captures images of hands and arms above the surface from video, and the display module visualizes hands and arms with built-in effects to identify them such as adding color and tattoo, and to show their height from surface by adding shadows or circle indicators, and to show hand and arm motion traces. The toolkit can greatly improve the expressiveness and usability of distributed tabletop groupware collaboration.

When hand gestures captured in a video-mediated remote collaboration system are displayed on a 2D screen, the spatial relationships between objects in the remote workspace cannot be understood well as in the real world. Amores et al. [20] developed ShowMe mobile collaboration system by using a head-mounted display and depth camera to share free hand 3D mesh gestures from the local person's view, which enables both users to present in the same physical environment and perceive real-time communication. Huang et al. developed HandsIn3D in a mixed reality application scenario using head-tracking, stereoscopic rendering, inter-occlusion handling and virtual shadowing for immersive remote collaboration [22]. Compared to their previous 2D system HandsInAir [62], HandsIn3D prototype system integrates the projection of instructive virtual hands into the 3D workspace from the local user's view point, enhances the user's ability to perceive spatial information and offers greater sense of copresence and interaction. Sharing natural hand gestures to each other was also supported in the mixed reality remote collaboration system Shared-Sphere [17, 63]. This system tracked user's hand motion with a hand tracking sensor, and showed the correspondent virtual representation in the other's VR or see-through HMD. Similarly, Cai and Tanaka reported that the participants could share video, audio, and 3D hand gestures in a mixed reality system to experience "go together" as in colocated collaboration [21]. The systems demonstrates flexibility in performing natural hand gestures in remote collaborations.

And it also suggests that natural hand gestures are also a favorable and effective communication method in remote collaboration systems.

Zhang et al. [64] developed a framework for hand gesture recognition based on the information fusion of a three-axis accelerometer (ACC) and multichannel electromyography (EMG) sensors. Their decision-tree and Markov model based classifier could interpret different hand gestures, as operation commands with over 90% accuracy to solve a virtual Rubik's cube game. In the MR collaboration system of Feick et al. [65], rendered hand gestures were specifically investigated in remote helper's view to operate virtual objects like flipping and rotating a virtual object along an arbitrary axis. Their studies and experiments suggested that hand gestures enabled remote collaborators to communicate with natural gestures, and to demonstrate very complicated 3D operations that require a lot of effort to be explained verbally. This is a very interesting practice presenting the potentials to combine gesture recognition and virtual object manipulation in AR/MR/VR-based collaboration. Compared with verbal and eye gaze communication, hand gestures demonstrate the capability to convey delicate and complicated information in an effective visual-based way, as in face-to-face communication.

Combination of Eye Gaze and Hand Gestures The combination of eye gaze and hand gestures is often applied for object manipulation in remote collaboration systems, where a target is selected by eye gaze and manipulated by hand gestures. Song et al. [46] developed an interface GaFinC for manipulating 3D models using eye gaze and hand gestures in a CAD system. Chatterjee et al. [47] compared the performance of three application scenarios under three conditions including gaze-alone, gesture-alone, and eye gaze with free-space hand gestures. Their experiment showed that the combination of eye gaze and hand gestures as input modalities led to rapid, precise, and expressive interaction comparable to the performance of "gold standard" input modalities such as mouse and track pad. Later, Deng et al. [66] investigated spatial mapping problems between the virtual space and the physical space when applying both mid-air hand gestures and eye gaze to select and place virtual 3D objects in virtual environment, where eye gaze functions as a pointer to select a virtual object and a location, and mid-air hand gestures work as operation commands to grab and release the selected virtual object to the designated location. Similarly, the CoVAR system [19] also combined eye gaze to select and hand gestures to manipulate an object in virtual reality. The comparison of eye gaze and hand gestures in a mixed reality remote collaboration of Bai et al. [3] also suggested that sharing both eye gaze and gestures from the remote user to the local can significantly reduce the task completion time and strengthen the copresence, and that hand gestures improve the system's usability and reduce the mental workload for the local workers than the eye gaze alone.

Head Gestures In some remote collaboration systems, head pose is tracked as an augmented pointing gesture [67–69], like hand and eye gaze pointers. When head pose functions as a pointing gesture, it can be regarded as a proxy of hand gesture or eye gaze. Head pose can also be tracked to control the rotation of a virtual camera, thus rotating the current point of view in VR systems [70]. Head and camera rotations are also tracked to detect precise eye-gaze direction in computer user interface systems, sports science, psychology, and biometrics as well. Moreover, as a part of body gestures, head gestures like nodding and shaking are commonly recognized as conveying emotional or social interaction information in addition to carrying linguistic content [71]. Terven et al. [72] developed a system based on Hidden Markov Models to recognize six head gestures of nodding, shaking, turning right, turning left, looking up, and looking down, which presented robustness to be integrated into a wearable device to improve the presence of ego-motion. When head gestures are embodied to avatars in virtual environments, it will significantly improve participants' identifiable self-embodiment and the social presence.

3.3.3.4 Eye Gaze
Intentional eye gaze can be regarded not only as a fast nonverbal and pointing gesture, or an indicator of focus of attention, or an operational gesture but also as means of social interaction [73].

Investigations suggest that eye gaze is a fast input modality which can significantly improve the collaboration by reducing task completion time [74] when compared to other modalities like hand gestures, since people response after seeing [66, 75]. Intentional eye gaze is also a behavior indicator to reveal interaction mechanism, which functions as an intention indicator to predict actions. As reported in some studies, participants can correctly predict other's intention and conduct the operation before conversation during on-site and remote collaborations [48]. Markov models [76] and support vector machine (SVM) [77] have been applied to predict intention of actions in collaborative works, and reported quite effective. These studies reveal that eye gaze information is a very important input modality supplementing other body gestures and speech, as well as a predictive indicator. In addition to the functionality as a pointing gesture and intention indicator, Piumsomboon et al. [78] indicated that eye gaze together with head movements can be used as operational gestures. In more details, by stabilizing eye gaze on a target which is called vestibulo-ocular reflex, the nodding, and rolling head movements can be interpreted as manipulation gestures of the target. As people's gazing practices and the visibility of these to others are important resources in social interaction, it is important to reproduce eye gaze in AR/VR/MR to improve immersive collaboration experiences [79–81].

With much more affordable eye-tracking hardware today than years ago, as well as the trend of computer vision-based eye-trackers instead of wearable equipment, eye-tracking technology is expected to be applied more often in the near future.

Eye gaze pointer has been most commonly visualized as a gaze cursor on a 2D scene screen or display. In addition, eye gaze hotspots [82], heatmaps [83, 84] or zoom focus [85] representing focus of attention, gaze trail [83, 85] demonstrating eye gaze travel history may also apply. In virtual environments eye gaze can be visualized as eye ray cast from avatars [81], and in real world it can be projected in the physical workspace [48]. It is also noted that some remote collaboration systems prefer visualizing eye movement instead of the eye gaze pointer such as the ThirdEye system [86] to guide the attention in videoconferencing. Visualization of eye gaze is supposed to strengthen mutual communication by sharing the visual awareness. However, an augmented collaboration system may suffer from visible pure-gaze input with many distractions and confusions, which has been noticed as *"Midas touch"* problem [87], due the lack of a natural input delimiter [47]. As a practice to alleviate the ambiguities, another input modality such as input from a mouse, keyboard, or touch device, or natural input from voice, body movements, or even facial expression, can be used as a trigger to visualize intentional eye gaze.

3.3.3.5 Virtual Replica of Objects

In object-focused remote collaborations, a virtual replica of the object of interest and manipulation of the virtual object will significantly improve the collaboration effectiveness and user experience.

The remote collaboration system of Yang et al. [88] virtualized the real world object located at the local side using computer vision, which was rendered in mixed reality, so that the remote user could manipulate the virtual object such as translating, rotating with or without hand adjustment, or interacting with a deformable object. Guided by the demo from the remote site, the local user could finalize the real object operations under further remote instructions when necessary. Similarly, in the POINT3D and DEMO3D systems presented by Oda et al. [89, 90], a remote expert could create virtual replica of physical objects in a virtual environment, with which the remote expert could share the professional knowledge and experience with a local user by pointing and annotating instructions to specific part of the virtual replica, as well as manipulating the virtual replica to demonstrate the physical operations. The results indicated that the virtual demonstration is very effective to improve the remote collaboration.

Instead of reproducing a virtual replica of physical object, Feick et al. [91] designed a remote manipulator (ReMa) to reproduce orientation manipulations on a physical proxy object at a remote site. With a physical proxy, participants used the conventional video chat systems to perform object manipulation more effectively. In addition, a shared perspective was reported more effective and preferred compared to the opposing perspective offered by the video chat systems.

Alternatively, a tangible remote collaboration system can not only share the workspace, participants could also remotely manipulate physical embodiment of objects in a video-mediated environment [92]. The physical embodiment and object manipulation of remote users at the local end enables the local user to experience the physical telepresence of the remote end, is very promising for collaborations on 3D models, remote assistance or communication.

The studies suggest that virtual demonstration significantly improves the remote collaboration despite the technical challenges including arbitrary object and workspace reconstruction, gesture tracking, effective spatial referencing, recognition and rendering of complicated object manipulation.

3.3.4 Haptic Communication Cues

In addition to visual and verbal communication, haptic communication cues are also employed in remote communication systems.

Gunn et al. [93] designed a remote haptic surgery training system by using a haptic virtual environment as an integral part of the learning process. The application offered haptic guiding hand function, with which the instructor can remotely grasp the student's tool to haptically guide it along with the student's hand to any point in the virtual scene. Both instructor and student can feel the pulling and resistance force from each other during the movement of the guiding hand. Both instructor and student can simultaneously push, grasp and stretch the body organs in the model, where the organs built up using a surface mesh are deformable, presenting diathermy, clipping, and cutting dust effect, according to different operations of the participants. User experience feedback of the application showed a 100% high sense of presence with their teacher and over 87% engagement with the scenario.

Yatani et al. [94] investigated the effects of visual and tactile feedback on spatial coordination in collaborative handheld systems. In visual feedback condition, visual annotations in different colors are triggered when one or both participants touch the same area, so that they can continue to collaborate to achieve good performance. In tactile feedback condition, periodical or continuous vibrations of the handheld device are triggered when one or both participants touch the same area. Their results showed that visual feedback can provide precise spatial information but may hamper collaboration when it is occluded and sometimes distracts the user's attention, and that spatial tactile feedback of the handheld device can reduce the cognitive workload as in visual feedback condition; however, improvements are necessary to convey precise spatial information. A combination of visual and spatial tactile feedback is a good solution due to the hardware limitation.

Recently, Rekimoto Lab [95] designed HapticPointer as a necklace-style device for haptic pointing of remote collaboration tasks.[1] The HapticPointer device has 16

---

1 https://lab.rekimoto.org/projects/hapticpointer.

vibration motors placed along a line of flexible string, which enables participants wearing it feel the position and intensity of each vibration to find a specific target. The system reported a 90.65% accuracy for successful trials and six seconds as average time spent to complete the search task. Their user study suggests that the device can simulate the sensation of walking together which is very promising to improve engagement between the local and remote users, and the portability of HapticPointer device also benefits participants conducting complicated tasks that need more visual or other forms of attention.

In the navigation study of Wang et al. [96] in virtual environments, all visible virtual targets are identified, and one of them, as the current virtual target is selected and mapped to its real prop. Then, the rotation and translation gains are computed based on the positions and the orientations of the current virtual target and its real prop, considering the avoidance of physical boundaries and obstacles. In case, there is no visible virtual target from the current viewpoint, the gains will be computed using a predicted direction. Finally, the walking path is redirected by the rotation and translation gains. Experiment results showed that passive haptic feedback built on this dynamic real-virtual target mapping method can enhance users' immersion in virtual environments, when compared to the conventional steer-to-center method path redirection method.

Wang et al. [97] conducted a within-subject user study to compare two remote collaboration interfaces between a local worker and a remote helper, one with mid-air free drawing and one with tangible physical drawing. The results showed no significant differences in task completion time and operation errors, however, users felt that the tangible interface supporting passive haptic feedback could significantly improve the presence and immersion of the remote helper in virtual reality.

The studies and investigations suggest that haptic collaboration systems improve user experience and immersion during collaboration, which also enable visually impaired users to be better involved. It is also seen that there is much more to research and test haptic remote communications due to the challenge to define the applicable physical sense of touch in untouchable remote conditions.

3.3.5 Empathic Communication Cues

The definition of the term *empathy* can be tracked back to the late eighteenth century, which referred to sympathy at the earliest time but has been extended to a number of related processes and cognitive capacities involved in empathic behaviors. Empathy has been studied in a number of fields of psychology, neuroscience and artificial intelligence [98]. In human-human collaboration, empathy is commonly referred to the ability to detect what others feel and build up an understanding of the emotion ourselves, which has always been regarded as an important physiological indicator reflecting group cohesion, self-esteem, and positive affect [99]. Communication competence, emotion regulation, and cognitive mechanisms are studied not only in social interactions but also in artificial intelligence for computational empathy modeling [98]. In line with this background, measuring, understanding, and sharing empathy based on the role of empathy are three crucial components in empathy computing[2] [78].

Hart et al. [100] proposed a concept and discussed technical issues of an emotion-sharing and augmentation prototype system, in which facial expression, eye gaze, voice with volume indicator, and physiological data such as heart rate could be collected and shared with remote collaborators by traditional videoconferencing or avatars in mixed reality. In addition, empathy regulation need to be considered to measure empathy level. Studies reveals that several factors such as valence, intensity, and saliency of emotions, social relations, context, as well as mood, personality, gender, age, and emotional repertoire of the virtual agent should be taken into account for empathy measurement in VR [101].

3.3.5.1 Facial Expression

Facial expression is the most significant component for empathy modeling, as facial expression has the richest communication competence in expressing emotions like happiness, sadness, anger, surprise, confusion, excitement, and many other implicit meanings. It has been studied in a few empathy computing models, as Yalçın and DiPaola summarized (see Table 3 in [98]). Technically, the Ekman and Friesen's facial action units and facial landmark positions are examined to detect emotional communication competence, in addition to other physiological cues such as blood pressure and respiratory rate [99], or gesture cues like eye gaze [102, 103] and head nod, or speech signals [104, 105]. The input data are collected and fed into empathy models to output real-time categorical empathy level or scaled empathy degree.

3.3.5.2 Eye Gaze and Body Gestures

Eye contact and specific body gestures are also significant empathic cues. Eye contact is a natural and important part of the communication process, in which people may observe other's eye movement patterns like whether they are making direct eye contact or averting their gaze, how much they are blinking, or if their pupils are dilated, as feedback for mutual communication and indicator for emotion expression [106]. Eye tracking technologies have been widely used in VR games to improve copresence and collaboration experience. However, it is a big challenge for empathic modeling, considering eye gaze as an input modality due

---

2 TED talk about empathic computing by Mark Billinghurst: https://youtube.com/watch?v=ybdcBrG0aPA.

to the versatility of its functionality. Compared to eye gaze, body gestures can be more explicit. For example, Kuzuoka et al. [57] controlled the GestureCam robot to take a Japanese bow and were surprised by the very positive response of the remote collaborator. The experiment suggested the importance of sharing sympathy in all way including body gestures to improve remote collaborations.

3.3.5.3 Paralinguistic Speech Signals

Paralinguistic speech signals as another source for empathy computing have been studied as well. Xiao et al. [107] investigated the paralinguistic speech characteristics including pitch, energy, jitter, shimmer, and utterance duration from the audio recordings of the psychotherapy interaction, to manually rank empathy levels according to motivational interviewing treatment integrity (MITI) system [108]. Based on the EmpatheticDialogues dataset containing 25k dialogues with empathic listener responses, Rashkin et al. [109] generated empathic responses to the utterances with a pretrained model. The data-driven approach is supposed to advance the empathic modeling and computing with more and more data sources available in the near future.

3.3.5.4 Physiological Cues

Physiological cues are also investigated to detect empathy level and their impact on interactions and collaborations, which include skin conductivity, electromyography, galvanic skin response, blood pressure, respiratory rate, heart rate, eye blinking, eye pupil size. Prendinger and Ishizuka [110] examined the physiological indicators in terms of skin conductivity and electromyography in a virtual job interview scenario, in order to detect the users' emotion and provide them empathic feedback to improve user experience. Emotional speech based on prosody changes, eye blinking and breathing frequencies in addition to facial expression was implemented to virtual avatars to improve human user's experience [111]. Tan et al. [99] investigated remote collaboration performance in terms of empathy levels under several conditions of voice only, video, or visual representation informed by physiological cues measured by galvanic skin response, blood pressure, and respiratory rate. Their user study suggested that compared to voice-only interactions, using visualization of physiological cues would significantly improve the empathy levels in terms of group cohesion and positive affect during remote collaboration, which was comparable to the usage of videoconference. Later, Australian and Japanese researchers developed Empathy Glasses [78, 112–114] for empathy computing in mixed reality. Their CoVAR system allows a local user wearing Empathy Glasses to share with a remote collaborator, their point of view, eye gaze information, and emotion detected from facial expressions, galvanic skin response, and heart rate. Shen et al. [115] used a Pepper robot to extract the nonverbal features from each human participant's habitual behavior using its on-board sensors, the collected information including head motion, gaze, and body motion energy, and voice pitch, voice energy, and mel-frequency cepstral coefficient were fed into ridge regression and linear support vector machine classifiers, where the samples were labeled by evaluating personality questionnaires, showed promising binary classification performance on recognizing each of the Big Five personality traits.

Natural collaboration, experience capture, and implicit understanding are suggested to be combined to enable empathy computing thus improve user experience of remote collaboration.

3.4 Challenges

3.4.1 Multimodality Communication Modeling

Fussell et al. [52] identified the benefits of introducing gestures in remote collaboration based on videos, and argued that more complex forms of representational gesture "*may facilitate conversational grounding in collaborative physical tasks by allowing speakers to communicate multiple pieces of information simultaneously.*" It has been noticed that co-occurrence of body gestures and spoken language is a significant characteristic of communication, and Goldin-Meadow and Brent [116] suggested in their study that speech and gestures should be studied at the same time. Li et al. [54] reviewed hand gesture interfaces and summarized that the gesture interaction alone lacks the full information to be convoyed by other modalities during the movement, further research on multimodality interactions is needed. As a matter of fact, Monge and Day suggested as early as in 1976 that communication research should introduce multivariate analysis based on multiple dependent variables [117].

Based on the fundamental verbal, text and video-mediated communications, the AR/VR/MR technologies have advanced remote collaboration by introducing a variety of natural communication cues such as haptic cues, augmented hand gestures and eye/head pose, as well as facial expressions and other physiological cues. As there are quite a few communication cues available with less cost than ever before, further investigation based on a multimodality communication model is necessary to identify what to be chosen in different situations to ensure a successful remote collaboration with as less cost as possible. Multimodality communication modeling is seen as a significant issue and challenge.

3.4.2 Ease of Use: Embrace New Technologies

Kirk et al. [118] argued one and half decades ago that "*understanding how best to deploy a technology is critical to its successful adoption and continued use.*" An example to illustrate the argument is the application of MR technology in a construction site safety inspection scenario recently, where MR technology was accepted by 1/3 of respondents vs. 17% not, due to *not easy to use* when compared with traditional communication channels like emails, phone calls, videoconferencing, and on-site walking through and talk [12].

Ease of use is one of the key evaluation standard that determines the acceptance of new technologies. Individual differences regarding age, education, experience, cultural background, and many other issues are some of the factors that determine how people accept new technologies. Despite the human factors, the concept of ease of use consists of user evaluation metrics such as less difficulties with technology, usability, portability, flexibility, physical comfort with the devices, less dependency on devices without losing reliability.

In earlier AR/VR/MR practices, users were required to wear a relatively heavy headset or equipped with sensors, as well as get trained to work in virtual environment, which seemed to add extra cognitive load to users in cases like remote collaborated field works, dance classes, or surgeries. Wille et al. [119] reported in year 2014 participants wearing HMDs for a 3.5-hour manufacturing task took about 20% more task execution time than with the Tablet-PC, 10% more time with the HMD in see-through mode and 18% more time with the HMD in look-around mode compared to the wall-mounted monitor. Responses of questionnaire about visual fatigue indicated prolonged work wearing HMDs are more likely to cause *heavy eyes, neck pain,* or *headache* due to the weight and discomfort of the HMD head carrier. In Adcock and Gunn's [36] experiment where users were required to wear a helmet with a projector, or they could choose a projector fixed on a tripod, participants showed a preference on the tripod that could set their hands free and without equipment stability issues caused by the users' postures, even though the quantitative results did not indicate an improvement. Similar findings were reported in the investigations [18, 44] as well. Moreover, Pejoska et al. [11] revealed that the usability of their remote collaboration AR app was restricted by many external conditions including building noise, screen visibility in the sunlight, rain, low temperatures, protection-ware, and mobile connectivity problems, such that the app could only provide participants with an alternative solution for assisting processes and monitoring status of field works, they concluded that *it was clear that an app would not replace face-to-face communication entirely.* Drogemuller et al. [120] also indicated that portability of the AR system plays an important role for a successful deployment.

To the highest standard of ease of use, new technologies will enable mutual communication between participants as in a face-to-face mode that demonstrates reliability, flexibility, and physical comfort under all conditions, with benefits from AR/VR/MR technologies. The previous investigations and experiences will provide guidelines for user-friendly remote collaboration system design, which are quite challenging, as the problems and limitations are to be studied not only from a technical but from a social science point of view.

3.5 Future Directions

3.5.1 Natural Communication-Based Remote Collaboration

Natural communication happens without any aid of technology or tools. People talk in different tones and loudness using the mouth, sign using the hands and other body parts. All gesturing, body posture, and movements, as well as facial expression, are natural.

In the future, the traditional user interfaces can be enriched and replaced by natural communication cues tracked and recognized by computer vision-based technologies [54], which enable remote participants communicate in less device-dependent and more natural ways. Deep learning has shown its capability to deliver high accurate gesture recognition based on computer vision. Chai et al. [121] recognized continuous hand gestures with a two streams Recurrent Neural Networks (2S-RNN) for the RGB-D data input as Continuous Gesture Dataset (ConGD). Tsironi et al. [122] employed the CNNLSTM model as a combination of convolutional neural networks and Long Short-Term Memory Recurrent Neural Network (CNNLSTM) to detect the most intense body motion as meaningful dynamic gestures, with over 90% recognition accuracy and F1 measurement. It is quite promising that the design of future remote collaboration systems will benefit from the technologies that keep moving forward.

Natural communication cues in augmented remote collaboration systems will be adopted with the advances of technologies to ensure and enhance the effectiveness and flexibility of remote collaboration, such as virtual training and education. In the study of Mikropoulos et al. [123], where AR/VR technology was applied in primary school teaching, the students presented much more interest in participation with natural hand gestures than a mouse despite the difficulty to control the geometric figure movements by hand gestures. Natural communication-based virtual training and education such as virtual sports education [4] and virtual patient-based medical/surgery training [13] will not only allow participants receiving didactic education, but also virtually recreate the "hands-on" training sessions with live and haptic feedback to acquire and improve competence without extra travel cost. It is very promising to conduct immersive virtual education as applicable, flexible, and reliable as in face-to-face mode, which will greatly benefit students distributed remotely. The state-of-the-art remote collaboration systems have demonstrated significant positive social-economic impact during the pandemic, and will continue to contribute in a long term thereafter.

3.5.2 Cloud-Based Remote Collaboration

As portability and mobility of tools are very important to some tasks like field works, where the observation of working environment is essential to conduct remote collaboration, augmented remote collaboration based on cloud computing is emerging as a prominent trend.

With the advances in AR technologies and cloud computing, the U.S. Army had already tested the integrated visual augmentation system (IVAS) based on Microsoft HoloLens and Azure cloud services in the past years and started to provide soldiers with IVAS headsets with a see-through display that shows real-time scene mapping, aided target acquisition, and identification without disrupting their field of view, and even with low-light and thermal sensors for night vision.[3] From a technical point of view, the cloud-based augmented remote collaboration technologies are believed to significantly enhance the power of army, survival chance of soldiers and the groupware remote collaboration quality.

3.5.3 Empathic Remote Collaboration

Recent advances in machine learning, robotics, and AR/VR/MR technologies also enable empathic remote collaborations in the future as well, promising to make remote collaboration systems smarter.

Smart remote collaboration systems can not only assist interpreting gesture meaning but also respond under different situations to improve remote collaboration. For instance, the system can switch to a dependent view if a user asks for help, so that the response from other remote collaborators can be more customized as in a face-to-face mode.

In addition, as explored earlier, quite a few state-of-the-art data-driven models for empathy computing are based on a single or a few modalities such as speech, text, and facial expression to train their models. The current progress in multi-modal emotion recognition techniques based on machine learning, coupled with the widespread adoption of remote AR/VR/MR-based collaboration environments, is likely to significantly advance research and applications in this area [98].

These developments indicate great potential to design smart and empathic remote collaboration systems which can help smoothing as much as possible the communication barriers and allow effective collaboration regardless of individuals' locations.

3.6 Conclusion

This chapter presents a survey of communication cues used in augmented remote collaboration. We discussed the communication context in remote collaboration systems, categorized communication cues as verbal, visual, haptic, and empathic

---

3 https://blogs.microsoft.com/blog/2021/03/31/army-moves-microsoft-hololens-based-headset-from-prototyping-to-production-phase.

cues, and reviewed the details of the state-of-the-art models and systems that applied these communication cues. We further identified and summarized the challenges and indicated the future research directions.

References

1 Inside Robotics (2020). COVID-19 social distancing will drive remote collaboration innovation.

2 Wolff, R., Roberts, D.J., Steed, A., and Otto, O. (2007). A review of telecollaboration technologies with respect to closely coupled collaboration. *International Journal of Computer Applications in Technology* 29 (1): 11–26.

3 Bai, H., Sasikumar, P., Yang, J., and Billinghurst, M. (2020). A user study on mixed reality remote collaboration with eye gaze and hand gesture sharing. In: *Proceedings of the 2020 CHI Conference on Human Factors in Computing Systems*, CHI '20, 1–13. New York, NY, USA: Association for Computing Machinery.

4 Schroeder, A.N. and Kruse, R.C. (2021). The future of virtual sports ultrasound education and collaboration. *Current Sports Medicine Reports* 20: 57–61.

5 Schwab, M., Saffo, D., Zhang, Y. et al. (2021). VisConnect: Distributed event synchronization for collaborative visualization. *IEEE Transactions on Visualization and Computer Graphics* 27 (2): 347–357.

6 Luxenburger, A., Prange, A., Moniri, M.M., and Sonntag, D. (2016). MedicalVR: Towards medical remote collaboration using virtual reality. In: *Proceedings of the 2016 ACM International Joint Conference on Pervasive and Ubiquitous Computing: Adjunct*, UbiComp '16, 321–324. New York, NY, USA: Association for Computing Machinery.

7 Jalaliniya, S. and Pederson, T. (2017). Qualitative study of surgeons using a wearable personal assistant in surgeries and ward rounds. In: *eHealth 360°*, Lecture Notes of the Institute for Computer Sciences, Social Informatics and Telecommunications Engineering, vol. 181 (ed. K. Giokas, L. Bokor, and F. Hopfgartner), 208–219. Cham: Springer.

8 Barros, R., Borst, J., Kleynenberg, S. et al. (2015). Remote collaboration, decision support, and on-demand medical image analysis for acute stroke care. In: *Proceedings of European Conference on Service-Oriented and Cloud Computing (ESOCC 2015)*, vol. 9306, 214–225. Cham: Springer.

9 Sohrabi, C., Mathew, G., Franchi, T. et al. (2021). Impact of the coronavirus (COVID-19) pandemic on scientific research and implications for clinical academic training –a review. *International Journal of Surgery* 86: 57–63.

10 Wang, X., Dunston, P., and Skibniewski, M. (2004). Mixed reality technology applications in construction equipment operator training. In: *Proceedings of the 21st International Symposium on Automation and Robotics in Construction (ISARC 2004)*, 21–25, International. The International Association for Automation and Robotics in Construction.

11 Pejoska-Laajola, J., Reponen, S., Virnes, M., and Leinonen, T. (2017). Mobile augmented communication for remote collaboration in a physical work context. *Australasian Journal of Educational Technology* 33: 11–26.

12 Dai, F., Olorunfemi, A., Peng, W. et al. (2021). Can mixed reality enhance safety communication on construction sites? An industry perspective. *Safety Science* 133: 105009

13 De Ponti, R., Marazzato, J., Maresca, A.M. et al. (2020). Pre-graduation medical training including virtual reality during COVID-19 pandemic: a report on students' perception. *BMC Medical Education* 20: 332.

14 Potts, J., Sookdeo, T., Westerheide, J., and Sharber, D. (2020). Enhanced augmented/mixed reality and process safety applications. In: *Proceedings of SPE International Conference and Exhibition on Health, Safety, Environment, and Sustainability 2020, HSE and Sustainability 2020*, vol. 86. Manama, Bahrain: Society of Petroleum Engineers.

15 Nguyen, T.T.H. and Duval, T. (2014). A survey of communication and awareness in collaborative virtual environments. In: *2014 International Workshop on Collaborative Virtual Environments (3DCVE)*, 1–8. USA: IEEE.

16 McLean, S. (2005). Basics of interpersonal communication. Pearson/A and B, UK.

17 Lee, G.A., Teo, T., Kim, S., and Billinghurst, M. (2017). SharedSphere: MR collaboration through shared live panorama. In: *SIGGRAPH Asia 2017 Emerging Technologies*, SA '17. New York, NY, USA: Association for Computing Machinery.

18 Gauglitz, S., Nuernberger, B., Turk, M., and Höllerer, T. (2014). World-stabilized annotations and virtual scene navigation for remote collaboration. In: *Proceedings of the 27th Annual ACM Symposium on User Interface Software and Technology*, UIST '14, 449–459. New York, NY, USA: Association for Computing Machinery.

19 Piumsomboon, T., Dey, A., Ens, B. et al. (2017). [POSTER] CoVAR: Mixed-platform remote collaborative augmented and virtual realities system with shared collaboration cues. In: *2017 IEEE International Symposium on Mixed and Augmented Reality (ISMAR-Adjunct)*, ISMAR '17, 218–219. USA: IEEE.

20 Amores, J., Benavides, X., and Maes, P. (2015). ShowMe: A remote collaboration system that supports immersive gestural communication. In: *Proceedings of the 33rd Annual ACM Conference Extended Abstracts on Human Factors in*

Computing Systems, CHI EA '15, 1343–1348. New York, NY, USA: Association for Computing Machinery.

21 Cai, M. and Tanaka, J. (2019). Go together: providing nonverbal awareness cues to enhance co-located sensation in remote communication. *Human-centric Computing and Information Sciences* 9: 19.

22 Huang, W., Alem, L., Tecchia, F., and Duh, H. (2017). Augmented 3D hands: a gesture-based mixed reality system for distributed collaboration. *Journal on Multimodal User Interfaces* 12: 77–89.

23 Tait, M. and Billinghurst, M. (2015). The effect of view independence in a collaborative AR system. *Computer Supported Cooperative Work (CSCW)* 24: 563–589.

24 Kütt, G.H., Tanprasert, T., Rodolitz, J. et al. (2020). Effects of shared gaze on audio- versus text-based remote collaborations. *Proceedings of the ACM on Human-Computer Interaction* 4 (CSCW2): 1–25.

25 Kim, S., Lee, G.A., Billinghurst, M., and Huang, W. (2020). The combination of visual communication cues in mixed reality remote collaboration. *Journal on Multimodal User Interfaces* 14: 321–335.

26 Karis, D., Wildman, D., and Mané, A. (2016). Improving remote collaboration with video conferencing and video portals. *Human–Computer Interaction* 31 (1): 1–58.

27 Reinhardt, J., Hillen, L., and Wolf, K. (2020). Embedding conversational agents into AR: invisible or with a realistic human body? In: *Proceedings of the Fourteenth International Conference on Tangible, Embedded, and Embodied Interaction*, TEI '20, 299–310. New York, NY, USA: Association for Computing Machinery.

28 Fort, A., Delpuech, C., Pernier, J., and Giard, M.-H. (2002). Dynamics of cortico-subcortical cross-modal operations involved in audio-visual object detection in humans. *Cerebral Cortex (New York, NY: 1991)* 12: 1031–1039.

29 Bauer, M., Kortuem, G., and Segall, Z. (1999). "Where are you pointing at?" A study of remote collaboration in a wearable videoconference system. In: *Proceedings of the 3rd IEEE International Symposium on Wearable Computers*, ISWC '99, 151–158. USA: IEEE.

30 Wu, F., Thomas, J., Chinnola, S., and Rosenberg, E. (2020). Exploring communication modalities to support collaborative guidance in virtual reality. In: *2020 IEEE Conference on Virtual Reality and 3D User Interfaces Abstracts and Workshops (VRW)*, 79–86. USA: IEEE.

31 Piumsomboon, T., Dey, A., Ens, B. et al. (2019). The effects of sharing awareness cues in collaborative mixed reality. *Frontiers in Robotics and AI* 6: 5.

32 Waldow, K. and Fuhrmann, A. (2020). Addressing deaf or hard-of-hearing people in avatar-based mixed reality collaboration systems. In: *Proceedings* of 2020 IEEE Conference on Virtual Reality and 3D User Interfaces, VR '20, 595–596. USA: IEEE.

33 Yang, J., Sasikumar, P., Bai, H. et al. (2020). The combination of visual communication cues in mixed reality remote collaboration. *Journal on Multimodal User Interfaces* 14: 337–352.

34 Kim, S., Lee, G.A., and Sakata, N. (2013). Comparing pointing and drawing for remote collaboration. In: *2013 IEEE International Symposium on Mixed and Augmented Reality (ISMAR)*, 1–6. USA: IEEE.

35 Kim, S., Lee, G.A., Sakata, N. et al. (2013). Study of augmented gesture communication cues and view sharing in remote collaboration. In: *2013 IEEE International Symposium on Mixed and Augmented Reality*, ISMAR 2013, 261–262. USA: IEEE.

36 Adcock, M. and Gunn, C. (2015). Using projected light for mobile remote guidance. *Computer Supported Cooperative Work (CSCW)* 24: 591–611.

37 Teo, T., Lee, G.A., Billinghurst, M., and Adcock, M. (2018). Hand gestures and visual annotation in live 360 panorama-based mixed reality remote collaboration. In: *Proceedings of the 30th Australian Conference on Computer-Human Interaction*, OzCHI '18, 406–410. New York, NY, USA: Association for Computing Machinery.

38 Madeira, T., Marques, B., Alves, J. et al. (2021). Exploring annotations and hand tracking in augmented reality for remote collaboration. In: *Human Systems Engineering and Design III*. IHSED 2020, Advances in Intelligent Systems and Computing, vol. 1269 (ed. W. Karwowski, T. Ahram, D. Etinger, et al.), 83–89. Cham: Springer.

39 Günther, S., Müller, F., Schmitz, M. et al. (2018). CheckMate: Exploring a tangible augmented reality interface for remote interaction. In: *Extended Abstracts of the 2018 CHI Conference on Human Factors in Computing Systems*, CHI EA '18, 1–6. New York, NY, USA: Association for Computing Machinery.

40 Chen, H., Lee, A.S., Swift, M., and Tang, J.C. (2015). 3D collaboration method over HoloLens™ and Skype™ end points. In: *Proceedings of the 3rd International Workshop on Immersive Media Experiences*, ImmersiveME '15, 27–30. New York, NY, USA: Association for Computing Machinery.

41 Fussell, S.R., Setlock, L.D., Parker, E.M., and Yang, J. (2003). Assessing the value of a cursor pointing device for remote collaboration on physical tasks. In: *CHI '03 Extended Abstracts on Human Factors in Computing Systems*, CHI EA '03, 788–789. New York, NY, USA: Association for Computing Machinery.

42 Sakata, N., Kurata, N., Kato, T. et al. (2003). WACL: Supporting telecommunications using - wearable active camera with laser pointer. In: *Proceedings of the 7th IEEE International Symposium on Wearable Computers*, 53–56. USA: IEEE.

43 Kurata, T., Sakata, N., Kourogi, M. et al. (2004). Remote collaboration using a shoulder-worn active camera/laser. In: *8th International Symposium on Wearable Computers, ISWC '04*, 62–69. Washington, DC, USA: IEEE.

44 Zhou, J., Lee, I., Thomas, B.H. et al. (2011). Facilitating collaboration with laser projector-based spatial augmented reality in industrial applications. In: *Recent Trends of Mobile Collaborative Augmented Reality Systems*, 161–173. New York, NY, USA: Springer.

45 Yamamoto, T., Otsuki, M., Kuzuoka, H., and Suzuki, Y. (2018). Tele-guidance system to support anticipation during communication. *Special issue "Spatial Augmented Reality" of Multimodal Technologies and Interact* 3: 55

46 Song, J., Cho, S., Baek, S.-Y. et al. (2014). GaFinC: Gaze and finger control interface for 3D model manipulation in CAD application. *Computer-Aided Design* 46: 239–245.

47 Chatterjee, I., Xiao, R., and Harrison, C. (2015). Gaze+Gesture: expressive, precise and targeted free-space interactions. In: *Proceedings of the 2015 ACM on International Conference on Multimodal Interaction*, ICMI '15, 131–138. New York, NY, USA: Association for Computing Machinery.

48 Higuch, K., Yonetani, R., and Sato, Y. (2016). Can eye help you? Effects of visualizing eye fixations on remote collaboration scenarios for physical tasks. In: *Proceedings of the 2016 CHI Conference on Human Factors in Computing Systems*, CHI '16, 5180–5190. New York, NY, USA: Association for Computing Machinery.

49 Akkil, D. and Isokoski, P. (2019). Comparison of gaze and mouse pointers for video-based collaborative physical task. *Interacting with Computers* 30: 524–542.

50 Kendon, A. (2013). *Exploring the Utterance Roles of Visible Body Action: A Personal Account*, 7–27. Berlin/Boston, MA: Walter de Gruyter GmbH.

51 Pavlovic, V.I., Sharma, R., and Huang, T.S. (1997). Visual interpretation of hand gestures for human-computer interaction: a review. *IEEE Transactions on Pattern Analysis and Machine Intelligence* 19 (7): 677–695.

52 Fussell, S., Setlock, L., Ou, J. et al. (2004). Gestures over video streams to support remote collaboration on physical tasks. *Human-Computer Interaction* 19: 273–309.

53 Kirk, D., Crabtree, A., and Rodden, T. (2005). Ways of the hands. In: *Proceedings of ECSCW 2005*, 1–21. Dordrecht: Springer.

54 Li, Y., Huang, J., Tian, F. et al. (2019). Gesture interaction in virtual reality. *Virtual Reality & Intelligent Hardware* 1: 84–112.

55 Tang, J.C. and Minneman, S.L. (1991). VideoDraw: A video interface for collaborative drawing. *ACM Transactions on Information Systems* 9 (2): 170–184.

56 Tang, J.C. and Minneman, S. (1991). VideoWhiteboard: Video shadows to support remote collaboration. In: *Proceedings of the SIGCHI Conference on*

Human Factors in Computing Systems, CHI '91, 315–322. New York, NY, USA: Association for Computing Machinery.

57 Kuzuoka, H., Kosuge, T., and Tanaka, M. (1994). GestureCam: A video communication system for sympathetic remote collaboration. In: *Proceedings of the 1994 ACM Conference on Computer Supported Cooperative Work*, CSCW '94, 35–43. New York, NY, USA: Association for Computing Machinery.

58 Kuzuoka, H., Oyama, S., Yamazaki, K. et al. (2000). GestureMan: A mobile robot that embodies a remote instructor's actions. In: *Proceedings of the 2000 ACM Conference on Computer Supported Cooperative Work*, CSCW '00, 155–162. New York, NY, USA: Association for Computing Machinery.

59 Tang, A., Boyle, M., and Greenberg, S. (2005). Understanding and mitigating display and presence disparity in mixed presence groupware. *Journal of Research and Practice in Information Technology - ACJ* 37 (2): 71–88.

60 Yamashita, N., Kaji, K., Kuzuoka, H., and Hirata, K. (2011). Improving visibility of remote gestures in distributed tabletop collaboration. In: *Proceedings of the ACM 2011 Conference on Computer Supported Cooperative Work*, CSCW '11, 95–104. New York, NY, USA: Association for Computing Machinery.

61 Genest, A.M., Gutwin, C., Tang, A. et al. (2013). KinectArms: A toolkit for capturing and displaying arm embodiments in distributed tabletop groupware. In: *Proceedings of the 2013 Conference on Computer Supported Cooperative Work*, CSCW '13, 157–166. New York, NY, USA: Association for Computing Machinery.

62 Huang, W. and Alem, L. (2013). Gesturing in the air: supporting full mobility in remote collaboration on physical tasks. *Journal of Universal Computer Science* 19: 1158–1174.

63 Lee, G.A., Teo, T., Kim, S., and Billinghurst, M. (2017). Mixed reality collaboration through sharing a live panorama. In: *SIGGRAPH Asia 2017 Mobile Graphics & Interactive Applications*, SA '17. New York, NY, USA: Association for Computing Machinery.

64 Zhang, X., Chen, X., Li, Y. et al. (2011). A framework for hand gesture recognition based on accelerometer and EMG sensors. *IEEE Transactions on Systems, Man, and Cybernetics - Part A: Systems and Humans* 41: 1064–1076.

65 Feick, M., Tang, A., and Bateman, S. (2018). Mixed-reality for object-focused remote collaboration. In: *The 31st Annual ACM Symposium on User Interface Software and Technology Adjunct Proceedings*, UIST '18 Adjunct, 63–65. New York, NY, USA: Association for Computing Machinery.

66 Deng, S., Jiang, N., Chang, J. et al. (2017). Understanding the impact of multimodal interaction using gaze informed mid-air gesture control in 3D virtual objects manipulation. *International Journal of Human-Computer Studies* 105: 68–80.

67 Wang, P., Zhang, S., Bai, X. et al. (2019). Head pointer or eye gaze: which helps more in MR remote collaboration? In: *2019 IEEE Conference on Virtual Reality and 3D User Interfaces (VR)*, VR '19, 1219–1220. USA: IEEE.

68 Wang, P., Bai, X., Billinghurst, M. et al. (2020). Using a head pointer or eye gaze: the effect of gaze on spatial ar remote collaboration for physical tasks. *Interacting with Computers* 32: 153–169.

69 Jones, B., Zhang, Y., Wong, P.N.Y., and Rintel, S. (2020). VROOM: Virtual robot overlay for online meetings. In: *Extended Abstracts of the 2020 CHI Conference on Human Factors in Computing Systems*, CHI EA '20, 1–10. New York, NY, USA: Association for Computing Machinery.

70 Calandra, D.M., Di Mauro, D., Cutugno, F., and Di Martino, S. (2016). Navigating wall-sized displays with the gaze: a proposal for cultural heritage. *Proceedings of the 1st Workshop on Advanced Visual Interfaces for Cultural Heritage Co-Located with the International Working Conference on Advanced Visual Interfaces (AVI 2016)*, Bari, Italy (7–10 June 2016), Volume 1621 of *CEUR Workshop Proceedings*, 36–43. Aachen University, Germany. CEUR-WS.org.

71 Roter, D.L., Frankel, R.M., Hall, J.A., and Sluyter, D. (2006). The expression of emotion through nonverbal behavior in medical visits. *Journal of General Internal Medicine* 21: 28–34.

72 Terven, J.R., Salas, J., and Raducanu, B. (2014). Robust head gestures recognition for assistive technology. In: *Pattern Recognition. Mexican Conference on Pattern Recognition (MCPR 2014)*, Lecture Notes in Computer Science, vol. 8495 (ed. J.F. Martínez-Trinidad, J.A. Carrasco-Ochoa, J.A. Olvera-Lopez, et al.), 152–161. Cham: Springer.

73 Xiao, C., Huang, W., and Billinghurst, M. (2020). Usage and effect of eye tracking in remote guidance. In: *32nd Australia Conference on Human-Computer Interaction (OzCHI '20)*. New York, NY, USA: Association for Computing Machinery.

74 Wang, H. and Shi, B.E. (2019). Gaze awareness improves collaboration efficiency in a collaborative assembly task. In: *Proceedings of the 11th ACM Symposium on Eye Tracking Research & Applications*, ETRA '19. New York, NY, USA: Association for Computing Machinery.

75 Müller, R., Helmert, J., and Pannasch, S. (2014). Limitations of gaze transfer: without visual context, eye movements do not to help to coordinate joint action, whereas mouse movements do. *Acta Psychologica* 152: 19–28.

76 Ou, J., Oh, L., Fussell, S.R. et al. (2008). Predicting visual focus of attention from intention in remote collaborative tasks. *IEEE Transactions on Multimedia* 10: 1034–1045.

77 Huang, C.-M., Andrist, S., Sauppé, A., and Mutlu, B. (2015). Using gaze patterns to predict task intent in collaboration. *Frontiers in Psychology* 6: 1049

78 Piumsomboon, T., Lee, Y., Lee, G. et al. (2017). Empathic mixed reality: sharing what you feel and interacting with what you see. In: *2017 International Symposium on Ubiquitous Virtual Reality (ISUVR)*, 38–41. USA: IEEE.

79 Steptoe, W., Oyekoya, O., Murgia, A. et al. (2009). Eye tracking for avatar eye gaze control during object-focused multiparty interaction in immersive collaborative virtual environments. In: *Proceedings of IEEE Virtual Reality 2009*, VR '09, 83–90. USA: IEEE.

80 Andrist, S., Gleicher, M., and Mutlu, B. (2017). Looking coordinated: bidirectional gaze mechanisms for collaborative interaction with virtual characters. In: *Proceedings of the 2017 CHI Conference on Human Factors in Computing Systems*, CHI '17, 2571–2582. New York, NY, USA: Association for Computing Machinery.

81 Piumsomboon, T., Lee, G.A., Hart, J.D. et al. (2018). Mini-me: an adaptive avatar for mixed reality remote collaboration. In: *Proceedings of the 2018 CHI Conference on Human Factors in Computing Systems*, CHI '18 (ed. A. Cox and M. Perry). New York, NY, USA: Association for Computing Machinery. *International Conference on Human Factors in Computing Systems 2018*, CHI 2018; Conference date: 21-04-2018 Through 26-04-2018.

82 Chetwood, A., Kwok, K.-W., Sun, L.-W. et al. (2012). Collaborative eye tracking: a potential training tool in laparoscopic surgery. *Surgical Endoscopy* 26: 2003–2009.

83 Zhang, Y., Pfeuffer, K., Chong, M.K. et al. (2016). Look together: using gaze for assisting co-located collaborative search. *Personal and Ubiquitous Computing* 21: 173–186.

84 D'Angelo, S. and Gergle, D. (2018). An eye for design: gaze visualizations for remote collaborative work. In: *Proceedings of the 2018 CHI Conference on Human Factors in Computing Systems*, CHI '18, 1–12. New York, NY, USA: Association for Computing Machinery.

85 Li, J., Manavalan, M., D'Angelo, S., and Gergle, D. (2016). Designing shared gaze awareness for remote collaboration. In: *Proceedings of the 19th ACM Conference on Computer Supported Cooperative Work and Social Computing Companion*, CSCW '16 Companion, 325–328. New York, NY, USA: Association for Computing Machinery.

86 Otsuki, M., Maruyama, K., Kuzuoka, H., and Suzuki, Y. (2018). Effects of enhanced gaze presentation on gaze leading in remote collaborative physical tasks. In: *Proceedings of the 2018 CHI Conference on Human Factors in Computing Systems*, CHI '18, 1–11. New York, NY, USA: Association for Computing Machinery.

87 Jacob, R. and Stellmach, S. (2016). What you look at is what you get: gaze-based user interfaces. *Interactions* 23 (5): 62–65.

88 Yang, P., Kitahara, I., and Ohta, Y. (2015). [Poster] Remote mixed reality system supporting interactions with virtualized objects. In: *2015 IEEE International Symposium on Mixed and Augmented Reality*, ISMAR '15, 64–67. USA: IEEE.

89 Oda, O., Elvezio, C., Sukan, M. et al. (2015). Virtual replicas for remote assistance in virtual and augmented reality. In: *Proceedings of the 28th ACM Symposium on User Interface Software and Technology (UIST)*, 405–415. New York, NY, USA: Association for Computing Machinery.

90 Elvezio, C., Sukan, M., Oda, O. et al. (2017). Remote collaboration in AR and VR using virtual replicas. In: *ACM SIGGRAPH 2017 VR Village*, SIGGRAPH '17. New York, NY, USA: Association for Computing Machinery.

91 Feick, M., Mok, T., Tang, A. et al. (2018). Perspective on and re-orientation of physical proxies in object-focused remote collaboration. In: *Proceedings of the 2018 CHI Conference on Human Factors in Computing Systems*, CHI '18, 1–13. New York, NY, USA: Association for Computing Machinery.

92 Leithinger, D., Follmer, S., Olwal, A., and Ishii, H. (2014). Physical telepresence: shape capture and display for embodied, computer-mediated remote collaboration. In: *Proceedings of the 27th Annual ACM Symposium on User Interface Software and Technology*, UIST '14, 461–470. New York, NY, USA: Association for Computing Machinery.

93 Gunn, C., Hutchins, M., Stevenson, D. et al. (2005). Using collaborative haptics in remote surgical training. *1st Joint Eurohaptics Conference and Symposium on Haptic Interfaces for Virtual Environment and Teleoperator Systems. World Haptics Conference*, 481–482.

94 Yatani, K., Gergle, D., and Truong, K. (2012). Investigating effects of visual and tactile feedback on spatial coordination in collaborative handheld systems. In: *Proceedings of the ACM 2012 Conference on Computer Supported Cooperative Work*, CSCW '12, 661–670. New York, NY, USA: Association for Computing Machinery.

95 Matsuda, A., Nozawa, K., Takata, K. et al. (2020). HapticPointer: A neck-worn device that presents direction by vibrotactile feedback for remote collaboration tasks. In: *Proceedings of the Augmented Humans International Conference*, AHs '20. New York, NY, USA: Association for Computing Machinery.

96 Wang, L., Zhao, Z., Yang, X. et al. (2020). A constrained path redirection for passive haptics. In: *2020 IEEE Conference on Virtual Reality and 3D User Interfaces Abstracts and Workshops (VRW)*, 650–651. USA: IEEE.

97 Wang, P., Bai, X., Billinghurst, M. et al. (2020). Haptic feedback helps me? A VR-SAR remote collaborative system with tangible interaction. *International Journal of Human-Computer Interaction* 36: 1–16.

98 Yalçın, Ö.N. and DiPaola, S. (2020). Modeling empathy: building a link between affective and cognitive processes. *Artificial Intelligence Review* 53 (4): 2983–3006.

99 Tan, C.S.S., Luyten, K., Van den Bergh, J. et al. (2014). The role of physiological cues during remote collaboration. *Presence: Teleoperators and Virtual Environments* 23 (1): 90–107.

100 Hart, J., Piumsomboon, T., Lee, G., and Billinghurst, M. (2018). Sharing and augmenting emotion in collaborative mixed reality. In: *2018 IEEE International Symposium on Mixed and Augmented Reality Adjunct (ISMAR-Adjunct)*, ISMAR '18, 212–213. USA: IEEE.

101 Paiva, A., Leite, I., Boukricha, H., and Wachsmuth, I. (2017). Empathy in virtual agents and robots: a survey. *ACM Transactions on Interactive Intelligent Systems* 7 (3): 1–40.

102 Leite, I. (2015). Long-term interactions with empathic social robots. *AI Matters* 1 (3): 13–15.

103 Kumano, S., Otsuka, K., Mikami, D. et al. (2015). Analyzing interpersonal empathy via collective impressions. *IEEE Transactions on Affective Computing* 6 (4): 324–336.

104 Yalçın, Ö.N. and DiPaola, S. (2019). Evaluating levels of emotional contagion with an embodied conversational agent. In: *Proceedings of the 41st Annual Conference of the Cognitive Science Society*, CogSci 2019, 3143–3149. Canada. cognitivesciencesociety.org.

105 Yalçın, Ö.N. (2020). Empathy framework for embodied conversational agents. *Cognitive Systems Research* 59: 123–132.

106 D'Agostino, T. and Bylund, C. (2014). Nonverbal accommodation in health care communication. *Health Communication* 29: 563–573.

107 Xiao, B., Bone, D., Van Segbroeck, M. et al. (2014). Modeling therapist empathy through prosody in drug addiction counseling. In: *Proceedings of the Annual Conference of the International Speech Communication Association*, 213–217. INTERSPEECH. https://www.isca-speech.org (accessed 18 January 2024).

108 Moyers, T., Rowell, L., Manuel, J. et al. (2016). The motivational interviewing treatment integrity code (MITI 4): rationale, preliminary reliability and validity. *Journal of Substance Abuse Treatment* 65: 36–42.

109 Rashkin, H., Smith, E., Li, M., and Boureau, Y.-L. (2019). Towards empathetic open-domain conversation models: a new benchmark and dataset. In: *Proceedings of the 57th Annual Meeting of the Association for Computational Linguistics*, 5370–5381. ACL. https://www.aclweb.org/.

110 Prendinger, H. and Ishizuka, M. (2005). The empathic companion: a character-based interface that addresses users' affective states. *Applied Artificial Intelligence* 19: 267–285.

111 Boukricha, H., Wachsmuth, I., Carminati, M., and Knoeferle, P. (2013). A computational model of empathy: empirical evaluation. In: *Proceedings - 2013 Humaine Association Conference on Affective Computing and Intelligent Interaction*, ACII 2013, 1–6. USA: IEEE.

112 Billinghurst, M., Gupta, K., Katsutoshi, M. et al. (2016). Is it in your eyes? Explorations in using gaze cues for remote collaboration. In: *Collaboration Meets Interactive Spaces* (ed. C. Anslow, P. Campos, and J. Jorge), 177–199. Switzerland: Springer.

113 Masai, K., Kunze, K., Sugimoto, M., and Billinghurst, M. (2016). Empathy glasses. In: *Proceedings of the 2016 CHI Conference Extended Abstracts on Human Factors in Computing Systems*, CHI EA '16, 1257–1263. New York, NY, USA: Association for Computing Machinery.

114 Lee, Y., Masai, K., Kunze, K. et al. (2016). A remote collaboration system with empathy glasses. In: *2016 IEEE International Symposium on Mixed and Augmented Reality Adjunct (ISMAR-Adjunct)*, ISMAR '16, 342–343. USA: IEEE.

115 Shen, Z., Elibol, A., and Chong, N. (2020). Understanding nonverbal communication cues of human personality traits in human-robot interaction. *IEEE/CAA Journal of Automatica Sinica* 7 (6): 1465–1477.

116 Goldin-Meadow, S. and Brentari, D. (2017). Gesture, sign and language: the coming of age of sign language and gesture studies. *The Behavioral and Brain Sciences* 40: 1–17.

117 Monge, P.R. and Day, P.D. (1976). Multivariate analysis in communication research. *Human Communication Research* 2: 207–220.

118 Kirk, D., Rodden, T., and Fraser, D.S. (2007). Turn it this way: grounding collaborative action with remote gestures. In: *Proceedings of the SIGCHI Conference on Human Factors in Computing Systems*, CHI '07, 1039–1048. New York, NY, USA: Association for Computing Machinery.

119 Wille, M., Adolph, L., Grauel, B. et al. (2014). Prolonged work with head mounted displays. In: *Proceedings of the 2014 ACM International Symposium on Wearable Computers: Adjunct Program*, ISWC '14 Adjunct, 221–224. New York, NY, USA: Association for Computing Machinery.

120 Drogemuller, A., Walsh, J., Smith, R.T. et al. (2021). Turning everyday objects into passive tangible controllers. In: *Proceedings of the 15th International Conference on Tangible, Embedded, and Embodied Interaction*, TEI '21. USA: Association for Computing Machinery.

121 Chai, X., Liu, Z., Yin, F. et al. (2016). Two streams recurrent neural networks for large-scale continuous gesture recognition. In: *2016 23rd International Conference on Pattern Recognition (ICPR)*, 31–36. USA: IEEE.

122 Tsironi, E., Barros, P., Weber, C., and Wermter, S. (2017). An analysis of convolutional long short-term memory recurrent neural networks for gesture recognition. *Neurocomputing* 268: 76–86.

123 Mikropoulos, T., Vrellis, I., and Moutsiolis, A. (2014). Primary school students' attitude towards gesture based interaction: a Comparison between Microsoft Kinect and mouse. In: *Proceedings - IEEE 14th International Conference on Advanced Learning Technologies*, ICALT 2014, 678–682. USA: IEEE.

4

Communication Cues for Remote Guidance

4.1 Introduction

Many real-world situations require a local worker to work under guidance from a remote helper to manipulate physical objects for a collaborative task (or physical task) [1]. This can be seen in our daily life. For example, a technician in a call center helps a user to fix an internet connection issue in which connection of wires and inspection of network devices are required. This can also be seen in professional workplaces of industry sections. For example in the mining industry, more and more complex technologies and machines are nowadays being introduced into remote mine sites to improve productivity. However, such advanced machines often require specific knowledge and expertise to maintain, and personnel with such expertise are often in high demand and not locally available. Flying an expert onsite is not cost-effective when a machine breaks down, particularly if it takes the helper only a few minutes or hours to fix and bring the machine back running again. Therefore, there is a strong need for systems that enable helpers to respond to requests in real time and be able to fix the machine remotely without having to travel to the site [2]. Another example of remote assistance on physical tasks is a typical scenario in the medical and health domain, in which a local doctor receives guidance from a remote specialist to perform emergency surgery on a patient [3]. Similar applications in this area also include a nursing student receiving guidance from a university tutor to perform an assigned health procedure while on placement in a hospital and a home-care professional performing rescue procedures such as Cardiopulmonary resuscitation (CPR) on a suddenly sick person with assistance from a remote ambulance service provider [4].

A typical setup of systems supporting this type of collaboration is illustrated in Figure 4.1. The system has two units: one for helper and the other for worker, which are connected via the internet or a communication network. The local worker unit has a head-mounted near-eye display (HMD) and a front-facing camera, while the helper unit has a camera mounted over a large display.

Computer-Supported Collaboration: Theory and Practice, First Edition.
Weidong Huang, Mark Billinghurst, Leila Alem, Chun Xiao, and Troels Rasmussen.

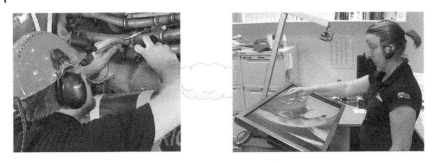

Figure 4.1 Local worker unit with a camera attached to the HMD (left) and the remote helper unit with a touch screen and a camera (right).

Both sides can communicate with each other verbally via audio headsets. Once the connection is established, the local camera captures the workspace, and the video is sent to the helper side, where the helper can see the workspace video on the display and perform hand gestures when needed while speaking with the worker. The hand gestures of the helper will be captured by the helper camera, and the hand gesture video is then combined with the workspace video and displayed on the HMD for the worker.

According to Kim et al. [5], there are five components that are involved with remote guidance on physical tasks. These components are tasks, local users, remote users, communication, and tools. A task is what the collaborators aim to complete together and should involve physical objects to be manipulated, such as a machine that is no longer working and tools that are used to fix the machine. A local user is a person who is located within the workspace and has direct access to the physical objects but does not have knowledge of how to manipulate the objects, thus requiring guidance from a remote user. The remote user, on the other hand, is a person who is located away from the workspace and does not have direct access to the objects but has the expert knowledge, thus being able to tell and show the local user how to complete the task. Communication is the activity that the collaborators use to understand each other for task performance. For this purpose, in addition to verbal communication, a range of other communication cues can be used to share their thoughts and intention to reach a consensus and establish the common ground between the collaborators. Tools are mechanisms, shared spaces or systems that are used to for users to show the current status and progress of the task, share each other's views and display the communication cues from both sides. Tools are critical for effective support of remote collaboration as they provide foundations for collaborators to work together on the task at hand while physically distributed [6].

Therefore, a typical feature of remote guidance on physical tasks is that the local worker has access to physical objects but does not know how to operate on or manipulate them, while the remote helper knows how but does not have direct access to the physical objects [7, 8]. As a result, in a conventional setting of remote guidance, the local user is expected to follow the instructions of the remote helper to manipulate objects in performing the collaborative task. Taking the above-mentioned five components into consideration, attempts have been made to study on a range of research aspects for remote guidance in the past few decades [5]. These include how to convey and validate multimodal communication cues, promote situation awareness, understand their effects and share the cues in a shared visual space. Systems that support remote guidance are usually designed and developed in such a way that collaborators using the system feel like being together and behave like being together [9]. In 1990s and early 2000s, most systems were proof of concept in laboratory, and much of the research was to test theories. Due to recent rapid advances in networking, mobile, and wearable technologies, many systems have been developed based on theoretical foundations to meet the demand from industrial application scenarios and to conduct research for better support of remote guidance [10].

More specifically, with remote collaboration, the collaborators are separated in different physical environments. They no longer have common ground to communicate with each other. To support collaboration, shared visual spaces have been used to achieve common ground [8, 11]. A shared space is a place where the collaborators can see, talk about, and manipulate the objects and where the collaborators can see each other's actions and monitor the progress of the task. Further, in face-to-face communication, while verbally communicating with each other, collaboration partners also perform gestures (hand gestures and other body actions), and these gestures are visually available to all collaborators. And this has been found to be one of the main reasons why face-to-face communication is more effective than computer-medicated remote communication [12]. As a result, in developing remote guidance systems, the focus has been on providing communication cues and realistic collaborative environments for better user experience and task performance. Examples of communication cues include pointers, hand gestures, sketches, and eye gazes. Also, depending on the specific objectives of the system and hardware devices used by the system (wearable, mobile, or desktop-based), these communication cues can be implemented in different ways. For example, hand gestures can be unmediated video-based hands, projected hands, or recreated digital hands [2]. Pointers can be hand pointing, gaze pointer, or delivered by a stick mounted in worker's workspace [13–15]. Despite extensive research on how remote guidance should be supported, the research results are reported in different venues.

Given the importance and variety of communication cues, we conducted a survey to summarize the communication cues being used, approaches that implement the cues, and their effects on remote guidance. In this chapter, we categorize the communication cues into explicit and implicit ones and report our findings. In the context of remote guidance on physical tasks, we define communication cues as explicit if they are directly and specifically used for guidance and conversational grounding purposes. On the other hand, implicit cues are those that are not specifically mentioned, are generated as a result of interactions between collaborators, are used to seek confirmation for what is perceived consciously and unconsciously by users, or are to provide situation awareness for users. It should be noted that whether a communication cue is explicit or implicit should be considered in the context of the collaboration. For example, if eye gaze is used as a pointer, then it is explicit. But if it is used to convey information about workload or fatigue, then it is implicit. At the end of the chapter, we also discuss challenges and envision possible future work. This survey is based on papers that the authors collected during their more than ten years' extensive research on the topic. It is a review of selective and representative studies that together represent the current state of art of the research and development.

4.2   Explicit Communication Cues

Gestural information has been found to be important for common ground in remote collaboration (e.g. [16]) compared with its absence (e.g. [7]). This is especially the case when actions and objects are difficult to describe verbally (e.g. [17]). Fussell et al. [18] set out four types of gesture that may be important: deictic (pointing), representational, spatial, and motion. Deictic are gestures that facilitate object and location references. Representational are gestures that indicate literal shapes or orientations for things. Spatial gestures show distances and sizes, and motion gestures indicate a required action or movement. These kinds of gestures may be achieved "directly" by making a view of hands available or via a "surrogate" such as a cursor or drawing.

The conveyance of gestures may be particularly important for novice users and users who are unfamiliar with each other. Experimental work has found that the advantages that gestures bring to verbal communication are largely when helpers and workers are performing a task for the first time [12]. Once common ground between the communicators has been established for the task; there is little difference between voice only and voice with gesture communication methods. That experimental work indicated that this was in part due to reducing the burden of communicative effort in establishing common ground by the worker.

A variety of methods have been devised to share gestural and other nonverbal communication cues in remote guidance settings. The type of technology used may limit or facilitate different types of gestural information. Like display media, the gestural affordances of a remote guidance system may be asymmetric between helper and worker. It is often the case that worker's gestures are more accessible and effective than helpers due to their place within the workspace context.

4.2.1 Pointer

To enable deictic communication from the helper to the worker, it is necessary to provide some means of pointing. Usually, workers can use deictic communication easily since the shared visual space is often focused on their context (i.e., the workspace). Since the need to share visual information about the helper context is often not necessary, simply providing a pointer that the helper can use in the shared visual space or workspace can suffice to enable deictic communication.

4.2.1.1 Cursor

A very simple means of providing a pointer to the helper is a cursor within the shared visual space. The cursor is usually operated by a mouse onto a video feed, which is then fed back to the worker [14, 19]. Cursors can also be displayed on other shared visual media, such as reconstructed models of the workspace [20]. It can also be projected directly onto the physical space via a standard or laser projection system [21]. Such pointers are also possible on other visual media, such as screen sharing [22]. When a cursor is available to a helper, it is used extensively and rated favorably [14, 19]. A cursor alone, however, does not appear to facilitate the efficiency of task performance when compared to a system without one [19].

4.2.1.2 Laser Pointer

Laser pointers provide a means by which the helper can point to real physical objects without the use of a medium. This may carry advantages for mobile workers who need to maintain an unobstructed field of vision. Typically, laser pointers are mounted with cameras and oriented to the center of their field of view, the camera itself being controlled remotely by the helper [13, 23]. It is also possible for the laser pointer to be controllable independently from the camera such that the helper can select a point within the camera's view, and the laser pointer orients to that specific location [24].

A good example of a laser-pointed system is the "Wearable Active Camera with Laser Pointer" (WACL) system described by Sakata et al. [23]. This system involved a device that a worker wore on their shoulder. A camera with a laser pointer was placed on a remote controllable actuator that was mounted on the worker's shoulder. This setup allowed the helper to have an independent view of the workspace and point to objects and locations in the physical space.

In the case that a camera and laser pointer are mounted on the worker, it is necessary to provide some means of stabilization and directional correction to account for the movements of the worker [23]. This may be especially the case when there are multiple workers, as when the worker wearing the device is working, their body movement makes it difficult for the helper to point at objects for other workers [25]. Without such corrections, the pointer would require too many manual adjustments by the helper to be useful.

The use of laser pointers is a simple means of allowing helpers to use deictic communication. In remote guidance scenarios that do not require complex gesturing, a laser pointer can replace the need for the worker to have any display medium (i.e. no shared visual space). While this could be especially useful for worker mobility, the need to stabilize the pointing location is a technical challenge if the pointer is worn or carried by the worker.

4.2.1.3 Stick Mounted on a Robot

One means of ensuring laser pointer stability and mobility is to mount it on a helper-controlled device such as a robot. The small dot created by a laser pointer, however, may be difficult to track and locate by the remote worker. Thus, the inclusion of a physical pointer to indicate the laser's orientation may be important for the worker to efficiently detect the object or location in question.

A remote-controlled camera and laser system [13] as well as a remote-controlled robot with laser system [24] are examples of systems that include a physical pointer. In the former system, a static finger-like object was mounted on the camera along with a laser pointer to make the system orientation more salient to the worker. The latter system used a dynamic physical pointer that raised up in the direction of the laser pointer when the laser was activated. The authors note, however, that getting the pointer to raise up before the laser was activated would facilitate communication efficiency by leveraging preparatory actions taken by the helper.

4.2.1.4 Eye-Gaze pointer and Head Pointer

Eye gaze, or eye fixation, is an important communication cue in collaboration. Users can use their eyes as a pointer to draw the attention of their opener to the target objects. Higuch et al. [26] developed systems in which eye fixations of the helper were visualized and displayed to the worker both by fixed devices and by mobile devices. With fixed devices, eye gaze of the helper was captured by an off-the-shelf eye tracker and then projected to the worker's workspace by a projector indicated as a green circle together with a polyline representing five preceding gazes. With mobile devices, the captured eye gaze of the helper was shown as a green square in a light-weighted optical see-through head-mounted display worn by the worker. The authors conducted a series of user studies to investigate how collaboration behaviors and interaction patterns between a local worker and a remote helper change if the helper's eye fixations are presented to the worker. And it was found that in the context of remote guidance, eye fixations can be used as a fast and accurate pointer to the object of interest and that when used together with verbal instructions, eyes can be used as explicit communication cues.

Akkil et al. [27] presented a novel remote guidance interface in which the helper's eye gaze is projected onto the physical task space of the worker and displayed as a spotlight. This effectively transfers gaze information into a pointer, guiding the worker into the object or an area that the gaze is pointing to. The authors conducted a user study investigating the usefulness of the gaze interface for remote collaboration in a circuit-building task and found that the gaze pointer helped make referencing and object identification easier and improved the worker's confidence in task performance.

In order to explore what effects different sharing methods of gaze information have in collaborative augmented reality applications, Weng et al. [15] developed a remote collaboration platform that supports the sharing of both head pointer and gaze pointer cues of the remote helper. The authors conducted a user study comparing eye-gaze pointer with head pointer and cursor pointer and found that eye-gaze pointer helped improve situational awareness and facilitate conversational grounding, and that head pointer can be a more accessible alternative to the more expensive and unreliable eye-gaze pointer.

4.2.1.5 Tracked Marker Annotations

Digital location markers can achieve a similar effect as a laser pointer if the worker is wearing or carrying a display. Much like laser pointers, which can lose their orientation when a worker moves, digital location markers arguably also need to be stabilized to their location.

A prototype system of an augmented reality framework designed to provide view independence and spatial tracking of annotations was described by Gauglitz et al. [28]. The prototype allowed a remote helper to take a snapshot of a worker-controlled video feed (a tablet with screen and camera). This snapshot did not interrupt the worker's view of the video feed (c.f. [14]). When a shot was taken, the live stream of the video was superimposed on it in greyscale so that the helper did not lose sight of the current worker's focus. The snapshot allowed the helper to mark numbered annotations ("X" markers) onto workspace objects and locations. These markers were tracked to their physical locations by a software representation of the visual space that was made up of images taken from the video feed. As such, even if the annotations were outside the field of view of the worker's camera, they would appear in the correct positions if the worker panned or tilted the camera to bring them into view. When outside the worker's view, an arrow would point in the direction of the annotation so that the worker could locate it. Evaluation of the prototype found that the annotation system led to better task completion performance than video alone. However, there was no statistically significant improvement over a video annotation system that involved no tracking or stabilization of the marker location. Users, however, showed a strong subjective preference for the annotation tracking system. A later iteration of this prototype allowed for helper view independence such that the snapshot feature was unnecessary [29]. The evaluation of that version produced similar results.

Another system that involved the spatial tracking of marker annotations is described by Chang et al. [30]. This system used object detection via image segmentation to draw bounding boxes around objects. Live video was streamed from the worker to the helper. The worker used a smartphone with a tripod mount, while the helper used a tablet. The helper could scribble over the region of the image that contained an object to annotate it. The system then detected the object and placed a bounding box around it with a label (i.e. "A"). The object was then tracked so long as it remained within the view of the camera. The labeled bounding boxes were designed to facilitate object referencing in conversation. An evaluation of the system found that labeled annotations with tracking produced the fastest task completion times and conversational efficiency. Consistent with other tracking systems, no reliable differences were found for task performance between tracked and nontracked labeled annotations overall (however, the combination of tracking and label was still best). User ratings also indicated that annotation labels were appreciated by workers compared to when they were absent, but the presence or absence of this feature made no difference to helper ratings.

Techniques used to stabilize location annotations on video feeds generally appear to be more successful than attempts to stabilize laser pointers. While users generally prefer such stabilization, it does not appear to convey any advantages to task performance. This is likely due to deictic gestures generally being transient and short-lived within conversations [31]. Since location indications are only needed briefly during communication, their need to be stabilized to a location is also only brief.

4.2.2 Sketches and Annotations

Sketches and annotations can come in handy when explaining a complex task procedure in relation to physical objects. The ability of the communicator to sketch or draw into the shared visual space increases the amount of gestural information that they can convey compared to simple pointers. Drawing allows for deictic, representational, spatial, and, to some extent, motion gestures to be conveyed [18]. The ability of a helper to send drawn information has been found to increase the efficiency of remote guidance tasks compared to pointers or verbal descriptions [16, 18].

Fakourfar et al. [31] conducted research into drawing annotation use during remote collaboration tasks. These authors identified three types of annotation: referential, procedural, and deictic. Referential were commonly used to depict end-states of object configurations or legends to indicate annotation types (e.g. what different color annotations meant) and were retained over the course of a task. Procedural was used to indicate transformations to be enacted on task objects to change their state and were removed once the action was performed. Deictic were more transient and included dots, arrows, circles, and scribbles to indicate objects. The extent to which these types of annotations were used varied with the task demands. For example, in a tangram puzzle task, participants commonly drew the end-state of the puzzle as a reference; in contrast, in a brick model building task, participants relied on deictic annotations to indicate pieces and locations. For an origami task, procedural annotations were used to indicate how to fold the paper.

4.2.2.1 Draw on Video

Drawing annotations into the shared visual space is dependent on the nature of the shared visual space. Live video is a very common form of shared visual space, and thus, the earliest drawing techniques were onto a live video stream. Naturally, this technique requires that the worker also has a view of the video stream on a screen or head-mounted display.

Ou et al. [32] described a prime example of system that relied on helper sketches onto a live video feed which is often cited in the literature. The system involved a fixed scene camera at the worker's site. The helper viewed this feed on a tablet computer, which afforded the drawing of annotations on top of the video feed using a stylus pen. The video feed and helper drawings were displayed on a desktop screen at the worker's site. Research using this system found that automatically erasing the helper drawings after a few seconds facilitated task efficiency [16] toward that observed when the helper and worker are together in the same place, working side-by-side [18]. The system also involved a feature to "normalize" drawings (e.g. to make curves smoother and lines straighter) while this feature may reduce the amount of data to transmit and facilitate the use of drawings as computer commands [32], users did not rate this feature as useful over simple freehand drawings [18]. Taking a slightly different approach, Chantziaras et al. [33] developed a system in which prespecified digital annotations, such as arrow, text, and 3D hand annotation, can be sent and imposed on videos of live streams of the workspace.

Drawing directly onto a live video feed is a robust means of conveying gestural information to a helper. Its usefulness, however, is constrained to situations in which the helper has a view of the video feed, which is not too obstructive to their ability to work.

4.2.2.2 Draw into Workspace

Drawing annotations can be projected directly into the workspace by means of a projection system. Much like the advantages of laser pointer systems, this approach can remove the need for the worker to have a display medium that can obstruct their view of the workspace.

Palmer et al. [21] describe a laser projection system that allows a remote helper to draw directly into a workspace. The system was developed as part of a telehealth system. A laser projector (like those found in entertainment stage events) was mounted on a pole in the workspace and could display vector graphics on any surface. A variety of interchangeable, remotely controlled cameras displayed a video feed onto tablets at the helper and worker sites (creating a shared visual space). Both helper and worker could use a tablet stylus to control the cameras and draw predefined shapes and freeform drawings as well as to control a visible cursor. The system used an alignment algorithm to correct for any differences between the camera's and laser projector's position and orientation. In this way, the image is shown in the intended physical location.

Another system that allows for drawing directly into the physical workspace is described by Gurevich et al. [34]. This system involved a portable device that could be placed in the target workspace. The device consisted of a robotic arm with a camera and a portable digital projector. The helper could control the robotic arm to change the view of the workspace and had a feature that allowed them to "bookmark" views, allowing automatic reorientation back to that perspective at a later point. The projector displayed a green tracing square around the drawable space. This square allowed for the system to automatically adjust the angle of projection to account for changes in planes. It also served to provide awareness to the helper about the drawable area, as well as providing workers with information about the helper's current focus of attention. The helper could use the projector for a cursor, free-drawing, placing predefined shapes, and icons as well as presenting text in a range of colors. The authors note projector and camera alignment issues that could arise due to object depth differences. In their prototype system, they allow helpers to adjust their annotations by small increments to account for any such distortions; however, they also suggest that depth-sensing cameras could be used. A qualitative user evaluation showed that the system supported grounding and had good user acceptance. Users, however, did not use the text or icons features. Free draw was used most, and the cursor was used for more transient gesturing.

Drawing annotations directly into the workspace via a projection system appears to be a robust method for gesturing. Considerations regarding the brightness and power consumption of projectors relative to the lighting conditions of the workspace need to be taken into account. Additionally, image corrections may need to be applied to account for uneven surfaces and orientation angles.

4.2.2.3 Tracked Drawing Annotations

Drawing onto virtual representations of a workspace for augmented reality is another possibility. Systems that use this approach will usually involve helper view independence. For visual augmentation to work effectively, these annotations need to be tracked and stabilized to their intended locations.

Gauglitz et al. [35] extended their previous work on tracked marker annotations in augmented reality [28, 29] to also allow for drawing. The system allowed helpers to draw annotations into any visible region within a 3D reconstructed view of the workspace. The reconstructed view was limited to points of view that had been captured by a worker-controlled camera. The helper drew annotations using a touchscreen tablet, and these annotations were tracked to their coordinates in the workspace. Thus, when the worker pointed their camera and screen (i.e., a smartphone) at the location, the annotation was viewable as superimposed over the video feed. The authors note challenges associated with resolving drawings over 3D space. Since the drawing is done in 2D, a change in camera orientation (i.e., to view the same location from the side) affects the interpretability of the drawing. The authors tested some solutions to this problem with users and found that anchoring them to a dominant plane was most favorable. They note, however, that this causes extreme distortions when viewed from other angles.

Adcock and Gunn [36] also present a tracked drawing annotation system. The system involves the use of a camera, a projector, and fiducial markers. Fiducial markers are objects, such as cardboard, or a piece of paper, with an image on them, which can be captured in a video by camera and recognized through digital image processing for tracking purposes. They are often used in Augmented Reality-based applications or virtual 3D representations of the environment. This representation was used to track locations and make alignment adjustments to any drawing that was to be presented on the projector. The system was designed to be mobile by mounting the camera and projector onto a helmet. It is, however, unclear how the use of fiducial markers can be resolved in a mobile environment. An evaluation of the system compared performance when the system was helmet-mounted versus mounted on a wheeled tripod. The tracked annotations were also compared to nontracked annotations (i.e., that were not reoriented or moved when tracked environmental objects were moved). Results indicated that nontracked annotations from a tripod mount performed best in terms of quantitative task performance. User ratings, however, indicated a preference for the tracked annotations over nontracked, as well as a preference for the tripod mount over helmet mounting. User comments noted that tracking was more comfortable when the device was head-mounted, as otherwise. they felt the need to keep their head still. The authors noted a latency issue with the annotation correction system, which led to users making their own adaptations before the correction was executed, thus undermining the purpose of automatic realignment.

Research by Fakourfar et al. [31] questions the utility of tracked annotations in many collaboration scenarios. This group's research noted that annotation use is generally very rough, transient, and used dynamically as part of conversation.

The need then to track annotations to locations and objects is often short-lived. Once the message has been conveyed at the moment, it is generally no longer needed. Referential annotations that are more long-lived are used to convey end-state information or color codes for annotations. Such annotations, however, also may not have much utility in remaining attached to locations or objects because the intention is to be constantly accessible. These authors found no quantitative benefit to tracked annotations over nontracked annotations; however, users rated a greater preference for tracking.

As with other tracked annotations, this technique does not appear to convey any particular quantitative advantages in task performance. Users, however, consistently favor alignment correction techniques in these circumstances. This pattern of results may suggest that helpers use drawing annotations only once a joint focus has been achieved with the worker. Stabilization may only be necessary to correct for camera angle changes due to such things as minor head movements, which may suggest the use of more simple techniques.

4.2.2.4 Combination

Systems are not necessarily constrained to a single drawing annotation technique. In some circumstances, it may be prudent to combine multiple mediums onto which annotations can be drawn.

Stevenson et al. [37] describe and pilot-tested a component of a telehealth system (see also [21]), which included both laser drawing and drawing over video. The system was designed to facilitate remote collaboration between medical experts and assistants in patient consultations. The subsystem involved video feeds of the helper and worker's faces and an overhead view of the worker's workspace. The workspace view was visible to the worker on a desktop screen and to the helper on a tablet. The helper was able to draw on the tablet with a stylus that could be displayed either on the video feed shown on the worker's display or directly onto the workspace via a laser projector. The authors rationalized that the two drawing mediums provided different strengths and weaknesses. Drawing on video was more stable and accurate but required the worker to resolve the disjunction in their perspective between the workspace and the screen. On the other hand, drawing directly onto the workspace via laser involved no disjunction but was less accurate as it required the helper to try and resolve any differences in 3D space (i.e., between the 2D display and the actual workspace). A qualitative assessment of the system found that users negotiated how to use the technologies for their purposes. The technology selected differed between pairs of collaborators and between different task demands.

4.2.3 Hand Gesture

The use of hands presents a naturalistic means of conveying gestural information. While the types of gestural information that may be conveyed are the same as for drawing, it is potentially more limited for conveying representational information (it is easier to draw some shapes than to gesture them with hands). For model-building tasks, however, hand gesture systems have been found to be more effective than drawing [38]. Thus, depending on the demands of the task [31], the natural intuitiveness of hand gestures makes them more helpful in conveying object manipulation information [26].

Some authors have created taxonomies of hand gestures that are used during remote instruction scenarios (primarily in model-building tasks). For example, Kirk et al. [39] described six types of hand gestures from their observations that were used to indicate task-specific actions (e.g. search for object and action to take) and to facilitate conversation (e.g. established shared visual perspective and conversational turn-taking). Wickey and Alem [40] added additional gestures to this list, as well as organized gestures into those that convey awareness, instructions, or manipulations. Gestures of these three types are often compositional, and the authors recommend that hand gesture systems have such affordances as well as suggesting that some gestures may be useful for controlling the view (e.g. a pointing gesture may pan/zoom the camera to focus on a located object).

Hand gestures may be conveyed to a remote location in a few ways. As with gestures in general, these methods use either direct or surrogate technologies.

4.2.3.1 Video Feed of Hand Gestures

A simple means of conveying hand gestures is to use a live video feed. This style of hand gesturing system is rudimentary and was primarily used in earlier work. It involved a camera feed of the helper's hands in front of a screen displaying another video feed of the workspace. Notably, it involved no video mixing and was less likely to preserve the orientation of both the worker and helper to the workspace.

Kato et al. [41] present some early work using this approach. In their system, the helper selects one of three video feeds from the worker's context and makes hand gestures in front of a screen that displays it. A camera pointing at the screen captures the helper's hand gestures and the screen's content and sends it to the worker's site. Experimentation with different locations for the worker-side display of the helper's gestures showed that placing the display in view of a context camera was best. Since the helper could see both the worker and their own gestures together, it improved their confidence that their gestures had been noticed.

Another system involving video of hand gestures is described by Alem et al. [10, 42]. This system was designed with worker mobility in mind, for industrial use for maintenance and repairs of equipment in mines. The system involved a head-mounted camera and near-eye display for the worker. The helper had access to a tabletop touchscreen display, which showed the worker's perspective view and a panoramic view of the workspace. A fixed camera oriented downwards also captured the portion of the display that included the worker's video feed, such that helper gestures toward that view were captured and sent back to the worker. The worker could then glance at the display in their upper periphery to see gestures while not having their main view occluded. This setup reduces discomfort and potential loss of situational awareness caused by displays that cover the eye that were used in early setups of this kind (e.g. [11]) while providing the advantages of natural hand gestures (e.g. [38]). Pilot testing with the system, however, indicated that care needs to be taken in aligning the near-eye display to ensure its full benefits are experienced in practice. Other issues were noted related to video lag and image quality. Otherwise, the system was generally well accepted.

4.2.3.2 Superimposed Hand Gesture

Superimposing the helper's hands onto the shared visual space is a means of creating a visual copresence. This is because, in systems such as these, the hand gestures of both the worker and helper become visible within the shared visual space.

A mixed reality system created by Kirk et al. [39] projected the hands of a remote helper directly into the workspace. A ceiling camera above the workspace captured video that was displayed on a screen for the remote helper. Another ceiling camera captured the helpers' hands, which were superimposed onto the workspace via a ceiling-mounted projector. Thus, the worker could see the helper's hands in the workspace, and the helper could see their own hands, the worker's hands, and task objects on the screen. For the worker, at least, this meant that there was no disjunction in their perspective of helper gestures and the task space. Evaluation of this system [43] indicated that the use of gestures improved the quality of communication over voice only in that workers showed better learning of the task when tested a day later. Conversely, however, workers provided ratings that suggested they felt that communication was more one-sided, with the workers simply doing what they were told.

Various other versions of the system introduced by Kirk et al. [39] have been reported on in the literature. Another version of this style of gesture system involved video mixing of the worker and helper's hands video feeds for display on a desktop monitor [9, 40] rather than projection onto the workspace. No evidence has been found for a performance advantage of either projection or desktop monitor display with this style of system for workers [12]. However, some evidence suggests that adaptation is required on the part of helpers to resolve the disjunction in their perspective. Helpers have been observed pointing at the desktop monitor instead of within the overhead camera field of view [40].

The authors who made that observation suggest that the projection of the shared visual space onto the helper's side may be important to make the system more intuitive. Given the aforementioned evidence, this suggests that the opposite display setup to the Kirk et al. [39] system such that the worker has a desktop monitor and the helper has a projection on top of their hands.

A system to superimpose hand gestures into the shared visual space while also allowing for helper mobility is presented by Huang and Alem [2]. This system distinguishes itself from previous remote guidance work in which helpers are typically stationary, providing guidance from a fixed workstation. The system involved a head-mounted camera and near-eye display for both the helper and worker. The worker's camera provided a view of the workspace, whereas the helper's view was oriented toward their hands. Software was used to filter out other visual information other than the hands (it detected skin), and then mixed this information with the view of the workspace. The workspace with the helper's hands superimposed was displayed on both helper's and worker's near-eye displays. Thus, both the worker and helper could attend to their environment without occlusion from the display, glancing upwards to view the shared visual space when necessary. A usability evaluation of the system showed positive acceptance by users. It was noted, however, that continual switching between the near-eye display and the workspace was tiring, prompting the authors to suggest the use of see-through displays. Similar work, called MobileHelper, was also presented by Robert et al. [44]. Their system uses a head-mounted display on the worker side and a mobile guidance deceive on the helper side, thus enabling both worker and helper to be mobile.

A system described by Higuch et al. [26] also presented a means of displaying helper hand gestures to a mobile worker. This system involved a worker with a see-through head-mounted display and head-mounted camera. The helper viewed the video feed on a desktop monitor. A gesture camera at the helper's site used a polarizing lens to filter out images from the monitor. This allowed the helper's gestures to be isolated from the screen images to be sent to the worker. The gestures were then superimposed into the worker's view via their head-mounted display.

Some work by Gao et al. [45] has also used depth-sensing cameras to capture hand gestures. The depth-sensing camera provides a 3D point cloud of the workspace, which is viewable in virtual reality. Depth sensing cameras can then also be used to capture a helper's hand gestures as a point cloud and superimpose them into the 3D virtual representation of the workspace.

Another example of a system set up for 3D representations of hand gestures from depth-sensing cameras is the system presented by Huang et al. [46] and Tecchia et al. [47]. This system used overhead depth sensing cameras above a tabletop workspace for both the worker and helper. On the worker's side, the camera was used to capture a 3D point cloud of the workspace. On the helper's side, the camera was used to capture a 3D point cloud of the helper's hands. The workspace and helper's hands data were then fused into the same virtual scene and displayed on a desktop monitor for the worker and a virtual reality head-mounted display for the helper. This system allowed the helper to have a real-time, independent view of the workspace while also conveying 3D hand gestures to the worker. The 3D aspect of the helper's hand allowed more complex spatial gesturing than is possible with 2D renditions. User evaluations of the system indicated that 3D representation of the workspace and gestures improved the sense of copresence, clarity of representational gestures, and spatial relations, as well as immersion and interaction.

Superimposing hand gestures into the shared visual space has proven to be a robust means of conveying gestural information. Hand gestures are intuitive and support conversational grounding well. Providing the helper with good alignment between their own hands and the visual information that they are gesturing about appears to be important. Workers, on the other hand, appear to be able to resolve a disjunction in perspective, perhaps because they only need to refer to helper gestures at critical moments during task instructions.

4.2.3.3 Combination

Systems that include hand gesturing techniques can also implement other gesture technologies. Providing multiple gesturing affordances can potentially increase the richness of information that can be conveyed while allowing users to choose the easiest method for their purposes.

A gesturing approach that involved a combination of both projected hand gestures and direct feeds of hand gestures was used in the immersive virtual workspace room of Yamashita et al. [17, 48, 49]. That system involved wall screens that showed the remote collaborators' upper bodies as well as a tabletop display which was the workspace. Hand and arm gestures and objects captured by the overhead camera were displayed remotely on the tabletop. While this was like a projection system, the image was presented on a screen on the table's surface, which produced issues of occlusion. To address this issue, a snapshot feature [48] was used in which an image of the current workspace could be captured and displayed on a wall screen. Evaluation of the system with and without the snapshot function found evidence to support the gesture system's efficacy. Analysis of collaborator communication indicated greater degrees of deixis, grounding, joint focus of attention, monitoring of comprehension, and communication efficiency (i.e., less utterances) with the snapshot gesturing feature.

Yamashita et al. [17] extended their work by investigating another method (instead of snapshot) for addressing the visual occlusion problem. For the tabletop display, a visual history was used such that gestures remained visible for some time. A greyscale version of gestures from 1.1 seconds in the past was displayed alongside the current state of gestures. Past gestures and current ones were connected by a "motion flow" effect, which shows wispy lines of motion between them. This method addressed the problem of visual occlusion by allowing remote workers to move objects to see the specific gesture underneath and provided a means of seeing gestures they may have missed due to looking elsewhere (i.e., while focusing on some other aspect of the task). Evidence was found that this system reduced the need for clarifications in instructions in terms of less questions and confirmations made by workers and less utterances by helpers. Workers also perceived the visual history technique as requiring less effort in the completion of the task. A subsequent iteration of the system involved a downward-facing projector, presumably to better circumvent occlusion problems by projecting gestures over the top of objects [49].

Huang et al. [50] presented a prototype system that used both hand gestures and drawing annotations. A live video feed from a worker's head-mounted camera was streamed to the helper. The helper used a large touchscreen display, which allowed for snapshots of the video feed to be taken. The snapshots or live video afforded both hand gestures (via an overhead camera with polarizing lens) and drawings (via the touchscreen). The helper's hand gestures and drawings were then sent back to the worker and displayed on a near-eye display. A user evaluation compared the use of hand gesture withdrawal to hand gestures alone for a brick model building task and a laptop repair task. No differences were found between the conditions for the brick task on task completion times or cognitive load. The combination of both hand gestures and drawings was, however, found to decrease task completion times for the repair task. However, it was also found to increase cognitive load. Hand gestures were used more often in general; however, when information was too complex to convey with hand gestures drawings, were used instead.

4.2.4 Object Models

Representational gestures, motion gestures, and spatial gestures may often be complex and difficult to convey with hands. In some kinds of remote guidance tasks, these gestures may be used to indicate specific task objects, how to orient task objects together, and actions to perform with those task objects. By providing a remote helper with a virtual object model of important task objects (e.g. engine parts), the helper may be able to more effectively demonstrate complex and intricate steps that are required.

A system described by Oda et al. [51] uses virtual 3D models of known task objects allowed for helper annotations and demos that were tracked to the physical objects themselves. One version of the system allowed helpers to annotate virtual replicas of task models with positions and alignments between objects. Points marked on one object were linked to points marked on another object.

These points were then augmented onto the physical task objects via the worker's head-mounted display with lines linking the points so that the worker could see how to position and align the pieces. Another version of the system allowed the helper to manipulate and position the models together virtually. This information was then augmented onto the physical objects the worker could see, such that a virtual model part would be shown in its required position on the physical object. The worker could then copy the demonstration with the real objects. An evaluation of the system found that virtual manipulation of objects led to better task performance than virtual annotations. The interface for manipulating virtual objects was more naturalistic and intuitive for users than pointing to specific locations in virtual space. This was especially the case for novice helpers (helpers who had experience with the system – i.e., the experimenters – had less trouble with the pointing system but were still slower with it than direct virtual object manipulation).

Another system that used known task objects that were represented virtually and tracked in 3D was described by Tait and Billinghurst [52]. This system used depth cameras to display a virtual representation of the workspace in 3D to the helper. The helper was then able to manipulate the positioning of several predefined task objects within the workspace via desktop software with a mouse. The placement of the object models was then augmented onto the physical workspace for the worker via a see-through head-mounted display. As the worker moved the known task object, its position was updated on the helper's virtual representation of the workspace.

In remote guidance scenarios that involve known task objects and potentially complex actions to be performed, virtual replicas of objects may facilitate helper instructions. Task domains such as machinery maintenance, repair, and construction would be examples. Naturalistic interaction with virtual objects can provide an intuitive way for helpers to demonstrate complex operations.

4.2.5 Comparison of Explicit Communication Cues

Some studies have formally compared the performance of different technologies used for conveying gestures in remote collaboration settings. Such comparisons provide some insight into the utility of different techniques in supporting remote guidance.

A study by Tajimi et al. [25] compared drawing on video with laser pointer technologies for remotely instructing multiple workers. The drawing on the video system involved a head-mounted camera and display, which was worn by one worker. The helper viewed the video stream and drew on the video, which was shown on the worker's head-mounted display. This setup meant that the worker wearing the display had to convey instructions to other workers as well as that the helper did not have direct control of the workspace view. The laser pointer setup was similar to the WACL by Sakata et al. [23]. A remote-control camera with a laser pointer was mounted on the worker to provide an independent view of the workspace and a means of pointing at physical objects and locations. While the study found no differences in task completion times, there were strong user preferences for the WACL-type system. In a task that required two workers to work in parallel, the WACL-type system led to fast completion times. This was because the worker wearing the device could work while the helper gave instructions to the other worker (in the head-mounted display setup, the worker had to relay their instructions to the other worker, interrupting their work). The authors did, however, note that worker movements during work caused problems with stabilization of the camera and pointer for providing instructions to another worker.

Kirk and Stanton Fraser [38] compared their hand projection system to drawing gesture systems (e.g. [32]). Video feeds of hands, hands with a whiteboard, and a digital-only drawing system were compared. Tabletop projection and screen displays of the shared visual space for the worker's view were also compared. It was found that hand gestures alone produced the best performance in a brick model-building task. They did not, however, find any advantage to projection versus screen display, which did not support any efficacy of mixed reality in such systems. This was contrary to previous arguments that disjunction in perspective may have impaired performance due to greater cognitive demands for coordination [39].

A hand projection system, similar to the Kirk et al. [39] system, has also been compared to a cursor pointer in an experiment by Li et al. [9]. In that experiment, a video feed of the helper's hands was mixed into a video feed of the worker's hands and workspace for the hand projection system. For the cursor condition, the helper controlled a mouse for the pointer. In both cases, the video feed of the worker's hands and workspace (with either superimposed hands or cursor) was displayed to both the worker and helper on a desktop monitor. While no statistically significant results were found, there was a strong trend for the hand gesture condition being faster for task completion of a block model-building task. Workers and helpers rated the use of hand gestures as more helpful than the cursor, and helpers perceived better interaction with hand gestures as well.

Kim et al. [53, 54] and Huang et al. [50] conducted a series of user studies comparing the effects of different combinations of gesturing cues, including hand gestures, pointing gestures, and sketches in different shared space settings such as 2D, 3D dependent and independent views of the local workspace, and it was found that each additional cue had a different level of effect depending on specific task situations. For example, additional sketch cues reduced the task time and increased usability when hand gesturing was limited. However, it may not be very useful if the object to be annotated is small.

As part of the evaluation of the Oda et al. [51] tracked 3D virtual model virtual reality and augmented reality annotation and demonstration system, a 2D drawing annotation system was compared. The 2D drawing system included the same 3D virtual models and augmented the annotations onto the worker's view. The main difference, however, was that it used a 2D tablet display with a touch interface for the helper, whereas the main system used a virtual reality display with tracking devices that were manipulated in 3D by the helper's hands. The virtual reality 3D system outperformed the 2D touchscreen system on the object assembly task used in the evaluation. However, participants found that pointing to specific locations in the virtual 3D environment is more difficult than pointing to the same objects on a standard 2D touchscreen interface.

In general, hand gesturing systems have been found to be superior to many other gesturing techniques. Gesturing techniques that allow for greater conveyance of information also appear to be better than simple pointing techniques. Specific task demands and interruptions caused by gesturing displays, however, may change what is effective in some circumstances.

4.3 Implicit Communication Cues

Much of the work in remote guidance has focused on means of conveying explicit gestures. In general, there has been a deliberate focus away from face-to-face visual feeds due to a perception that this information does little to improve conversational grounding when working on physical tasks (e.g. see [7]). However, more and more recent evidence has shown that additional communication cues could help improve collaborative experience such as copresence and situation awareness. For example, eye-tracking research, however, has found that helpers do attend to faces albeit infrequently when they are physically copresent [8]. Other studies have found that approximately half of users appreciate a view of their collaborator's face being available [9]. That same study found that helpers generally found facial information as important for solving the task, suggesting it may be used to help grounding by providing a backchannel that supports situational awareness (i.e., for helpers to monitor worker comprehension).

Research by Billinghurst et al. [55, 56] investigated how eye-gaze cues may affect remote guidance. The research used a system that included a head-mounted camera and display as well as eye-tracking of the worker. The helper viewed a live video feed of the worker's perspective, which also included the location of the worker's gaze. A cursor was also available for the helper to point within the video feed. A comparison of conditions in which eye-gaze location was present or absent, as well as when the helper cursor was present or absent, was studied. It was found that task performance and user ratings were best when both helper pointer and worker gaze features were present. The magnitude of this effect appeared greater than either the cursor or eye-gaze features on their own. This result may suggest that reciprocity of collaborators' focus of attention best supports task performance. A further pilot study by the group added a worker facial expression detection feature. While many of the results were equivocal due to small sample size, there was a nominal trend in user ratings for combination of features being better. Facial expression with gaze and pointing was generally rated best compared to other combinations of features. This may suggest that adding additional implicit communication cues enhances remote collaboration. The effect of eye gaze was also investigated by Higuch et al. [26]. These authors present a system that tracks the helper's eye gazes onto the video feed of the workspace via either a fixed scene camera or a worker head-mounted camera. User evaluations indicated that eye-gaze information combined with hand gestures led to a task performance benefit in terms of speed and accuracy over hand gestures alone and was rated more favorably by participants. Similar findings were also obtained from a user study by Bai et al. [57].

Research by Otsuki et al. [58] investigated eye-gaze cues via the use of a surrogate. A transparent hemisphere was attached to a tablet computer's screen. A black circle, representing the iris, was displayed underneath the hemisphere, giving the illusion of a 3D eyeball. By moving the iris behind the hemisphere, the orientation of the gaze could be gleaned. This surrogate was then used to indicate the eye-gaze location of a remote collaborator whose face and upper body were displayed below the surrogate on the tablet. A user evaluation found that the surrogate allowed participants to attain a joint focus of attention to objects in front of themselves faster than when the surrogate was absent.

Robot systems involve using a remotely controlled robot as a surrogate for the helper at the worker's site. Such surrogates allow for a considerable amount of freedom for the helper (i.e., of movement, field of view, etc.). Perhaps the largest advantage of robot systems, however, is their ability to convey body movements like gestures that may convey implicit communication cues.

In the robot system created by Kuzuoka et al. [24, 59], there was a considerable focus on implicit signals created by various parts of the body. The orientation of the helper's head was conveyed to the robot such that the worker could see which direction the instructor was looking. Research by the group led to this feature being controlled by 3D motion tracking of the helper's head [24] rather than manual control by hand [59], as it was more intuitive for the helper. The robot also included a physical pointing stick to support a laser pointer. The physical pointer raised up from pointing downwards to the direction of the laser beam after the laser was activated. The authors noted from an evaluation, however, that capturing preparatory actions by the helper to make the pointer raise up before the helper activated the laser may make communication more efficient by providing more implicit cues for orientation [24].

The orientation of collaborators' heads and fields of view can also be conveyed in virtual reality. A system by Gao et al. [45] constructed a 3D virtual representation of the workspace by stitching together point clouds generated from a depth-sensing camera mounted on the worker. Both workers and helpers wore virtual reality headsets, which tracked the orientation of their views within the virtual space. The collaborator's view orientation was then displayed in the virtual representation of the workspace to provide awareness of each other's focus of attention.

Body orientation can also be conveyed in immersive spaces. Yamashita et al. [49] explain that in fixed-position collaboration settings, the remote partner's body orientation is uninformative. They argue that this is part of the reason why upper-body views are often omitted in remote collaboration systems and research. However, in an immersive space such as their own system, where collaborators move around a central table, video projections of the remote partners' upper bodies convey information that allows for orientation information to be gleaned. Being able to see where a collaborator is in the room also provides expectations as to where their hand gestures and actions will be located on a tabletop display. Evaluation of their system with and without upper-body views showed faster task completion times when upper-body views were available. A view of the upper body also changed where users positioned themselves. Without a view of the upper body, collaborators stood in the same position, but with the view, they stood next to each other, such that they could glance at the remote partner.

A system by Yamamoto et al. [60] conveyed body gesture cues via the use of a virtual avatar. The system involved the helper wearing a virtual reality headset within a room with positional sensing cameras. The worker used a see-through head-mounted display within another room with positional sensing cameras. The workspace was set around a couple of whiteboards which had cameras capturing their surface. The helper could see the whiteboard surfaces in virtual reality along with a skeleton model of themselves within the space, which they viewed from a third-person perspective. This model helped them to orient their arm gestures within the space. The worker then saw the virtual avatar of the helper augmented into their view of the workspace. A comparison of the system, which showed a full-body avatar to one that only showed red spheres where the helper's hands were indicated that task performance was faster with the avatar. The presence of a virtual body avatar also reduced the cognitive load for the worker. Piumsomboon et al. [61] also proposed an adaptive avatar that represents a remote user's gaze direction and body gestures within the local user's field of view and found in a user study that the avatar improved social presence and could reduce the task difficulty and mental effort, leading to improved performance time. Wang et al. [62]

investigated the effect of the body representation level of the avatar and found that a whole-body representation avatar induces lower workload and higher efficiency.

Orientation cues can be conveyed by various means and may be used to guide workers to locations and objects. Research by Günther et al. [63] explored the use of three means of providing orientation information to workers: audio, visual, and tactile. For audio cues, stereo-tones of different intensities to indicate direction and distance were used to guide workers to task objects. While this technique led to the fastest task completion times, it was generally rated unfavorably by workers. Visual cues were conveyed by an augmenting label over the objects in the worker's view via a see-through display. While users rated this approach highly, it produced slower task completion times than live video with voice. Tactile cues were conveyed by vibrators on a glove, such that the vibrator closest to the target was active. This approach was rated quite favorably; however, it was not significantly faster than visual cues. The authors note from user comments that cues should only persist as long as necessary to convey the necessary information (in this case, finding the object). Cues in any modality that persist too long are distracting for workers and may impair task performance.

Effects of haptic feedback have also been investigated in a user study by Wang et al. [64]. The authors presented a tangible physical drawing interface, which provides haptic feedback with tangible interaction between a local worker and a remote expert helper. They compared it with another interface without haptic feedback and found that haptic feedback could significantly improve the remote expert's user experience.

Implicit cues need not only be used to convey information for the purposes of communication between helper and worker. Ou et al. [65–67] explore the possibility of using implicit cues to intelligently control what information is presented to the helper during remote collaboration. The authors rationalize that while multiple views of a workspace may carry advantages, limitations in data transmission and the ineffectiveness of manual switching between sources of information make it ineffective. Task actions and verbal utterances, however, may convey information about what helpers will likely attend to next and if captured, may be used to intelligently control information feeds without user intervention. The authors provide some initial proof of concept using eye-tracking and automatic classifiers. In circumstances where one helper is guiding multiple workers, however, linguistic cues appear to be more complex and less reliable for indicating changes in helper focus [68].

In summary, providing additional implicit communication cues can help support more effective conversation and task performance. With implicit cues such as facial expression, eye gaze, and body orientation, it appears that adding them helps to enhance the main channels of communication. These cues alone may be ineffective, but when combined with gesturing and speech, they may enhance things considerably. Due to the general asymmetry of information shared between helper and worker, it appears that providing additional implicit cues from the helper (i.e., orientation) is particularly effective for assisting in worker comprehension.

4.4 Challenges and Future Directions

4.4.1 Technical and Technological Limitations

Network performance for transmitting data can have a critical effect on the effectiveness of communication. For example, the introduction of delays to a shared visual space, as may be found in a poor network connection, can severely impact on its usefulness, even to the point where it no longer supports communication [8, 69]. Systems that involve drawing onto real-world objects also are affected by the round trip time of data transmission; the drawing must be transmitted to the remote site, rendered, captured, and returned to the user before they can see the result of their input [21]. There also seems to be a trend of making all communication cues that are available for colocated collaboration available for remote collaboration by reproducing colocated collaborative environment in remote guidance systems. This can be computationally expensive, causing delays in system response to users' interactions. The central question is whether current available AR/VR devices and technologies allow us to produce a new collaborative environment for remote guidance, which is cheaper to produce and set up and, with proper training if necessary, can lead to a similar level of user experience and task performance as colocated. For example, although eye contact and eye gaze can be important for guiding collaboration, Wang et al. [15] investigated the effect of head pointer and found that it can be an alternative to eye gaze with similar task performance but easier to set up. This research is promising as it shows that it is feasible to support remote guidance without having to reproduce everything that is available for colocated collaboration.

4.4.2 Empathic, Mental Effort, and Fatigue Cues

In remotely collaborating on physical tasks in safety-critical professional settings, such as fixing a machine in a remote mine site or a manufacturing factory or performing surgery on a patient in a remote clinic, it is important to monitor and understand the remote partner's physical and psychological status so that a proper guidance can be provided [70]. This type of information can be obtained through words they use, speech analysis, sensors attached to the collaborators' body, pupil size, and eye movement patterns for emotion status, stress level, cognitive load and fatigue status [70–72]. A positive side about this information is that tools that can be used to extract and process the information are already mature, and that the whole process can be done automatically and locally without human interference.

Thus, the exchange of information over the network between the two sides of the system can be kept to be minimum. The question is to what extent this information is helpful to justify the cost involved and how this information should be displayed to the collaboration partners in a proper way so that it is ubiquitous and easily accessible.

4.4.3 Evaluation Metrics and Methodologies

Evaluation is important in assessing and validating the effectiveness of communication cues. Current evaluations are mainly conducted in research laboratories, and these include controlled experiments and qualitative studies with researchers and students as participants. Also, in these studies, traditional task measures and questionnaire instruments, such as task completion times, self-reported Likert scale-based cognitive load, and usability questions. However, as remote guidance systems are normally developed for practical use, it would be more desirable to conduct longitudinal or field studies in real-world settings with real users for more realistic evaluation of and comprehensive insights into the value of the system in consideration and communication cues [10, 70]. Further, when it comes to task performance measures such as completion time or accuracy, it is often group-based; thus, the individual contribution and interaction during the task performance is unknown, and currently, we do not have quantitative measures for these for remote guidance. On the other hand, when it comes to usability measures, it is often individually measured, and again, we do not have group-based metrics for these. However, there have been well-established group-based metrics in other fields, such as psychological safety, group efficacy, and group cognitive load [73, 74], which we could borrow, adapt, and apply in the context of remote guidance. Clearly, more research is needed in this direction.

4.5 Conclusion

For remote guidance on physical tasks, in addition to verbal communications, how to convey other communication cues effectively such as hand gestures has been researched for many years. Our review indicates that a number of novel and effective methods have been developed and shown to be effective in improving system usability and helping collaborators achieve optimal user experience and task performance. More specifically, there is a growing interest in providing combination of multiple explicit communication cues to cater to the needs of different task purposes and in providing a combination of explicit and implicit communication cues. More and more research results have also been reported in the literature on their implications for system design and usage.

On the other hand, despite the progress being made, there are challenges ahead that require further investigations due to complexity and diversity of application requirements, limitations in technologies available, evaluation measurements, and our understanding of how to best support remote guidance. Additionally, it became evident to us during the surveying process that there are still limitations in current research and development practices that we should address in future work. These limitations include: (1) In testing and validating a communication cue or different combinations of cues, much effort has been focused on their effects on task performance. Relatively little attention has been paid to trying to understand how and why those effects happened from the end user's perspective. More specifically, how and why are those communication cues perceived and responded to mentally and behaviorally by users? How and why do different formats of the same cue affect the way that users perceive and respond to the cue? Answering these questions would be helpful for us to understand root causes of the effects and, as a result, enable us to design and develop more user-friendly and task-effective remote guidance tools. (2) Extensive development work has been done using a range of equipment and devices, different formats of a communication cue, and different combinations of communication cues for different tasks in various application domains. However, our understanding is still limited on what effects these different parameters, such as task, application domain, device type, and communication cue format, could have on the effectiveness of communication cues for task performance and collaborative user experience. Although some effort has been made to compare different communication modalities for their effects on task performance (e.g. [38, 54]), more systematic studies are needed. (3) Human factors have a significant role in Augmented Reality-based collaboration [75]. Le et al. [76] have reviewed and identified some factors that could have impact on remote collaboration, including culture, language, trust, and social status. Further investigations and user studies are required to validate their effects on remote guidance, particularly for their effects on user perception of communication cues.

It should be noted that our survey also has limitations. This survey is mostly based on the publications that we collected while conducting our extensive research in this field in the past years, and it is focused only on communication cues. Reviews and surveys on the research of other aspects of remote guidance, such as shared visual spaces or eye tracking, will also be valuable for the research community and application domains.

Acknowledgment

This chapter is a reprint of Huang et al. [1] with minor edits. Permission is granted from Springer.

References

1 Huang, W., Wakefield, M., Rasmussen, T.A. et al. (2022). A review on communication cues for augmented reality based remote guidance. *Journal on Multimodal User Interfaces* 16: 239–256. https://doi.org/10.1007/s12193-022-00387-1.

2 Huang, W. and Alem, L. (2013). Gesturing in the air: supporting full mobility in remote collaboration on physical tasks. *Journal of Universal Computer Science* 19 (8): 1158–1174.

3 Gasques, D., Johnson, J.G., Sharkey, T. et al. (2021). ARTEMIS: a collaborative mixed-reality system for immersive surgical telementoring. *Proceedings of the 2021 CHI Conference on Human Factors in Computing Systems*, Yokohama, Japan (08–13 May 2021). pp. 1–14. https://doi.org/10.1145/3411764.3445576.

4 Mather, C., Barnett, T., Broucek, V. et al. (2017). Helping hands: using augmented reality to provide remote guidance to health professionals. *Studies in Health Technology and Informatics* 241, Context Sensitive Health Informatics: Redesigning Healthcare Work: 57–62.

5 Kim, S., Billinghurst, M., and Kim, K. (2020). Multimodal interfaces and communication cues for remote collaboration. *Journal on Multimodal User Interfaces* 14 (4): 313–319. https://doi.org/10.1007/s12193-020-00346-8.

6 Kim, S., Billinghurst, M., Lee, C., and Lee, G. (2018). Using freeze frame and visual notifications in an annotation drawing interface for remote collaboration. *KSII Transactions on Internet and Information Systems* 12 (12): 6034–6056.

7 Fussell, S.R., Kraut, R.E., and Siegel, J. (2000). Coordination of communication: Effects of shared visual context on collaborative work. *Proceedings of the 2000 ACM conference on Computer Supported Cooperative Work*, Philadelphia, PA, USA (02-06 December 2000). ACM, pp. 21–30.

8 Kraut, R.E., Gergle, D., and Fussell, S.R. (2002). The use of visual information in shared visual spaces: informing the development of virtual co-presence. *Proceedings of the 2002 ACM Conference on Computer Supported Cooperative Work*, New Orleans, Louisiana, USA (16–20 November 2002). ACM, pp. 31–40.

9 Li, J., Wessels, A., Alem, L., and Stitzlein, C. (2007). Exploring interface with representation of gesture for remote collaboration. *Proceedings of the 19th Australasian Conference on Computer-Human Interaction: Entertaining User Interfaces*, Adelaide, Australia (28–30 November 2007), 2007. ACM, pp. 179–182. http://doi.acm.org/10.1145/1324892.1324926.

10 Alem, L., Huang, W., and Tecchia, F. (2011). Supporting the changing roles of maintenance operators in mining: a human factors perspective. *The Ergonomics Open Journal* 4 (1).

11 Kraut, R.E., Miller, M.D., and Siegel, J. (1996). Collaboration in performance of physical tasks: effects on outcomes and communication. *Proceedings of the 1996 ACM Conference on Computer Supported Cooperative Work*. ACM, pp. 57–66.

12 Kirk, D., Rodden, T., and Fraser, D.S. (2007). Turn it way: grounding collaborative action with remote gestures, *CHI '07: Proceedings of the SIGCHI Conference on Human Factors in Computing Systems*. ACM, pp. 1039–1048. http://portal.acm.org/citation.cfm?id=1240782#.

13 Kuzuoka, H., Kosuge, T., and Tanaka, M. (1994). GestureCam: a video communication system for sympathetic remote collaboration. *Proceedings of the 1994 ACM Conference on Computer Supported Cooperative Work*. ACM, pp. 35–43. http://doi.acm.org/10.1145/192844.192866.

14 Bauer, M., Kortuem, G., and Segall, Z. (1999). Where are you pointing at? A study of remote collaboration in a wearable videoconference system. *Proceedings of the 3rd IEEE International Symposium on Wearable Computers*. IEEE Computer Society, pp. 151–151. http://dl.acm.org/citation.cfm?id=519309 .856501.

15 Wang, P., Bai, X., Billinghurst, M. et al. (2020). Using a head pointer or eye gaze: the effect of gaze on spatial ar remote collaboration for physical tasks. *Interacting with Computers* 32 (2): 153–169. https://doi.org/10.1093/iwcomp/ iwaa012.

16 Ou, J., Fussell, S.R., Chen, X., et al. (2003). Gestural communication over video stream: supporting multimodal interaction for remote collaborative physical tasks. *Proceedings of the 5th International Conference on Multimodal Interfaces*, Vancouver, British Columbia, Canada (5–7 November 2003). ACM, pp. 242–249.

17 Yamashita, N., Kaji, K., Kuzuoka, H., and Hirata, K. (2011). Improving visibility of remote gestures in distributed tabletop collaboration. *Proceedings of the ACM 2011 Conference on Computer Supported Cooperative Work*. ACM, pp. 95–104. http://doi.acm.org/10.1145/1958824.1958839.

18 Fussell, S.R., Setlock, L.D., Yang, J. et al. (2004). Gestures over video streams to support remote collaboration on physical tasks. *Human–Computer Interaction* 19 (3): 273–309. https://doi.org/10.1207/s15327051hci1903_3.

19 Fussell, S.R., Setlock, L.D., Parker, E.M., and Yang, J. (2003). Assessing the value of a cursor pointing device for remote collaboration on physical tasks. *CHI'03 Extended Abstracts on Human Factors in Computing Systems*, Ft. Lauderdale Florida USA (05–10 April 2003). ACM, pp. 788–789.

20 O'Neill, J., Castellani, S., Roulland, F., et al. (2011). From ethnographic study to mixed reality: a remote collaborative troubleshooting system. *Proceedings of the ACM 2011 Conference on Computer Supported Cooperative Work*. ACM,

Hangzhou, China (19–23 March 2011), pp. 225–234. http://doi.acm.org/10 .1145/1958824.1958859.

21 Palmer, D., Adcock, M., Smith, J. et al. (2007). Annotating with light for remote guidance. *Proceedings of the 19th Australasian Conference on Computer-Human Interaction: Entertaining User Interfaces*, Adelaide, Australia (28–30 November 2007). ACM, pp. 103–110.

22 Karsenty, L. (1999). Cooperative work and shared visual context: an empirical study of comprehension problems in side-by-side and remote help dialogues. *Human–Computer Interaction* 14 (3): 283–315. https://doi.org/10 .1207/S15327051HCI1403_2.

23 Sakata, N., Kurata, T., Kato, T., et al. (2003). WACL: supporting telecommunications using - wearable active camera with laser pointer. *2003 Proceedings of the Seventh IEEE International Symposium on Wearable Computers* (21–23 October 2003), pp. 53–56. https://doi.org/10.1109/ISWC.2003.1241393.

24 H. Kuzuoka, Kosaka JI, Yamazaki K et al. (2004). Mediating dual ecologies. *Proceedings of the 2004 ACM Conference on Computer Supported Cooperative Work*. ACM, pp. 477–486.

25 Tajimi, K., Sakata, N., Uemura, K., and Nishida, S. (2010). Remote collaboration using real-world projection interface. *2010 IEEE International Conference on Systems Man and Cybernetics (SMC)*, Istanbul, Turkey (10–13 October 2010), pp. 3008–3013.

26 Higuch, K., Yonetani, R., and Sato, Y. (2016). Can eye help you?: effects of visualizing eye fixations on remote collaboration scenarios for physical tasks. *Proceedings of the 2016 CHI Conference on Human Factors in Computing Systems*. ACM, pp. 5180–5190. http://doi.acm.org/10.1145/2858036.2858438

27 Akkil, D., James, J.M., Isokoski, P., and Kangas, J. (2016). GazeTorch: enabling gaze awareness in collaborative physical tasks. *Proceedings of the 2016 CHI Conference Extended Abstracts on Human Factors in Computing Systems*. San Jose, California, USA. https://doi.org/10.1145/2851581.2892459.

28 Gauglitz, S., Lee, C., Turk, M., and Höllerer, T. (2012). Integrating the physical environment into mobile remote collaboration. *Proceedings of the 14th International Conference on Human-Computer Interaction with Mobile Devices and Services*. ACM, pp. 241–250. http://doi.acm.org/10.1145/2371574.2371610

29 Gauglitz, S., Nuernberger, B., Turk, M., and Höllerer, T. (2014). World-stabilized Annotations and Virtual Scene Navigation for Remote Collaboration. *Proceedings of the 27th Annual ACM Symposium on User Interface Software and Technology*. ACM, pp. 449–459. http://doi.acm.org/10.1145/ 2642918.2647372.

30 Chang, Y.-C., Wang, H.-C., Chu, H.-K., et al. (2017). AlphaRead: support unambiguous referencing in remote collaboration with readable object annotation. *Proceedings of the 2017 ACM Conference on Computer Supported*

Cooperative Work and Social Computing. ACM, pp. 2246–2259. http://doi .acm.org/10.1145/2998181.2998258.

31 Fakourfar, O., Ta, K., Tang, R., et al. (2016). Stabilized annotations for mobile remote assistance. *Proceedings of the 2016 CHI Conference on Human Factors in Computing Systems.* ACM, pp. 1548–1560. http://doi.acm.org/10.1145/ 2858036.2858171.

32 Ou, J., Chen, X., Fussell, S.R., and Yang, J. (2003). DOVE: drawing over video environment. *MULTIMEDIA '03 Proceedings of the Eleventh ACM International Conference on Multimedia.* ACM, pp. 100–101.

33 Chantziaras, G. et al. (2021). An augmented reality-based remote collaboration platform for worker assistance. In: *Pattern Recognition. ICPR International Workshops and Challenges. ICPR 2021*, Lecture Notes in Computer Science, vol. 12667 (ed. A. Del Bimbo et al.). Cham: Springer https://doi.org/10.1007/978-3-030-68787-8_30.

34 Gurevich, P., Lanir, J., Cohen, B., and Stone, R. (2012). TeleAdvisor: a versatile augmented reality tool for remote assistance. *Proceedings of the 2012 ACM Annual Conference on Human Factors in Computing Systems.* ACM, pp. 619–622. http://doi.acm.org/10.1145/2207676.2207763.

35 Gauglitz, S., Nuernberger, B., Turk, M., and Höllerer, T. (2014). In touch with the remote world: remote collaboration with augmented reality drawings and virtual navigation. *Proceedings of the 20th ACM Symposium on Virtual Reality Software and Technology.* ACM, pp. 197–205. http://doi.acm.org.ezproxy.lib .swin.edu.au/10.1145/2671015.2671016.

36 Adcock, M. and Gunn, C. (2015). Using projected light for mobile remote guidance. *Computer Supported Cooperative Work (CSCW)* 24 (6): 591–611. https:// doi.org/10.1007/s10606-015-9237-2.

37 Stevenson, D., Li, J., Smith, J., and Hutchins, M. (2008). A collaborative guidance case study. *Ninth Australasian User Interface Conference (AUIC 2008)*, B. Plimmer and G. Weber, Eds., Vol. 76: ACS, pp. 33–42.

38 Kirk, D. and Stanton Fraser, D. (2006). Comparing remote gesture technologies for supporting collaborative physical tasks. *Proceedings of the SIGCHI Conference on Human Factors in Computing Systems.* ACM, pp. 1191–1200.

39 Kirk, D., Crabtree, A., and Rodden, T. (2005). Ways of the hands. In: *ECSCW 2005*, 1–21. Springer.

40 Wickey, A. and Alem, L. (2007). Analysis of hand gestures in remote collaboration: some design recommendations. *Proceedings of the 19th Australasian Conference on Computer-Human Interaction: Entertaining User Interfaces*, Paris, France (18–22 September 2005). ACM, pp. 87–93. http://doi.acm.org/10.1145/ 1324892.1324909.

41 Kato, H., Yamazaki, K., Suzuki, H., et al. (1997). Designing a video-mediated collaboration system based on a body metaphor. *CSCL '97: Proceedings of the*

2nd International Conference on Computer Support for Collaborative Learning. International Society of the Learning Sciences, Toronto Ontario Canada (10–14 December 1997), pp. 148–156.

42 Alem, L., Tecchia, F., and Huang, W. (2011). HandsOnVideo: towards a gesture based mobile AR system for remote collaboration. In: *Recent Trends of Mobile Collaborative Augmented Reality Systems*, 135–148. New York: Springer.

43 Kirk, D.S. and Fraser, D.S. (2005). The effects of remote gesturing on distance instruction. *Proceedings of the 2005 Conference on Computer Support for Collaborative Learning: Learning 2005: The Next 10 Years!*

44 Robert, K., Zhu, D., Huang, W. et al. (2013). MobileHelper: remote guiding using smart mobile devices, hand gestures and augmented reality. *SIGGRAPH Asia 2013 Symposium on Mobile Graphics and Interactive Applications*, Hong Kong, Hong Kong (19 November 2013–22 November 2013). https://doi.org/10 .1145/2543651.2543664.

45 Gao, L., Bai, H., Lindeman, R., and Billinghurst, M. (2017). Static local environment capturing and sharing for MR remote collaboration. *SIGGRAPH Asia 2017 Mobile Graphics & Interactive Applications*, Bangkok, Thailand (27– 30 November 2017). ACM, pp. 17:1–17:6. http://doi.acm.org/10.1145/ 3132787.3139204.

46 Huang, W., Alem, L., Tecchia, F., and Duh, H.B.-L. (2018). Augmented 3D hands: a gesture-based mixed reality system for distributed collaboration. *Journal on Multimodal User Interfaces* 12 (2): 77–89.

47 Tecchia, F., Alem, L., and Huang, W. (2012). 3D helping hands: a gesture based MR system for remote collaboration. *Proceedings of the 11th ACM SIGGRAPH International Conference on Virtual-Reality Continuum and its Applications in Industry.* Singapore, Singapore (02–04 December 2012). pp. 323–328. https://doi.org/10.1145/2407516.2407590.

48 Yamashita, N., Hirata, K., Takada, T. et al. (2007). Effects of room-sized sharing on remote collaboration on physical tasks. *IPSJ Digital Courier* 3: 788–799.

49 Yamashita, N., Kuzuoka, H., Hirata, K. et al. (2011). Supporting fluid tabletop collaboration across distances. *Proceedings of the 2011 Annual Conference on Human Factors in Computing Systems.* ACM, pp. 2827–2836. http://doi.acm .org/10.1145/1978942.1979362.

50 Huang, W., Kim, S., Billinghurst, M., and Alem, L. (2019). Sharing hand gesture and sketch cues in remote collaboration. *Journal of Visual Communication and Image Representation* 58: 428–438. https://doi.org/10.1016/j.jvcir.2018.12 .010.

51 Oda, O., Elvezio, C., Sukan, M., et al. (2015). Virtual replicas for remote assistance in virtual and augmented reality. *Proceedings of the 28th Annual ACM Symposium on User Interface Software and Technology.* ACM, pp. 405–415. http://doi.acm.org/10.1145/2807442.2807497.

52 Tait, M. and Billinghurst, M. (2015). The effect of view independence in a collaborative AR system. *Computer Supported Cooperative Work (CSCW)* 1–27. http://dx.doi.org/10.1007/s10606-015-9231-8.

53 Kim, S., Lee, G., Huang, W., et al. (2019). Evaluating the combination of visual communication cues for HMD-based mixed reality remote collaboration. *Proceedings of the 2019 CHI Conference on Human Factors in Computing Systems.* Glasgow, Scotland UK (04–09 May 2019). pp. 01–13. https://doi.org/10.1145/3290605.3300403.

54 Kim, S., Lee, G., Billinghurst, M., and Huang, W. (2020). The combination of visual communication cues in mixed reality remote collaboration. *Journal on Multimodal User Interfaces* https://doi.org/10.1007/s12193-020-00335-x.

55 Billinghurst, M., Gupta, K., Katsutoshi, M. et al. (2016). Is it in your eyes? Explorations in using gaze cues for remote collaboration. In: *Collaboration Meets Interactive Spaces*, 177–199. Springer.

56 Gupta, K., Lee, G.A., and Billinghurst, M. (2016). Do you see what I see? The effect of gaze tracking on task space remote collaboration. *IEEE Transactions on Visualization and Computer Graphics* 22 (11): 2413–2422. https://doi.org/10.1109/TVCG.2016.2593778.

57 Bai, H., Sasikumar, P., Yang, J., and Billinghurst, M. (2020). A user study on mixed reality remote collaboration with eye gaze and hand gesture sharing. *Proceedings of the 2020 CHI Conference on Human Factors in Computing Systems.* Honolulu, HI, USA (25–30 April 2019). pp. 01–13. https://doi-org.ezproxy.lib.uts.edu.au/10.1145/3313831.3376550.

58 Otsuki, M., Maruyama, K., Kuzuoka, H., and Suzuki, Y. (2018). Effects of enhanced gaze presentation on gaze leading in remote collaborative physical tasks. *Proceedings of the 2018 CHI Conference on Human Factors in Computing Systems.* ACM, pp. 368:1–368:11. http://doi.acm.org/10.1145/3173574.3173942.

59 Kuzuoka, H., Yamazaki, K., Yamazaki, A., et al. (2004). Dual ecologies of robot as communication media: thoughts on coordinating orientations and projectability. *Proceedings of the SIGCHI Conference on Human Factors in Computing Systems*, 2004. ACM, pp. 183–190. http://doi.acm.org/10.1145/985692.985716.

60 Yamamoto, T., Otsuki, M., Kuzuoka, H., and Suzuki, Y. (2018). Tele-guidance system to support anticipation during communication. *Multimodal Technologies Interact* 2: 55–55.

61 Piumsomboon, T., Lee, G.A., Hart, J.D. et al. (2018). Mini-me: an adaptive avatar for mixed reality remote collaboration. *Proceedings of the 2018 CHI Conference on Human Factors in Computing Systems.* Montreal QC, Canada (21–26 April 2018). pp. 01–13. https://doi.org/10.1145/3173574.3173620.

62 Wang, T.-Y., Sato, Y., Otsuki, M. et al. (2020). Effect of body representation level of an avatar on quality of AR-based remote instruction. *Multimodal Technologies and Interaction* 4 (1): 3. https://www.mdpi.com/2414-4088/4/1/3.

63 Günther, S., Kratz, S., Avrahami, D., and Mühlhäuser, M. (2018). Exploring audio, visual, and tactile cues for synchronous remote assistance. *Proceedings of the 11th PErvasive Technologies Related to Assistive Environments Conference.* ACM, pp. 339–344. http://doi.acm.org.ezproxy.lib.swin.edu.au/10.1145/3197768 .3201568.

64 Wang, P., Bai, X., Billinghurst, M. et al. (2020). Haptic feedback helps me? A VR-SAR remote collaborative system with tangible interaction. *International Journal of Human–Computer Interaction* 36 (13): 1242–1257. https://doi.org/10 .1080/10447318.2020.1732140.

65 Ou, J., Oh, L.M., Fussell, S.R., et al. (2005). Analyzing and predicting focus of attention in remote collaborative tasks. *Proceedings of the 7th International Conference on Multimodal Interfaces*, Toronto Italy (4–6 October 2005). ACM, pp. 116–123.

66 Ou, J., Oh, L.M., Fussell, S.R. et al. (2008). Predicting visual focus of attention from intention in remote collaborative tasks. *IEEE Transactions on Multimedia* 10 (6): 1034–1045.

67 Ou, J., Oh, L.M., Yang, J., and Fussell, S.R. (2005). Effects of task properties, partner actions, and message content on eye gaze patterns in a collaborative task. *Proceedings of the SIGCHI Conference on Human Factors in Computing Systems*, Portland Oregon USA (2–7 April 2005). ACM, pp. 231–240.

68 Wong, J., Oh, L.M., Ou, J., et al. (2007). Sharing a single expert among multiple partners. *Proceedings of the SIGCHI Conference on Human Factors in Computing Systems*, San Jose California USA (28 April 2007): ACM, pp. 261–270.

69 Gergle, D., Kraut, R.E., and Fussell, S.R. (2006). The impact of delayed visual feedback on collaborative performance. *Proceedings of the SIGCHI Conference on Human Factors in Computing Systems*, Montréal Québec Canada (22–27 April 2006): ACM, pp. 1303–1312.

70 Huang, W., Alem, L., Nepal, S., and Thilakanathan, D. (2013). Supporting tele-assistance and tele-monitoring in safety-critical environments. *Proceedings of the 25th Australian Computer-Human Interaction Conference: Augmentation, Application, Innovation, Collaboration, Adelaide, Australia, 2013.* https://doi .org/10.1145/2541016.2541065.

71 Yap, T.F., Epps, J., Ambikairajah, E., and Choi, E.H.C. (2015). Voice source under cognitive load: Effects and classification. *Speech Communication* 72: 74–95. https://doi.org/10.1016/j.specom.2015.05.007.

72 Liebenthal, E., Silbersweig, D.A., and Stern, E. (2016). The Language, tone and prosody of emotions: neural substrates and dynamics of spoken-word emotion perception, (in English), *Frontiers in Neuroscience, Review* 10 (506): https://doi.org/10.3389/fnins.2016.00506.

73 Kirschner, F., Paas, F., and Kirschner, P.A. (2008). Individual versus group learning as a function of task complexity: an exploration into the measurement of group cognitive load. In: *Beyond Knowledge: The Legacy of Competence*, 21–28. Dordrecht, Netherlands: Springer.

74 Kim, S., Lee, H., and Connerton, T.P. (2020). How psychological safety affects team performance: mediating role of efficacy and learning behavior. *(in English), Frontiers in Psychology, Original Research* 11 (1581): https://doi.org/10.3389/fpsyg.2020.01581.

75 Huang, W., Alem, L., and Livingston, M.A. (2013). *Human Factors in Augmented Reality Environments*, 274. New York: Springer Science+Business Media.

76 Le, H.C., Huang, W., Billinghurst, M., and Yap, E.H. (2021). Identifying human factors for remote guidance on physical tasks. In: *Cooperative Design, Visualization, and Engineering*, 271–283. Cham: Springer International Publishing.

5

Communicating Eye Gaze Cues in Remote Collaboration on Physical Tasks

5.1 Introduction

A dramatic increase in remote collaboration in recent years is seen to have a significant socioeconomic impact globally over the long term [1]. Remote guidance on physical tasks is one type of remote collaboration that has many applications in our everyday lives and in industrial sectors such as remote maintenance. A real-life scenario for remote guidance could be a local worker performing manipulations of physical objects under the guidance of a remote helper, which can save time and money as the remote helper is no longer required to travel to the site [2]. However, a key problem arises for collaborations under the remote guidance when compared to the face-to-face mode is how to share real-time workspace awareness. More specifically, challenges in real-time remote collaboration include monitoring the distant workspace, building up communication ground, sharing a referential system, and understanding the remote activities and perceptions of the collaborators via the available communication channels [3–5].

As indicated by Gutwin and Greenberg [6], eye gaze is an indicator of workspace awareness sharing where a participant is looking during groupware collaborations. While eye gaze can be observed in face-to-face collaboration with less effort, gaze input was often missing due to hardware limitations or availability in earlier video-mediated remote collaboration. The value of eye tracking as a method for investigating collaborative physical tasks in medical and other domains was explored [3, 7] until the series of experimental studies conducted by Fussell's research team in early 2000s. One of their experiments studied a remote-guided collaborative physical task in which the helper was eye tracked and instructed the worker to construct a robot head in year 2003 [7]. The gaze cues from the helper were found to be used to monitor worker's actions, establish joint focus of attention (FOA), and formulate messages as a pointing gesture. Another experiment conducted by the Fussell's research group revealed that a head-mounted camera with eye-tracking capabilities worn by a worker provided little benefit to

Computer-Supported Collaboration: Theory and Practice, First Edition.
Weidong Huang, Mark Billinghurst, Leila Alem, Chun Xiao, and Troels Rasmussen.
© 2024 The Institute of Electrical and Electronics Engineers, Inc. Published 2024 by John Wiley & Sons, Inc.

improve the efficiency of remote-guided robot construction task when compared to a scene camera setup at the worker's end [3], although, in the former case, the helper had more detailed information on worker's focus of attention. Please note that in the early 2000s, eye-tracking technology was considerably less developed than today, which limited the studies on the value of gaze information for remote collaboration on physical tasks. As high-quality eye tracking equipment is becoming much affordable than ever before, more examples are seen to bring values of gaze awareness to remote collaboration system design and practices. Remote collaboration on physical tasks using augmented reality (AR) and mixed reality (MR) technologies has attracted increasing interest in the past decades [8, 9].

In human–human collaboration, eye gaze is an instant and effortless input modality in addition to touch, speech, hand, and other body postures and gestures [10]. Eye gaze functions as a quick and precise pointing gesture and is very effective when the FOA is too complicated to describe verbally or by other ways [11, 12]. In addition to designating and monitoring practices [13], interactions of eye gaze during collaboration also help in synchronizing actions, improving mutual understanding and emotional awareness of collaborative participants, and so on [14]. We are interested in the advantages and limitations of the collaborative systems that implemented eye-tracking technology under remote guidance. Would the introduction of eye tracking contribute to high remote collaborative performance? If yes, what kind of system configuration could best improve the performance? This leads to our motivation to review the current research status on gaze that is used in collaborative systems to conduct physical tasks, or tasks aiming to support object manipulations that include a phase of locating an object. We expect that the literature review on this topic will help us to better understand the usage and effect of implementation of eye tracking in these systems, and to identify the technical and social challenges when developing eye tracking-supported systems for collaborative physical works under remote guidance. Our end goal is to offer information and evidence to support remote collaborative system design with eye-tracking facilities for physical tasks and similar application scenarios in the future.

This chapter surveys the research on the use of eye tracking and sharing of gaze cues for remote collaboration on physical tasks. The main contributions are:

- First, we demonstrate a changing research landscape with respect to research topics related to using eye tracking to support remote physical task collaborations over the last two decades.
- Second, we categorize the eye tracking-supported prototypes and systems according to system setup and gaze visualization, summarize the gaze functionality, and analyze and discuss the current challenges of using tracked eye gaze information in remote guidance for physical works.

- Last, we discuss the potential future research directions in remote collaboration systems, aiming to improve the flexibility and affordability of such systems with modest hardware configurations without compromising accuracy, to address privacy concerns, and to leverage machine learning technologies for enhanced remote collaborations.

5.2 The Changing Research Landscape – Research Topic Trends and Teams over the Past Two Decades

5.2.1 Method of Data Collection from Scopus

We collected articles published after the year 2000 by searching four keyword combinations from *eye tracking* or *gaze*, and *remote*, in addition to *collaborative* or *collaboration* in Scopus, as we wanted to reach a broad coverage of related publications with less restriction of search keywords. In this way, a bibliometric dataset of 149 articles was collected after removing duplicates and items without article title or author names.

5.2.2 Bibliometric Dataset Cleaning for Topic and Coauthorship Network Visualization

The collected bibliometric dataset from Scopus is cleaned by the following process in order to visualize the research topics and academic coauthorship.

- *Deduplication*: DOI is the key for raw data deduplication. In case DOIs are not available, deduplication is conducted by matching article title and authors.
- *Author keyword normalization*: The author keywords of articles are cleaned by removing nonalphanumeric characters, and normalizing selected keywords/terms including mapping term abbreviations to full names. Table 5.1 lists some mapping examples. We believe that the author keywords are more customized to reflect the content of the publications.
- *Author name normalization*: Author name duplicates are often found in forms like *Doe J.* and *Doe J.M.* Fortunately, in Scopus, each author name can be mapped to an author id so that a dictionary of names and ids can be built. However, exceptions can still be found due to reasons such as encoding issues. In our case, 2.48% of the author names are duplicates, where an author name has several mappings to Scopus author id. Interestingly but not surprisingly, most name duplicates seem to be Asian names in this corpus. Name duplicates are distinguished by appending a suffix in the form of "$n.m$" to the name, where n is the size of the specific duplicate subset, and m stands for the unique order number in the duplicate subset.

Table 5.1 Normalization of selected keywords and terms.

Original keyword/term	Normalized keyword/term
Nonalphanumeric characters except "-"	" " (whitespace)
Computer-supported cooperative work	CSCW
Computer-supported collaborative work	CSCW
Computer-supported collaborative learning	CSCL
Human–computer interaction	HCI
Video conferencing	Video conferencing
Virtual environments	Virtual environment

5.2.3 An Overview of Research Topic Changes and Collaborations

We visualize the co-occurrence of author keywords (Figure 5.1) and coauthorship (Figure 5.2) network of the cleaned bibliometric collection with Python and VOSviewer,[1] thus presenting an overview of the research trends and research group collaborations over the past two decades.

Figure 5.1 is the co-occurrence network of author keywords. The node size represents the occurrence of a keyword or term in the corpus, while the node darkness represents the average year of the publications . This visualization can be interpreted as a network of the specific research topics over time in the research field. The figure shows that the research topics related to *eye tracking* had been the focus before 2010. It is represented by the dark nodes like *focus of attention*. In the next decade from 2010, the focus of research topics had moved to light gray nodes like *augmented reality* and *mixed reality*. We can also notice that some light gray nodes represent the most recent topics of *gaze visualization, facial expression, mixed reality remote collaboration*, and *hand gesture*. It indicates that in the earlier years, the investigation focus was theoretical mechanism of eye tracking in remote collaboration due to technology constraints; in the two decades between 2010 and 2020, eye tracking has contributed to pair programming and immersive collaboration in addition to remote collaboration. Recently, the research focus has moved to the fields of augmented/mixed realities with more details such as gaze interaction and hand gesture.

Figure 5.2 is a visualization of the coauthorship network which shows individual authors' collaboration networks over the past two decades. In this figure, node size represents the number of publications that an author has in the bibliometric dataset, and the node darkness represents the average of publication years. We can

---

1 www.vosviewer.com.

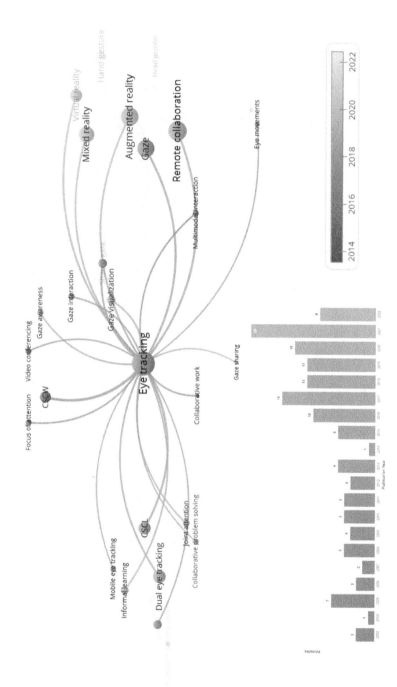

Figure 5.1 Author keyword co-occurrence network sourced from Scopus.

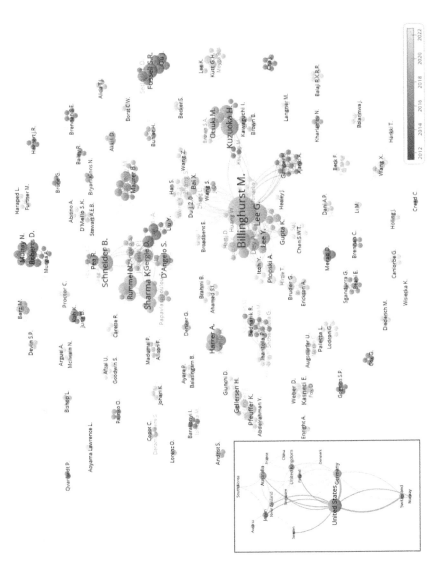

Figure 5.2 Coauthorship network, sourced from Scopus.

easily identify several research groups from this figure. For example, in the early 2000s, a group led by Fussell[2] was active in eye-tracking research, while in recent years, Billinghurst's group[3] and Kuzuoka's group[4] have been active in this space. Furthermore, a country-based coauthorship network shown at the bottom left corner of the figure indicates that authors from United States, Australia, and Japan significantly contributed to the research field.

5.2.4 Screening for Detailed Article Review

As shown in Figure 5.1, the collected dataset contains the research and experimental articles in a series of subtopics ranging from remote collaborative learning, pair programming, guided navigation, online cooperative gaming, to collaborative physical work. In this survey, we would like to focus on remote collaboration on physical tasks. However, we found that the keyword *physical* does not always appear in the metadata. Only 7 out of 141 articles contain *physical* in the author keywords, 10 and 36 articles contain this keyword in the index keywords and abstracts, respectively. It suggests that eye tracking has been deployed in a variety of remote collaborative systems. While all the subtopics share some common characteristics in terms of eye tracking, gaze information is collected and applied in human-to-human collaborations. We will narrow down the scope for detailed review by focusing on the systems used for physical task collaboration or part of a physical task collaboration, such as collaborative physical location search. A physical task can be specified as an object-oriented manipulation work. It could be a task in a real workspace scene, or a simulated scene shown on a screen or in a virtual environment, where the participants need to take operational actions on an object target. Physical task collaboration is a basic kind of human-to-human collaboration that we believe the studies on remote collaborative physical tasks would present the generalizability of use cases supported by eye tracking.

Due to the planned scope of this survey, a keyword-based screening procedure has been applied. Table 5.2 lists the keywords used for screening/filtering.

Table 5.2 Selected keywords for screening.

Applied to	Words for screening
Article titles	Shopping, writing, e-learning, drive
Author keywords	Robot, game, programming, education

---

2 http://sfussell.hci.cornell.edu/people/people.shtml.
3 http://empathiccomputing.org.
4 https://www.cyber.t.u-tokyo.ac.jp/ja/members.

Research topics like computer-supported collaborative learning (CSCL) will not be discussed specifically in this survey. The screening process is followed by a review process to select articles describing a collaborative physical task under remote guidance. As a result of the screening and review, 38 out of the 141 articles are identified. We examine again and find that there are 4, 6, and 19 out of the final selected 38 articles containing *physical* in author keywords, index keywords, and abstracts, respectively. The majority of articles in the screened set have been published in *ACM Conference series on Human Factors in Computing Systems (CHI)*, which is well known in the domain. Please note that there are overlaps in the subtopics, as the selected articles might cover several subtopics, even though the screening process has been applied. In the discussion of this survey, we also refer to several additional references from the reference lists of the selected articles to support the discussion.

5.3 Categorization of System Setup Based on the Screened Publications

In general, pair collaboration scenario applies to the screened publications. In this scenario, a remote participant named as helper collaborates with a local worker on an object manipulation task, in which most of the tasks have a procedure to locate an object. This pair collaboration scenario also applies to group collaboration where either the helper or worker could be replaced by multiple participants. In some cases, the work environment is virtual, or one or several of the participants are remote or virtual representation. Table 5.3 and Figure 5.3 summarize the typical setup of the screened publications with regards to eye-tracked subjects, display configuration, and the type of gaze awareness which eye tracking is applied for. Gaze awareness is specified by the applied subject, i.e. the helper's or worker's gaze in workplace, or full gaze awareness (FGA) specified by the eye-tracked subject, for example, the helper can see where the worker is looking when the worker is eye tracked, and/or vice versa, or mutual gaze with which both helper and worker can tell that they are looking at each other.

Basically, a remote physical task collaborative system consists of the setup of a local worker and a remote helper, who are separated by a physical distance but connected via audio and shared video. Figure 5.3 illustrates several typical setups from both ends. A local worker can be equipped with a wearable headset with display, camera, and eye tracker (as worker's setup 1(c)), or a desktop or tablet with a fixed or wearable eye tracker (as worker's setup 4). If the workspace is relatively small, a local worker can work with the monitor of a scene camera and guided by a helper's augmented gaze pointer from a projector (as worker's setup 2), or

Table 5.3 A summary of eye tracking-supported collaborative system setup based on the screened publications.

Publications	Year	Worker's setup	Helper's setup	Awareness type	Gaze visualization
Eye gaze observed but not visible to other participants					
[7]	2003	—	H2(a)	Helper's gaze	—
[3]	2003	W1(b)	—	Worker's gaze	—
[15–17]	2005–2008	—	H2(a)	Helper's gaze	—
Eye gaze overlaid on a fixed screen or HMD					
[18]	2010	W4(b)	H1	FGA[a)] of both	Cursor
[19]	2012	W4(b)	H1	Helper's FGA	Hotspot
[20]	2012	W4(b)	H1	Helper's FGA	Hotspot
[21]	2014	W4(b)	H1	FGA of both	Cursor
[22]	2016	W1(c)	H1 w/o ET	Worker's FGA	Cursor
[23]	2016	W1(c)	H1 w/o ET	Worker's FGA	Cursor
[24]	2016	W4(b)	H1	FGA of both	Cursor
[25]	2016	W4(b)	H1	FGA of both	Cursor or trail
[26]	2016	W1(c) w/o ET	H1	Helper's FGA	Cursor
[27]	2017	W1(c)	H1	FGA of both	Cursor
[28]	2017	W4(b)	H1	FGA of both	Cursor, trajectory highlight, spotlight
[29]	2018	W4(a)	H1	Helper's FGA	Cursor
[30]	2018	W4(b)	H1	FGA of both	No vis, heatmap shared area, pointer
[31]	2018	—	H1	Helper's FGA	Physical equipment ThirdEye
Augmented eye gaze projected on the workspace					
[32, 33]	2016–2019	W2	H1	Helper's FGA	Projected gaze
[26]	2016	W2	H1	Helper's FGA	Projected gaze
[34]	2017	W2	H2(a)	Helper's FGA	Projected gaze
[35–37]	2018–2022	W2	H2(b)	Helper's FGA	Projected gaze

(Continued)

Table 5.3 (Continued)

Publications	Year	Worker's setup	Helper's setup	Awareness type	Gaze visualization
Augmented eye gaze in augmented/virtual/mixed reality and virtual environment					
[38]	2004	W1(c)	H2(c)	FGA of both	Cursor
[39]	2009	W1(b)	H2(a)	FGA of both	Gaze ray
[40]	2016	W3	System response	Worker's FGA	Cursor
[8]	2017	W1(a) or W3	Virtual helper	Worker's FGA, mutual	Gaze ray
[9, 41, 42]	2017–2019	W3	H3	FGA of both	Gaze ray
[43]	2019	W3	H3	FGA of both	Gaze ray
[12]	2020	W3	H3	FGA of both	Gaze ray
[44–46]	2021–2022	W3	H3	FGA of both	Gaze ray

a) FGA: full gaze awareness.

a worker may also be equipped with an AR/VR headset to conduct the work in a relatively large work environment. The apparatus setup of a remote helper is quite similar to that of the above, but more often a desktop with an eye tracker is set up at the helper's end (as helper's setup 1). A remote helper can also wear a head-mounted display with eye tracker (as helper's setup 2(b)). In both cases, the hand gesture or head gaze, or facial expression of both the helper and worker can be tracked by optional equipment. A remote helper can also collaborate with a worker in AR/virtual reality (VR)/MR when wearing a VR headset (as helper's setup 3).

When considering the display equipment where the eye gaze is visualized, we can categorize the eye tracking-supported systems into three types, i.e. where the eye gaze is overlaid on the screen, displaying the shared video stream, or the remote helper's eye gaze is augmented and projected in a fixed workspace, or the eye gaze is augmented and visualized in AR/VR/MR or virtual environment. Please note that the studies in the early 2000s did not apply overlaid eye gaze due to technology constrains. However, their investigation had led to more experimental studies later, thus significantly supported and validated theoretical research in this field.

Different system configurations in the studies which are listed in Table 5.3 support the investigations on the usability of gaze information from the helper, the worker, or both. We will discuss the functionality of tracked eye gaze in Section 5.5 and the effect of different gaze visualizations in Section 5.4.5.

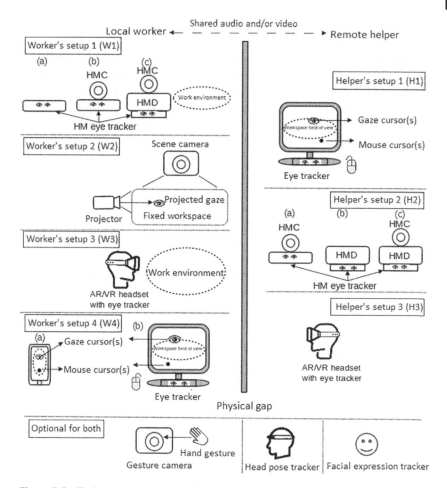

Figure 5.3 Typical system setups with eye tracker.

5.3.1 Eye Gaze Overlaid on a Fixed Screen or Head-Mounted Display

The system used by D'Angelo and Gergle [24] was a symmetric dual eye-tracking system with worker's setup 4(b) and helper's setup 1. In the collaborative puzzle assembly task, the eye gaze of one participant was visualized in the shared view as gaze cursor and monitored by the other. In the spatial search work later [30] and the work from Zhang et al. [28], eye gaze from both participants were overlaid on the screen. Another dual eye-tracking system as reported by Lee et al. [27] was set up using a fixed computer (helper's setup 1) and a wearable equipment (worker's setup 1(c)), where eye gaze of both participants were overlaid on the monitor in addition to the mouse cursors.

In asymmetric eye-tracking systems described in [22, 23, 26, 29, 47, 48], only one participant was eye tracked, and the gaze was visualized and monitored by the other. This type of systems were typically set up by equipping a remote helper with a fixed device (helper's setup 1) and a local worker with a wearable device (worker's setup 1), but only one end was eye tracked. A wearable device is much flexible and practical for a local worker in a real workspace.

In addition to eye gaze cues, some of the prototypes and systems applied other visual cues like augmented annotation, mouse cursor, or hand gesture to improve collaboration.

Most of the studies reported that shared eye gaze improves collaboration quality particularly when referring to linguistically complex objects, as well as significantly improves collaborative experience. However, continuous gaze visualization, lagged or inaccurate gaze signal cause distractions and confusions. The former case has been studied as *Midas touch* problem [10], which could be improved or solved by special system design (see Section 5.7). The latter case is believed to be eased and solved with the development of eye-tracking data processing, analytic techniques, and hardware design. In addition, the number of visible gaze and mouse cursors from both ends, in the case when two gaze cursors and two mouse cursors are all shown in the interface, could probably add extra cognitive load to the collaborators. The studies also suggest eye tracking-supported system design to take account of the burden on participants who carry or wear specific equipment. Gaze information needs to be considered to be visualized at a right time and right places to avoid distractions and additional cognitive load.

5.3.2 Augmented Eye Gaze Projected on a Fixed Workspace

In some studies, augmented eye gaze of a helper was projected into a relative small or fixed workspace. This kind of systems are usually set up with worker's setup 2 without eye tracker and helper's setup 1 or 2. Hand gesture or head pose of a helper can also be tracked to enforce remote guidance. In both cases, the remote helper can monitor the work status using the shared video on a screen or head mounted display (HMD).

Akkil et al. [32, 33] projected the helper's gaze as a torch/cursor into the worker's physical workspace to guide building blocks, as Higuch et al. [26] and Wang et al. [36, 49] did. van Rheden et al. [34] also projected the eye gaze of the helper onto a 6 × 6 grid whiteboard to guide the worker to draw a shape by connecting grid edges with lines. It should be noted that though Rheden's work is a located collaboration, their findings can be used as a reference for remote collaboration.

Compared to the case that a worker has to wear a head-mounted device or has to observe a display or a screen showing instructions from the helper, this relatively free-style work mode is much closer to a colocated one. The projected eye gaze in the real workspace strongly improves the efficiency. However, the limitation is also obvious. A gaze visual can be identified in a relatively small and plain workspace as seen in the studies. It will be challenging to identify a gaze visual in a big and complicated work environment, as noted in [50] that "monitoring and displaying 3D gaze in physical collaborative tasks is more challenging than tracking gaze on a 2D screen".

5.3.3 Augmented Eye Gaze in Virtual Environment

Immersive collaboration experience is one of the great advantages that augmented reality technologies have brought [51]. Avatars which were used to represent real-world participants in AR/VR/MR or virtual environment, when equipped with head pose and eye gaze, strongly improve the immersive experience as well as the collaboration efficiency.

In a remote collaboration supported by AR/VR/MR technologies, a typical scenario is that a local worker wears an AR headset with eye tracker (worker's setup 3) and a helper wears a VR headset with eye tracker (helper's setup 3), such as the CoVAR system [41]. Please note that some immersive video conferencing studied groupware collaborations, where all participants wear a VR headset [39]. Eye gaze from each participant is represented by avatars in a virtual environment.

5.4 Gaze Visualization

Visible gaze explicitly helps to attract awareness and guide attention. As there are different system configurations and applied environments, various gaze visualizations have been designed for different prototypes and systems.

5.4.1 Gaze Cursor in Two-Dimensional View

Eye gaze has been most commonly visualized as a cursor with a few variations such as a spot or eye-shaped icon visible on a 2D scene screen or display. In addition to gaze cursors, Chetwood et al. [19] used hotspots to represent the gaze of a supervisor over time in a simulated laparoscopic surgical equipment to guide surgeon training. The trainee's gaze was visualized the same way and monitored by the supervisor. Higuch et al. [26] visualized the helper's gaze in a combined way for an object manipulation task, where the gaze direction was represented as a piece of highlighted arc. When the gaze target was out of the limited HMD field of view of the worker, and the exact gaze focus in the worker's view was visualized as a square cursor. Li et al. [25] visualized gaze as gaze trail or zoom focus for a simulated collaborative puzzle assembly task and a bomb diffusion task.

During a joint spatial search task, D'Angelo and Gergle [30] studied and compared using heatmap, gaze path, or shared area as gaze visualizations for a joint on-screen pattern search task. They found that the collaborative visual search task was completed in less time when gaze was visualized as a shared area or gaze path, and the collaboration with gaze visual took less time than without gaze visual. Zhang et al. [28] studied four gaze visualizations including cursor, trajectory, highlight, and spotlight for a colocated map search task and reported that less explicit gaze visualization was preferred. The gaze path (trajectory) should be avoided, as high visibility of gaze path might cause more distraction.

The different findings of the two studies mentioned above reveal the fact that gaze visualization should be designed to fit different tasks, and more detailed gaze information is preferred in a pattern/shape search task, while explicit details of gaze movement do not help much for a target search task. The findings also suggest that displaying gaze indicators helps to establish joint attention using FGA so that the collaboration efficiency can be improved. Although Zhang et al.'s study focuses on colocated collaboration, their findings hold significance for remote collaborations as well. The insights gleaned from these investigations also inform further studies in gaze visualization and eye-tracking data processing and modeling.

5.4.2 Gaze Visible as a Projected Pointer in Physical Workspace

In Section 5.3.2, we discussed the projected gaze onto the workspace, where the helper's gaze was represented as either a torch [32], a cursor [26, 34, 36, 49], or gaze path [26] that preserves a brief history of gaze travel. This visualization proved informative guiding an object placement action from one place to another. However, it's important to note that the effectiveness of projecting gaze onto the workspace may be limited in large open spaces.

5.4.3 Gaze as a Cursor or Ray-Cast in AR/VR/MR Carried by Avatars

In a virtual environment, Duchowski et al. [38] made gaze visible as a lightspot to reference objects for both collaborators. The tracked eye gaze in MedicalVR system was visualized as a gaze cursor to trigger system response when it was overlaid on specific objects in virtual reality [40]. Steptoe et al. [39, 52] studied the difference among static gaze, tracked gaze, and modeled gaze pattern by head direction in an object-focused collaboration scenario in virtual environment. They reported that avatars with tracked or static gaze were seen to support significantly higher quality of communication, while tracked or static gaze did not show a significant difference. Andrist et al. [8] studied bidirectional gaze model in which tracked gaze was used to monitor the physical task status and reference objects. Their system enabled the virtual agent to interact with human collaborator in sharing mutual gaze, which significantly improves the immersive experience. Eye gaze of both AR and VR users were tracked and visualized as virtual ray-cast to instruct object manipulations in a real environment in system CoVAR [41] and MiniMe [9], where both worker's and helper's avatars were shown side-by-side in a mixed-reality environment. The studies and experiments show that tracked gaze can be used in AR/VR/MR not only to make avatars more alive as a significant channel for mutual communication, but also can function as gaze awareness pointer to support object manipulations.

5.4.4 Gaze Visualization with Controlled Visibility

To avoid the distractions caused by nonfiltered gaze visuals, some systems are designed to be able to visualize gaze in a controlled manner. The AR videoconferencing interface designed by Barakonyi et al. [47] set the eye cursor hidden by default, which could be selectively turned on by a keyboard shortcut to avoid meaningless gaze cues for object selection. Speicher et al. [53] also reported several cases in which the remote collaborative participants turned off the gaze awareness component showing the participant's viewpoint when audio and other communication options were available. The ThirdEye developed by Otsuki et al. [31] guided the attention in a video conference, where the gaze was not explicitly overlaid in the workspace. Instead, an anchored eye-shaped button kept showing eye movements like a real eye ball. As demonstrated in their work, the ThirdEye successfully led the attention.

5.4.5 Effectiveness of Gaze Visibility and Visualization

As shared gaze is reported to effectively reduce remote collaboration time and simplify conversations during collaboration in most study cases, we believe that the visibility of gaze cues is practical to most of the system designs, and an appropriate form of gaze visualization could improve the collaboration even further. Gaze visualizations vary with respect to different types of tasks and system setups. Basically, if the collaborative task is monitored and presented on a screen, a gaze cursor is the common choice with quite a few variations. In particular, gaze visualized as a gaze path could be applied when moving an object from one place to another, which is believed to guide object-moving actions even more effectively. However, a gaze path was reported to cause distractions and should be avoided in collaborative target search tasks [28], as the target but not the search history is of more interest to a locating task.

The visible gaze in a physical workspace could also improve collaboration, as the on-site worker would not need to pay additional attention to a shared view screen. The augmented gaze pointer can also improve the social copresence although a privacy issue may arise. For example, the on-site worker might feel "being watched" [32], or the helper might not want to keep sharing the gaze awareness for some reason [34].

5.5 Functionality of Tracked Gaze in Remote Guidance on Physical Tasks

The key role that tracked gaze information plays in the remote guidance on physical tasks is to lead gaze awareness for joint attention identification. Once a participant's gaze awareness attracts the attention of the other, it will help to formulate messages for common understanding or provide visual reference for object manipulation [7]. The functionality of tracked gaze information can be interpreted from different perspectives.

As a fast and effective nonverbal input modality: The experiments reported in [21–24, 26, 32, 54] compared the collaborative efficiency under the conditions when gaze was visible and invisible during a remote collaboration. The results suggest that visible gaze improves the collaborative efficiency, as "eye fixations visualized by both a projector and an HMD show a fast and precise pointing capability over hand gestures" [26]. This is consistent with the finding that shared gaze significantly reduced completion time in a face-to-face collaborative construction task, as Lischke et al. [55] reported. Chetwood et al. [19] compared verbal-based and gaze-based instructions in a collaborative surgeon training. They found that eye gaze is especially useful in the collaboration between clinicians from different countries or background who spoke different languages and were under different physical conditions. In addition, face masks wore by the clinicians make verbal communication less effective during the collaboration. Similarly, Kwok et al. [20] indicated that eye gaze is more advantageous to carry information on the focus of the surgeon's attention than other input modalities in a collaborative surgery because during the surgery, the hands of the surgeons tend to be fully occupied with surgical instruments and verbal communication is limited when wearing masks. In particular, eye gaze cues are of great value to participants with speech and hearing issues. These investigations reveal that gaze information is a very important input modality supplementing body gesture and speech in collaborative physical tasks under remote guidance.

As a visual indicator of gaze awareness: Gaze information indicates what and where people are looking. Detection of collaborator's gaze awareness helps improving collaboration. For example, the surgery training system tested by Kwok et al. [20] enables a trainee to follow the supervisor's visible gaze to locate the tissue which needs to be removed. A visible gaze significantly improves the collaboration efficiency by resolving ambiguities, thus reducing the conversation time.

As an indicator to predict intentions: People's gaze also reflects what is going on inside of them. The studies of Ou et al. [15, 16] identified that the helper's FOA for physical tasks were the workspace, the jigsaw piece hub, and target puzzle when monitoring the helper's eye movement. Based on their observations, a Markov model to predict the helper's FOA was proposed [17]. Results showed that the amount of time a helper looking at different targets could be accurately predicted, given certain types of tasks and repeated trails. It indicates that it is possible to automate the scene camera to zoom in what the helpers might be interested in. Akkil and Isokoski [56] examined the gaze augmented in a video recorded from a head mounted camera (HMC) in a simulated driving task, and their results suggested that videos with augmented gaze, along with the available contextual information, could efficiently trigger the viewers' action to predict the observed people's intention. Higuch et al. [26] also reported that the worker could predict the intention of the helper when the helper's gaze was visible even without speech in a specific context. In addition to Markov model, a support vector machine (SVM) model was also applied to predict intended spoken requests in a sandwich assembly task [57], and the predictor was reported 1.8 seconds faster than the spoken requests on average. Smart collaborative system design has a promising future with reliable intention prediction.

As a trigger for system responses: Eye gaze can be used as a trigger for conversations and physical actions between collaborators and system responses. In MedicalVR system developed by Luxenburger et al. [40], the VR user's eye gaze could trigger system annotations to guide further physical operations when the user's eye gaze was overlaid on the objects of interest. In the bidirectional gaze VR model of Andrist et al. [8], the avatar could respond differently according to the AR user's eye gaze. For example, the avatar could wait until the AR user looked at the right object, or looked back to the AR user when being watched for social interaction purpose. In the CoVAR system developed by Piumsomboon et al. [41], when eye gaze of both AR and VR users spotted an object longer enough, the color of the targeted object could change as designed.

As a communication channel to improve social interactions: Eye-tracking techniques have been applied in AR/VR/MR with avatars as participant embodies [8, 9, 12, 38, 39, 41]. This was often done to support social interactions among participants in face-to-face collaborations, where eye gaze not only functions as a gesture and indicator of awareness to support physical operations, but also conveys mutual awareness to improve immersive copresence during remote collaboration.

5.6 Challenges of Utilizing Eye Tracking in Remote Collaboration

Gaze cues are usually applied in addition to other cues such as verbal conversation [19, 26, 33], head direction [8, 36, 49], hand gesture [12, 26, 47], facial expression [23, 58], cursor pointer [22], or a combination of several of these [12, 47]. The effectiveness of gaze cues is often investigated in user studies and evaluated by metrics such as task completion time, number of mistakes, number of words or phrases used in conversation, accompanied by subjective user questionnaires [26]. In most study cases, the shared gaze cues are reported to be able to improve the performance of collaborative tasks under certain conditions when compared to a specific baseline condition. Meanwhile, shared gaze is also found to cause distractions and confusions [12, 26, 33, 59]; due to the ambiguity between intended and unintended gaze, a person could be thinking when staring at somewhere. Identification of true gaze awareness is a challenging task that requires reliable eye tracking, a robust modeling that could extract intention from eye-tracking data with certain contextual information, and an optimum design of gaze visualization.

As mentioned above, gaze shown as a gaze pointer or a referential gesture is very effective when the referenced object is complicated to be described verbally or by other input modalities. However, it is most powerful when it is used in combination with other input modalities such as conversation and body gestures [59, 60]. Investigation of Higuch et al. [26] indicated that gaze cues can provide explicit instruction when they are combined with speech. Akkil and Isokoski [33] reported a similar case in which the gaze pointer needs more verbal instructions to accomplish a physical task under remote guidance. Consistently, gaze cues in AR/VR/MR were found to work together with other gesture cues better than to work alone for remote collaborative physical tasks, based on a verbal condition. The gaze cue alone works better only when referring to a complicated spatial layout and self-location awareness scenario than the combined cues [12]. The studies suggest that additional input modalities need to be considered for eye tracking-supported system design. However, it does not mean that the more facilities and indicator visualizations, the better. For example, it could be investigated if the mutual visible gaze pattern (the gaze cursor is only visible to the other side) or mirror visible gaze pattern (both gaze cursors are visible in the shared view) can function the same in dual eye-tracking systems, as the mirror visible gaze pattern may cause distractions and raise the collaborators cognitive load.

In addition to the technical challenges to track eye gaze in a 3D physical work environment, another practical challenge for eye tracking-supported system is the selection criteria of participants as they need to wear VR headsets with eye tracking as reported in [12]. Unfortunately, candidates with large head size or frame glasses, or having a wide distance between the eyes, may not be able to use eye-tracking in a wearable device.

When comparing mouse cursor pointer, head pointer, and eye gaze pointer, Wang et al. [49] found that the head pointer and eye gaze led to better collaboration quality, and the head pointer could be considered as a low-cost substitute for the eye gaze pointer [36]. Andrist et al. [8] also investigated the possibility of using head pose as a substitute for eye gaze. In addition to physical tasks, head pose has also been considered as a substitute for eye gaze in other collaborative tasks such as online gaming. However, it was found that shared eye gaze reacted faster than head gaze, as well as led to better immersive collaborative game experience, as reported by Špakov et al. [61]. There are also systems with no eye tracker or head tracker to capture gaze awareness, such as the live mobile remote physical task collaborative system developed by Gauglitz et al. [62], where the workspace scene was tracked and modeled incrementally in the real time and in 3D, using monocular vision-based simultaneous localization and mapping (SLAM) and subsequent surface modeling. Anchoring of annotations, virtual navigation, and synthesis of novel views were supported by the emerging model to enable a remote helper to effectively share gaze awareness with the on-site worker. The studies suggest that head pose could be considered as a much more affordable substitute for eye gaze, and alternative systems instead of eye tracking-supported systems could also work well for collaborative task that has no extreme time and precision requirements for eye tracking.

In addition to the physical functional usage of eye gaze as a pointer or action indicator in a collaborative work, mediating eye gaze is essential to improve social copresence for collaborative tasks. However, observing or visualizing a user's gaze may cause privacy concerns. In gaze torch study [32], feeling of *being watched* was reported by the on-site worker when the helper's non-filtered gaze visual was projected in the workspace at the worker's end. The eye-tracked collaborator of LaserViz prototype [34] was observed to interplay between intended and unintended gaze. This was called "attention dilemma," which the prototype wearer changed his behavior to avoid confusion or cover true gaze awareness. The studies suggest that eye tracking-supported systems could be designed to meet privacy needs, such as visualizing gaze only in workspace and the objects of interest, giving the user the flexibility to control the gaze visibility with minimum effort, which could be very challenging as well.

5.7 Future Directions

Gaze information acts as a subtle indicator of people's thinking behavior, as well as a physical pointing gesture. If only considering pointing purpose, a mouse, hand gesture, or other annotating input can work well or even outperform eye tracking, which leads to the question that in what scenarios eye tracking is necessary. The answer is many, as the prototypes and systems reviewed in this survey, remote surgery, navigation assistant system, and the tasks that have strict time limit or are time-sensitive, or restricted by physical conditions where verbal communication and body gesture might be unavailable, which could be benefit from this powerful eye-tracking input.

In response to the challenges that exist in the current systems, promising future research directions include identification of true gaze awareness and intention prediction, design of smart collaborative systems with privacy concerns. As presented in the survey, identification of joint attention is a key phase in collaborative tasks. However, in eye tracking-supported systems, not all overlapped gaze awareness leads to a common ground and joint attention. If the joint attention can be properly identified and predicted, correspondent annotations could be designed in the system to better support remote collaboration, so that the systems could respond smartly to better control the collaboration quality and reduce collaborators' cognitive load.

Machine learning is believed to support modeling. Silva et al. [63] also noted that machine learning could be used to model higher-level intents using eye-tracking data. Given the prototypes and systems surveyed in this survey are set up under different assumptions and validation conditions, shared datasets for machine learning could be built by collecting data for eye tracking-supported collaborative physical works under remote guidance. Building annotated shared datasets from collaborative scenarios will contribute to the setup of common criteria for system evaluation in the research community. One of the challenges is data collection which can be a complicated process. It is very challenging to collect a large volume of qualified data, as both quantity and quality of training data are very important for machine learning. In some applications such as rule extraction by machine learning, take reinforcement learning for example, building a dataset in a limited scale is possible. It even worked on a subset of the data [64], which leads to promising opportunities for more applications. The other challenge is data annotation. Collected data needs to be annotated by well-designed algorithms and/or well-trained *professionals* [65], which is both time-consuming and expensive. Despite the challenges, SVMs have been tested to achieve 76% accuracy in intention prediction from gaze cues alone in a collaborative assembly task using a dataset collected from 13 dyads of participants [57]. Most recently, Xu and Song [66] conducted experiment on a 20 volunteer group using Gaussian Auto-regression Hidden Markov modeling to predict eye gaze movement, the accuracy was reported to be 73.7%. Transfer learning has been seen to contribute significantly to gaze intention prediction as well. As a data-driven framework utilizing clustering, SVM, and transfer learning have been reported to achieve an average classification accuracy of 97.42% for gaze intention prediction [67]. A Microsoft research team developed a data-driven model for eye tracking using webcam, where human intention can be predicted using deep neural networks [68]. The model is built up based on an eye-tracking dataset collected by Krafka et al. [69] from 1474 people. More datasets are expected to be built and hopefully to be shared in the near future.

5.8 Conclusion

In this survey, we presented an overview of the state-of-the-art research on eye-tracking supported collaborative physical works under remote guidance. After studying the changing research landscape over the past two decades, we summarized and discussed typical prototypes and systems setups, eye gaze visualization, and gaze functionality in a remote physical task collaboration scenario. Our findings from most case studies indicate that collaboration performance and user experience have been improved when applying tracked gaze information, though there are technical and social challenges with respect to critical application issues. We suggest potential future research directions which include but are not limited to smart system design with gaze intention prediction, improving system flexibility and affordability with the modest hardware configuration without losing accuracy, taking care of privacy, and embracing machine learning to improve remote collaboration which can benefit the design of eye-tracking supported collaborative systems in the future.

References

1 Marks, P. (2020). Virtual collaboration in the age of the coronavirus. *Communications of the ACM* 63: 21–23.

2 Huang, W. and Alem, L. (2013). Gesturing in the air: supporting full mobility in remote collaboration on physical tasks. *Journal of Universal Computer Science* 19: 1158–1174.

3 Fussell, S.R., Setlock, L.D., and Kraut, R.E. (2003). Effects of head-mounted and scene-oriented video systems on remote collaboration on physical tasks. In: *Proceedings of the SIGCHI Conference on Human Factors in Computing Systems*, CHI '03, 513–520. New York, NY, USA: Association for Computing Machinery.

4 Le Chénéchal, M., Duval, T., Gouranton, V. et al. (2019). Help! i need a remote guide in my mixed reality collaborative environment. *Frontiers in Robotics and AI* 6: 106

5 Kim, S., Lee, G.A., Huang, W. et al. (2019). Evaluating the combination of visual communication cues for HMD-based mixed reality remote collaboration. In: *Proceedings of the 2019 CHI Conference on Human Factors in Computing Systems*, CHI '19, 1–13. New York, NY, USA: Association for Computing Machinery.

6 Gutwin, C. and Greenberg, S. (2002). A descriptive framework of workspace awareness for real-time groupware. *Computer Supported Cooperative Work (CSCW)* 11 (1): 411–446.

7 Fussell, S.R., Setlock, L.D., and Parker, E.M. (2003). Where do helpers look? Gaze targets during collaborative physical tasks. In: *Extended Abstracts of the 2003 Conference on Human Factors in Computing Systems*, CHI '03, 768–769. New York, NY, USA: Association for Computing Machinery.

8 Andrist, S., Gleicher, M., and Mutlu, B. (2017). Looking coordinated: bidirectional gaze mechanisms for collaborative interaction with virtual characters. In: *Proceedings of the 2017 CHI Conference on Human Factors in Computing Systems*, CHI '17, 2571–2582. New York, NY, USA: Association for Computing Machinery.

9 Piumsomboon, T., Lee, G.A., Hart, J.D. et al. (2018). Mini-me: an adaptive avatar for mixed reality remote collaboration. In: *Proceedings of the 2018 CHI Conference on Human Factors in Computing Systems*, CHI '18 (ed. A. Cox and M. Perry). New York, NY, USA: Association for Computing Machinery. *International Conference on Human Factors in Computing Systems 2018*, CHI 2018; Conference date: 21-04-2018 Through 26-04-2018.

10 Jacob, R. and Stellmach, S. (2016). What you look at is what you get: gaze-based user interfaces. *Interactions* 23 (5): 62–65.

11 Müller, R., Helmert, J., and Pannasch, S. (2014). Limitations of gaze transfer: without visual context, eye movements do not to help to coordinate joint action, whereas mouse movements do. *Acta Psychologica* 152: 19–28.

12 Bai, H., Sasikumar, P., Yang, J., and Billinghurst, M. (2020). A user study on mixed reality remote collaboration with eye gaze and hand gesture sharing. In: *Proceedings of the 2020 CHI Conference on Human Factors in Computing Systems*, CHI '20, 1–13. New York, NY, USA: Association for Computing Machinery.

13 Rae, J., Steptoe, W., and Roberts, D. (2011). Some implications of eye gaze behavior and perception for the design of immersive telecommunication systems. In: *Proceedings - IEEE International Symposium on Distributed Simulation and Real-Time Applications*, 108–114. USA: IEEE.

14 Chanel, G., Bétrancourt, M., Pun, T. et al. (2013). Assessment of computer-supported collaborative processes using interpersonal physiological and eye-movement coupling. In: *2013 Humaine Association Conference on Affective Computing and Intelligent Interaction*, 116–122. USA: IEEE.

15 Ou, J., Oh, L.M., Fussell, S.R. et al. (2005). Analyzing and predicting focus of attention in remote collaborative tasks. In: *Proceedings of the 7th International Conference on Multimodal Interfaces*, ICMI '05, 116–123. New York, NY, USA: Association for Computing Machinery.

16 Ou, J., Oh, L.M., Yang, J., and Fussell, S.R. (2005). Effects of task properties, partner actions, and message content on eye gaze patterns in a collaborative task. In: *Proceedings of the 2005 CHI Conference*, CHI '05, 231–240. New York, NY, USA: Association for Computing Machinery.

17 Ou, J., Oh, L., Fussell, S.R. et al. (2008). Predicting visual focus of attention from intention in remote collaborative tasks. *IEEE Transactions on Multimedia* 10: 1034–1045.

18 Carletta, J., Hill, R., Nicol, C. et al. (2010). Eyetracking for two-person tasks with manipulation of a virtual world. *Behavior Research Methods* 42 (1): 254–265.

19 Chetwood, A., Kwok, K.-W., Sun, L.-W. et al. (2012). Collaborative eye tracking: a potential training tool in laparoscopic surgery. *Surgical endoscopy* 26: 2003–2009.

20 Kwok, K.-W., Sun, L.-W., Mylonas, G. et al. (2012). Collaborative gaze channelling for improved cooperation during robotic assisted surgery. *Annals of Biomedical Engineering* 40: 2156–2167.

21 Bard, E.G., Hill, R.L., Foster, M.E., and Arai, M. (2014). Tuning accessibility of referring expressions in situated dialogue. *Language, Cognition and Neuroscience* 29 (8): 928–949.

22 Gupta, K., Lee, G.A., and Billinghurst, M. (2016). Do you see what I see? The effect of gaze tracking on task space remote collaboration. *IEEE Transactions on Visualization and Computer Graphics* 22 (11): 2413–2422.

23 Billinghurst, M., Gupta, K., Katsutoshi, M. et al. (2016). *Is It in Your Eyes? Explorations in Using Gaze Cues for Remote Collaboration*, 177–199. Switzerland: Springer.

24 D'Angelo, S. and Gergle, D. (2016). Gazed and confused: understanding and designing shared gaze for remote collaboration. In: *Proceedings of the 2016 CHI Conference on Human Factors in Computing Systems*, CHI '16, 2492–2496. New York, NY, USA: Association for Computing Machinery.

25 Li, J., Manavalan, M., D'Angelo, S., and Gergle, D. (2016). Designing shared gaze awareness for remote collaboration. In: *Proceedings of the 19th ACM Conference on Computer Supported Cooperative Work and Social Computing Companion*, CSCW '16 Companion, 325–328. New York, NY, USA: Association for Computing Machinery.

26 Higuch, K., Yonetani, R., and Sato, Y. (2016). Can eye help you? Effects of visualizing eye fixations on remote collaboration scenarios for physical tasks. In: *Proceedings of the 2016 CHI Conference on Human Factors in Computing*

Systems, CHI '16, 5180–5190. New York, NY, USA: Association for Computing Machinery.

27 Lee, G., Kim, S., Lee, Y. et al. (2017). [POSTER] Mutually shared gaze in augmented video conference. In: *2017 IEEE International Symposium on Mixed and Augmented Reality (ISMAR-Adjunct)*, ISMAR '17, 79–80. USA: IEEE.

28 Zhang, Y., Pfeuffer, K., Chong, M.K. et al. (2016). Look together: using gaze for assisting Co-located collaborative search. *Personal and Ubiquitous Computing* 21: 173–186.

29 Akkil, D., Thankachan, B., and Isokoski, P. (2018). I see what you see: gaze awareness in mobile video collaboration. In: *Proceedings of the 2018 ACM Symposium on Eye Tracking Research & Applications*, ETRA '18. New York, NY, USA: Association for Computing Machinery.

30 D'Angelo, S. and Gergle, D. (2018). An eye for design: gaze visualizations for remote collaborative work. In: *Proceedings of the 2018 CHI Conference on Human Factors in Computing Systems*, CHI '18, 1–12. New York, NY, USA: Association for Computing Machinery.

31 Otsuki, M., Maruyama, K., Kuzuoka, H., and Suzuki, Y. (2018). Effects of enhanced gaze presentation on gaze leading in remote collaborative physical tasks. In: *Proceedings of the 2018 CHI Conference on Human Factors in Computing Systems*, CHI '18, 1–11. New York, NY, USA: Association for Computing Machinery.

32 Akkil, D., James, J.M., Isokoski, P., and Kangas, J. (2016). GazeTorch: Enabling gaze awareness in collaborative physical tasks. In: *Proceedings of the 2016 CHI Conference Extended Abstracts on Human Factors in Computing Systems*, CHI EA '16, 1151–1158. New York, NY, USA: Association for Computing Machinery.

33 Akkil, D. and Isokoski, P. (2019). Comparison of gaze and mouse pointers for video-based collaborative physical task. *Interacting with Computers* 30: 524–542.

34 van Rheden, V., Maurer, B., Smit, D. et al. (2017). LaserViz: Shared gaze in the Co-located physical world. In: *Proceedings of the 11th International Conference on Tangible, Embedded, and Embodied Interaction*, TEI '17, 191–196. New York, NY, USA: Association for Computing Machinery.

35 Wang, P., Zhang, S., Bai, X. et al. (2018). Do you know what I mean? An MR-based collaborative platform. In: *2018 IEEE International Symposium on Mixed and Augmented Reality Adjunct (ISMAR-Adjunct)*, ISMAR '18, 77–78. USA: IEEE.

36 Wang, P., Zhang, S., Bai, X. et al. (2019). Head pointer or eye gaze: which helps more in mr remote collaboration? In: *2019 IEEE Conference on Virtual Reality and 3D User Interfaces (VR)*, VR '19, 1219–1220. USA: IEEE.

37 Wang, Y., Wang, P., Luo, Z., and Yan, Y. (2022). A novel AR remote collaborative platform for sharing 2.5DHANDS gestures and gaze. *The International Journal of Advanced Manufacturing Technology* 119: 6413–6421.

38 Duchowski, A.T., Cournia, N., Cumming, B. et al. (2004). Visual deictic reference in a collaborative virtual environment. In: *Proceedings of the 2004 Symposium on Eye Tracking Research & Applications*, ETRA '04, 35–40. New York, NY, USA: Association for Computing Machinery.

39 Steptoe, W., Oyekoya, O., Murgia, A. et al. (2009). Eye tracking for avatar eye gaze control during object-focused multiparty interaction in immersive collaborative virtual environments. In: *Proceedings of IEEE Virtual Reality 2009*, VR '09, 83–90. USA: IEEE.

40 Luxenburger, A., Prange, A., Moniri, M.M., and Sonntag, D. (2016). MedicaLVR: Towards medical remote collaboration using virtual reality. In: *Proceedings of the 2016 ACM International Joint Conference on Pervasive and Ubiquitous Computing: Adjunct*, UbiComp '16, 321–324. New York, NY, USA: Association for Computing Machinery.

41 Piumsomboon, T., Dey, A., Ens, B. et al. (2017). [POSTER] CoVAR: Mixed-platform remote collaborative augmented and virtual realities system with shared collaboration cues. In: *2017 IEEE International Symposium on Mixed and Augmented Reality (ISMAR-Adjunct)*, ISMAR '17, 218–219. USA: IEEE.

42 Piumsomboon, T., Dey, A., Ens, B. et al. (2019). The effects of sharing awareness cues in collaborative mixed reality. *Frontiers Robotics AI* 6: 5.

43 Sasikumar, P., Gao, L., Bai, H., and Billinghurst, M. (2019). Wearable RemoteFusion: a mixed reality remote collaboration system with local eye gaze and remote hand gesture sharing. In: *2019 IEEE International Symposium on Mixed and Augmented Reality Adjunct (ISMAR-Adjunct)*, ISMAR '19, 393–394. USA: IEEE.

44 Jing, A., May, K.W., Naeem, M. et al. (2021). eyemR-Vis: A mixed reality system to visualise bi-directional gaze behavioural cues between remote collaborators. In: *Extended Abstracts of the 2021 CHI Conference on Human Factors in Computing Systems*, CHI EA '21. New York, NY, USA: Association for Computing Machinery.

45 Jing, A., May, K.W., Naeem, M. et al. (2021). eyemR-Vis: Using bi-directional gaze behavioural cues to improve mixed reality remote collaboration. In: *Extended Abstracts of the 2021 CHI Conference on Human Factors in Computing Systems*, CHI EA '21. New York, NY, USA: Association for Computing Machinery.

46 Jing, A., Gupta, K., McDade, J. et al. (2022). Near-gaze visualisations of empathic communication cues in mixed reality collaboration. In: *ACM*

SIGGRAPH 2022 Posters, SIGGRAPH '22. New York, NY, USA: Association for Computing Machinery.

47 Barakonyi, I., Prendinger, H., Schmalstieg, D., and Ishizuka, M. (2007). Cascading hand and eye movement for augmented reality videoconferencing. In: *2007 IEEE Symposium on 3D User Interfaces*, 71–78. USA: IEEE.

48 Masai, K., Kunze, K., Sugimoto, M., and Billinghurst, M. (2016). Empathy glasses. In: *Proceedings of the 2016 CHI Conference Extended Abstracts on Human Factors in Computing Systems*, CHI EA '16, 1257–1263. New York, NY, USA: Association for Computing Machinery.

49 Wang, P., Bai, X., Billinghurst, M. et al. (2020). Using a head pointer or eye gaze: the effect of gaze on spatial AR remote collaboration for physical tasks. *Interacting with Computers* 32: 153–169.

50 Wang, H. and Shi, B.E. (2019). Gaze awareness improves collaboration efficiency in a collaborative assembly task. In: *Proceedings of the 11th ACM Symposium on Eye Tracking Research & Applications*, ETRA '19. New York, NY, USA: Association for Computing Machinery.

51 Billinghurst, M., Cordeil, M., Bezerianos, A., and Margolis, T. (2018). *Collaborative Immersive Analytics*, 221–257. Switzerland: Springer Nature.

52 Steptoe, W., Wolff, R., Murgia, A. et al. (2008). Eye-tracking for avatar eye-gaze and interactional analysis in immersive collaborative virtual environments. In: *Proceedings of the 2008 ACM Conference on Computer Supported Cooperative Work*, CSCW '08, 197–200. New York, NY, USA: Association for Computing Machinery.

53 Speicher, M., Cao, J., Yu, A. et al. (2018). 360anywhere: Mobile ad-hoc collaboration in any environment using 360 video and augmented reality. In: *Proceedings of the ACM on Human-Computer Interaction*, vol. 2. New York, NY, USA: Association for Computing Machinery.

54 Brennan, S., Chen, X., Dickinson, C. et al. (2008). Coordinating cognition: the costs and benefits of shared gaze during collaborative search. *Cognition* 106: 1465–1477.

55 Lischke, L., Schwind, V., Schweigert, R. et al. (2019). Understanding pointing for workspace tasks on large high-resolution displays. In: *Proceedings of the 18th International Conference on Mobile and Ubiquitous Multimedia*, MUM '19. New York, NY, USA: Association for Computing Machinery.

56 Akkil, D. and Isokoski, P. (2016). Gaze augmentation in egocentric video improves awareness of intention. In: *Proceedings of the 2016 CHI Conference on Human Factors in Computing Systems*, CHI '16, 1573–1584. New York, NY, USA: Association for Computing Machinery.

57 Huang, C.-M., Andrist, S., Sauppé, A., and Mutlu, B. (2015). Using gaze patterns to predict task intent in collaboration. *Frontiers in Psychology* 6: 1049

58 Piumsomboon, T., Lee, Y., Lee, G. et al. (2017). Empathic mixed reality: sharing what you feel and interacting with what you see. In: *2017 International Symposium on Ubiquitous Virtual Reality (ISUVR)*, 38–41. USA: IEEE.

59 Müller, R., Helmert, J., Pannasch, S., and Velichkovsky, B. (2013). Gaze transfer in remote cooperation: is it always helpful to see what your partner is attending to? *Quarterly Journal of Experimental Psychology (2006)* 66 (7): 1302–1316.

60 Bauer, M., Kortuem, G., and Segall, Z. (1999). "Where are you pointing at?" A study of remote collaboration in a wearable videoconference system. In: *Proceedings of the 3rd IEEE International Symposium on Wearable Computers*, ISWC '99, 151–158. USA: IEEE.

61 Špakov, O., Istance, H., Kari-Jouko, R. et al. (2019). Eye gaze and head gaze in collaborative games. In: *Proceedings of the 11th ACM Symposium on Eye Tracking Research & Applications*, ETRA '19. New York, NY, USA: Association for Computing Machinery.

62 Gauglitz, S., Nuernberger, B., Turk, M., and Höllerer, T. (2014). In touch with the remote world: remote collaboration with augmented reality drawings and virtual navigation. In: *Proceedings of the ACM Symposium on Virtual Reality Software and Technology, VRST*, 197–205. New York, NY, USA: Association for Computing Machinery.

63 Silva, N., Blascheck, T., Jianu, R. et al. (2019). Eye tracking support for visual analytics systems: foundations, current applications, and research challenges. In: *Proceedings of the 11th ACM Symposium on Eye Tracking Research & Applications*, ETRA '19. New York, NY, USA: Association for Computing Machinery.

64 Vogiatzis, D. and Stafylopatis, A. (2002). Reinforcement learning for rule extraction from a labeled dataset. *Cognitive Systems Research* 3 (2): 237–253. Integration of Symbolic and Connectionist Systems.

65 Reidsma, D. and op den Akker, R. (2008). Exploiting "subjective" annotations. In: *Proceedings of the Workshop on Human Judgements in Computational Linguistics*, HumanJudge '08, 8–16. USA: Association for Computing Machinery.

66 Xu, B. and Song, A. (2022). Modelling eye-gaze movement using Gaussian auto-regression hidden Markov. In: *AI 2021: Advances in Artificial Intelligence. AI 2022, Lecture Notes in Computer Science*, vol. 13151 (ed. G. Long, X. Yu, and S. Wang), 190–202. Springer.

67 Koochaki, F. and Najafizadeh, L. (2021). A data-driven framework for intention prediction via eye movement with applications to assistive systems. *IEEE Transactions on Neural Systems and Rehabilitation Engineering* 29: 974–984.

68 Sharma, J., Campbell, J., Ansell, P. et al. (2020). Towards hardware-agnostic gaze-trackers. arXiv pre-print, October 2020.

69 Krafka, K., Khosla, A., Kellnhofer, P. et al. (2016). Eye tracking for everyone. In: *2016 IEEE Conference on Computer Vision and Pattern Recognition (CVPR)*, 2176–2184. USA: IEEE.

6

Evaluating Augmented Reality Remote Guidance Systems

6.1 Introduction

As previously discussed in this book, communication is an essential part of remote collaboration, and many technologies have been developed to enable people to better connect and communicate with one another. However, the impact of these technologies can only be measured through conducting evaluation studies and measuring how the technologies change communication behavior between real people. In this chapter, we discuss methods for evaluating systems for remote guidance systems, particularly systems that use AR technology to improve collaboration on physical tasks.

There is a long history of performing evaluation studies on communication technologies. In the 1950s and 1960s, researchers began evaluating the signal quality of audio calls. For example, Fletcher [1] conducted early speech intelligibility and articulation tests over telephone lines in the 1950s, and similar tests were conducted by Draegert [2], Pickett [3], and others. In these tests, it was typical to play prerecorded sounds (e.g. spoken numbers) with different noise effects and have a listener repeat the information shared. In the 1960s, Riesz and Klemmer [4] explored the impact of audio delay on user satisfaction and call rejection rate for telephone calls. Early researchers also began exploring with ranking the perceived speech quality of audio signals, showing that listeners could reliably distinguish between different qualities of signals [5].

Similar studies were also performed with video conferencing from the 1960s and especially comparing audio-only communication to audio-video communication. For example, in the 1970s Chapanis [6] compared performance on a number of tasks using face-to-face, audio only, handwriting, and typed communication. The tasks included equipment-assembly, information retrieval, and geographic orientation. Surprisingly they found no difference in performance time between audio

Computer-Supported Collaboration: Theory and Practice, First Edition.
Weidong Huang, Mark Billinghurst, Leila Alem, Chun Xiao, and Troels Rasmussen.
© 2024 The Institute of Electrical and Electronics Engineers, Inc. Published 2024 by John Wiley & Sons, Inc.

and visual conditions. A similar result was found by Reid [7] and Willams [8] around the same time. However other experiments at the same time found that video did indeed impact task outcome [8, 9]. O'Conaill et al. [10] point out that the important lesson to be learned from this is that performance measures can be a poor measure of communication quality, and process measures, such as number of phrases spoken, are a better evaluation technique.

By the 1980s and 1990s well-established methods were used in the evaluation of video conferencing techniques. During this time were many experiments conducted comparing face-to-face, audio and video, and audio-only communication. Sellen provides a good summary [11]. While people generally did not prefer audio only, they were often able to perform tasks as effectively as in the video conditions. Both the audio and video, and audio-only cases typically produced poorer communication than face-to-face collaboration, so Sellen reports that the main effect on collaborative performance is due to whether the collaboration was technologically mediated or not, not on the type of technology mediation used. Naturally, this varies somewhat according to task. While face-to-face interaction was found to be no better than speech-only communication for cognitive problem-solving tasks [8], visual cues can be important in tasks requiring negotiation [6].

Researchers found that although the outcome may be the same, the process of communication can be affected by the presence or absence of visual cues [12]. This is because video can transmit social cues and effective information, although not as effectively as face-to-face interaction [13]. However, the usefulness of video for transmitting nonverbal cues may be overestimated, and video may be better used to show the communication availability of others or views of shared workspaces [14]. So even when users attempt nonverbal communication in a video conferencing environment, their gestures must be wildly exaggerated to be recognized as the equivalent face-to-face gesture [13].

These results imply that in collaborative AR experiments, process measures and subjective measures may be more important than quantitative outcome measures. Process measures are typically gathered by transcribing the speech and gesture interaction between the subjects and performing a conversational analysis. Measures that are often collected include the number of words spoken, the average number of works per phrase, number, and type of gestures, number of interruptions, number of questions, and the total speaking time. Although time-consuming, this type of fine-grained analysis often reveals differences in communication patterns between experimental conditions.

One of the difficulties with collecting process measures is that of deciding which metrics to use in developing a data coding technique. Transcribing audio and video tapes is a very time-consuming process and can be unfruitful if the wrong metrics are used. Nyerges et al. provide a good introduction to the art of coding groupware interactions and give guidance on good metrics [15]. Measures that were been found to be significantly different in these early video conferencing studies included:

- Frequency of conversational turns [16, 17]
- Conversational Handovers [16]
- Incidence/duration of overlapping speech [11, 17]
- Use of pronouns [18]
- Number of interruptions [16, 19]
- Turn completions [20]
- Dialogue length [12, 16, 19, 21]
- Dialogue structure [12, 19, 21]
- Backchannels [16]

Gesture and nonverbal behaviors can also be analyzed for characteristic features. Generally, these behaviors are first classified according to type and the occurrences of each type and then counted. Bekker et al. describe an observational study they performed on groups of subjects engaged in a face-to-face design task [22]. From video of the subject groups, four categories of gesture, *kinetic, spatial, pointing,* and *other,* were identified. They were then able to calculate the average number of gestures per minute for each of the different stages in the design task. These four categories were based on the more complex coding categories used by Ekman and Friesen [23].

However, despite this wealth of earlier experience in evaluating collaborative systems, the first collaborative AR systems either did not have any user evaluation or only used relatively primitive methods. Projects such as Studierstube [24] and Transvision [25] allowed users to see each other as well as 3D virtual objects in the space between them. Users could interact with the real world at the same time as the virtual images, supporting spatial cues and facilitating very natural collaboration. However, although these projects successfully demonstrated collaborative AR experiences there were no formal user studies conducted. These systems were also face-to-face collaborative AR experiences rather than remote guidance systems.

Kuzoka's SharedView system [26] from 1992 was one of the first to use an HMD and head-worn camera to enable a remote expert to help a user complete a remote guidance task. An experiment was conducted, and performance time and communication measures were recorded. However, there was no statistical analysis completed of the data, so not definitive conclusions could be drawn. Similarly, CamNet [27] was one of the first remote expert systems developed in 1992. It used a head-worn camera and monocular AR display to enable a medical worker to show the scene of an accident to a remote doctor. The first paper described the system only, and the follow-on paper [28] just had a few subjective questions to ask doctors their impression of the system. However, there was no conversational analysis, performance measures, or behavioral analysis.

Kraut et al. provide one of the first examples of using communication measures with a wearable interface and remote collaboration [29]. They were interested in how the presence or absence of a remote expert might help a subject repair a bicycle and what differences in communication patterns may result with and without shared video. Subjects wore an HMD that allowed them to see video of the remote expert or images of a repair manual. Subjects could complete the repairs in half the time with a remote expert and produce significantly higher-quality work. When video was used, they found that the experts were more proactive with help and that subjects did not need to be as explicit in describing their tasks. In a follow-up experiment, Fussell et al. added a condition where the expert was in the same room as the subject [30]. The same metrics were used (performance time and quality and conversational analysis of speech), and they found that the task was completed significantly faster face-to-face. This time, they found that speech patterns were significantly different between face-to-face and mediated conditions; experts in the face-to-face condition used significantly more deictic references and shorter phrases and were more efficient in their utterances.

The work of Kraut and Fussell just used simple pointing cues to support remote collaboration. However, more complicated AR content could be used. For example, in 1997, Billinghurst et al. [31] presented one of the first remote collaborative systems, where a user in a wearable AR display collaborated with a second user in a similar display on an object arranging task. The users completed the task using voice only, voice and gesture, and voice and graphics in each of the three display conditions (face-to-face, HMD to desktop, and HMD to HMD). The experimental measures were total task completion time and the number of words spoken, but these measures were so simple that nothing substantial could be learned from the experiment.

Several early collaborative AR experiments used a wider variety of experience measures, as summarized in Table 6.1. This table shows that performance time was a common measure used among early collaborative AR systems, but other process measures were being used, such as conversational analysis.

Since that time, methods have improved, but there is still a relative lack of research in collaborative AR systems and methods for evaluation. In 2008, Dünser et al. [32] provided a survey of evaluation techniques used in Augmented Reality, identifying 71 AR research papers from 2007 and earlier that had a user evaluation. Of these, there were only 10 papers that were collaborative AR systems, only four of which were remote collaboration systems. In 2018, Dey et al. [33] published a more extensive study, surveying papers from 2005 to 2014. They identified 15 user studies from 12 papers in the collaborative AR application area, more than half of which were papers about remote collaboration. These 12 papers

Table 6.1 Early collaborative AR experiments and measures used.

Paper	Task	Conditions	Measures used	Outcome
[5]	Pointing Colocated collaboration	AR vs. VR	Performance time Subjective ease of use survey	AR faster AR rated as easier to use
[6]	Face-to-face collaborative game	AR vs. VR	Subjective survey Game scores	Mixed
[7]	Machine operation Remote viewing Remote pointing	AR vs. FtF vs. Fixed Camera	Performance time Speech classification	FtF faster than AR and faster than fixed camera. Fixed camera and AR have similar speech patterns
[8]	Remote pointing Remote collaboration	AR pointing vs. no pointing	Gesture count Speech classification Subjective survey	Gestures used more than speech Deictic speech most common type of speech act
[9]	Bike repair Remote collaboration	Single user vs. remote expert	Performance time Performance quality Coding of speech acts	Subjects faster and produce better quality work with a remote expert
[10]	Bike repair Remote collaboration	Colocated expert vs. audio-video vs. audio only	Performance time Performance quality Conversational analysis	Performance and quality best in FtF condition Significant differences in conversational coding

represent only 4% of the total 291 research papers identified with user evaluation studies, so there are many more evaluation studies that could be conducted.

In Section 6.2, we will summarize some of the systematic reviews that have been conducted on collaborative AR systems, and the evaluation measures used. In Section 6.3, we will present some systems in depth, and describe the experimental methods and data collection methods used. Next, in Section 6.4, we will present some guidelines and lessons learned from previous research on how to conduct evaluation of AR systems for remote guidance. In Section 6.5, we highlight directions for future research and new methods that could be used for evaluation. Finally, we summarize the chapter in Section 6.6.

6.2 Evaluation Methods for Collaborative AR

As mentioned above, there have been relatively few papers published reviewing evaluation methods specifically for collaboration AR. Swan and Gabbard [34] published one of the first reviews of usability studies in AR, identifying 21 papers from before 2005, of which 3 were collaborative AR systems, but only one supporting remote collaboration [35]. Dünser et al.'s work in 2008 [32] identified half a dozen AR papers published before 2008 that describe a system for remote collaboration and have a user study. Similarly, from Dey's review from 2014 [33] there were only six AR remote collaboration papers with user evaluations in them.

These papers are summarized in Table 6.2 with a number of key features highlighted, including the number of subjects in the user study, roles of subjects, displays used by the remote and local workers, and type of subjective and objective measures used.

From this table, there are a number of key characteristics of these user studies. Most of the studies (3 from 14 studies) focus on asymmetric experiences, where a remote expert guides the local worker on some tasks. The remote expert normally was on a desktop interface (12 out of 14 studies), and in only half of the studies was the local worker using an AR HMD interface. A wide variety of tasks were explored, although remote maintenance/instruction, and assembly tasks were particularly popular. Most user studies used some type of survey as a subjective measure, with Likert scale usability surveys being particularly popular. However, a surprisingly large number of studies (5 of 14) did not have any type of objective measures. Also, there were only two studies that captured any communication measures, such as number of words spoken [31] and referential speech acts [36]. This suggests that there is a need for more communication measures to be captured in AR remote guidance studies.

There are a range of different evaluation methods that are typically used for AR experiences, and collaborative AR in particular. In their survey of evaluation

Table 6.2 Remote collaboration papers before 2014.

Paper	Year	Subjects	Display (worker)	Display (remote)	Roles	Task	Subjective measures	Objective measures	Interview
[35]	1999	12 pairs	Audio, Desktop	Audio, Desktop, HMD	Symmetric	Picture selection	Surveys on communication, copresence, awareness	None	No
[35]	1999	8 pairs	HMD	Desktop	Asymmetric	Assembly	Surveys on copresence, communication	Number of task steps	No
[31]	1997	3 pairs	Audio, HMD	Audio, HMD, Desktop	Symmetric	Object arranging	None	Performance time, words used	No
[36]	1999	9 pairs	HMD	Desktop	Asymmetric	Circuit board wiring	Usability survey	Referential speech acts, pointing acts, image freezing, task completion time	No
[37]	2005	9 pairs	VR HMD	AR HMD	Asymmetric	Maze Navigation	Survey on awareness, ease of communication, usability	Task completion time, path error,	Yes
[37]	2005	10 pairs	VR HMD	Desktop, AR HMD, Handheld	Asymmetric	Maze Navigation	Survey the satisfaction, preferred device	Task completion time, behavior observation	Yes
[38]	2012	5 pairs	Dekstop	Desktop	Symmetric, asymmetric	Game, puzzle, training	Survey enjoyment, ease of use, communication, social presence	None	No
[39]	2013	8 pairs	HHD	Desktop	Asymmetric	Remote maintenance	Survey on usability	Task completion time,	No
[40]	2012	24 pairs	HHD	Desktop	Asymmetric	Remote instruction	Interface ranking	Number tasks completed	No
[41]	2014	11 users	HHD	Desktop	Asymmetric	Remote instruction	User preference for interface elements	None	No
[42]	2014	30 pairs	HHD	Desktop	Asymmetric	Remote instruction	Subjective surveys for usability, interface elements	Task completion time, Errors	Yes
[43]	2014	5 pairs	HMD	Desktop	Asymmetric	Assembly	Surveys on ease of use, understanding	Camera movement	No
[44]	2012	5 users	HMD	Desktop	Asymmetric	Crime scene investigation	None	None	Yes
[45]	2013	4 pairs	HHD	HHD	Asymmetric	Assembly	Survey on ease of user, fatigue, satisfaction	None	Yes

techniques for AR, Dünser et al. [46] identified the following five main types of evaluation techniques used:

(1) *Objective measurements*: measures such as task completion times and accuracy/error rates, scores, positions, movements, number of actions, etc.

(2) *Subjective measurements*: measures such as user questionnaires, subjective user ratings, or judgments.

(3) *Qualitative analysis*: measures such as formal user observations, formal interviews, or classification or coding of user behavior (e.g. speech or gesture coding).

(4) *Usability evaluation techniques*: evaluation techniques that are often used in interface usability evaluations such as heuristic evaluation, expert-based evaluation, task analysis, think-aloud methods, or Wizard of OZ methods.

(5) *Informal evaluations*: evaluations such as informal user observations or informal collection of user feedback.

In this work they classified AR papers according to the main type of user evaluation performed. However it is common for researchers to combine several evaluation methods together to explore different aspects of the user experience. Gabbard et al. [47] provides a design methodology for combining together several different evaluation techniques in virtual environment system design. Their model iteratively moves through the stages of (1) User Task Analysis, (2) Expert Guidelines-Based Evaluation, (3) Formative User-Centered Evaluation, and (4) Summative Comparative Evaluation (see Figure 6.1). Outputs at each of these stages help progress the system design to a usable prototype.

In the design process, user task analysis involves identifying all of the tasks and methods needed to use a system. It is the process by which the interface designer gets deep insights into the needs of the user and the context of use for the system. User task analysis is a well-established process in interface design, using a formal approach such as that described by Norman and Draper [48], and Hix and Hartson [49]. The main output of this step is a set of detailed task descriptions that can be used by the evaluation steps (2, 3, and 4) to design the evaluation methods for the prototype system being tested.

Expert guidelines-based evaluation (often called heuristic evaluation) involves using domain experts to find potential usability issues. In this case, several domain experts with interaction design expertise would examine a prototype interface and provide recommendations for design improvements. These recommendations are typically based on well-established usability guidelines, such as those developed by Shneiderman et al. [50], and domain-specific guidelines, such as for Virtual Reality [51]. Nielsen provides an excellent overview of how to conduct a heuristic evaluation [52].

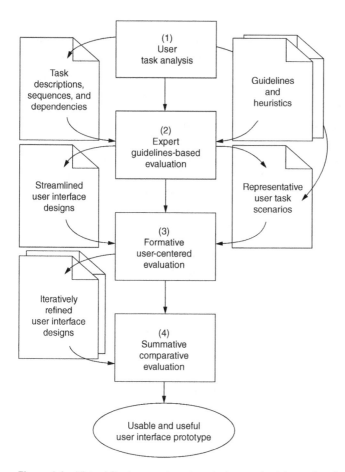

Figure 6.1 Virtual Environment system design methodology, showing evaluation steps. Source: Gabbard et al. [47]/IEEE.

Formative User-Centered Evaluation [53] is an evaluation process where the end users are heavily involved in early evaluation and throughout the entire design process. Gabbard [47] describes how it is an iterative, cyclical process through the steps of designers and evaluators developing user task scenarios, collecting qualitative and quantitative usability data, improving user interaction, and refining the task scenarios. The data collected is from representative users testing out the system in the task scenarios, and it is important that both qualitative and quantitative data are collected to get both subjective and objective measures of the system. Quantitative data is typically performance-based, such as time taken to complete a task, while qualitative is typically survey based, such as the user's impression of the usability of a system.

Summative Comparative Evaluation [53] is a way of evaluating and comparing different interface designs or systems for performing the same task. For example, comparing different ways of navigating through a virtual environment. In a summative evaluation, there is often statistical testing performed to compare interface conditions and see if one performs significantly better than another or provides a significantly more usable experience. In this case, the task scenario is the same, but there are different comparative systems being evaluated against each other. Users complete the task scenario with each of the different systems, and a variety of qualitative and quantitative measures can be used to collect their feedback.

Gabbard's [47] design methodology was introduced in the context of Virtual Environments but Hix et al. [54] demonstrate how this same methodology can be applied to the design of AR systems. In this case, they use the example of designing the Battlefield Augmented Reality System (BARS), an outdoor AR interface for information presentation and navigation for soldiers. In this case, the design of the system began with user needs and domain analysis to understand the requirements of soldiers in the field. This was followed by an expert evaluation and user-centered formative and summative evaluations. In all, there were six evaluation cycles conducted over a 2-month period with nearly 100 mockup prototypes created.

They reported that applying Gabbard's methodology and progressing from expert evaluation to user-based summative evaluation was an efficient and cost-effective strategy for assessing and improving a user interface design. In particular, the expert evaluations of BARS before conducting any user evaluations enabled the identification of any obvious usability problems or missing functionality early in the BARS development life cycle. This allowed improvements to be made to the user interface prior to performing user-based statistical and formative evaluations.

The same design methodology can be applied to AR remote guidance systems. The four stages of (1) User Task Analysis, (2) Expert Guidelines-Based Evaluation, (3) Formative User-Centered Evaluation, and (4) Summative Comparative Evaluation can be completed in a remote task guidance scenario. This is an iterative process so that learnings from each stage can be applied to the next, and can continue to drive improvement in the collaborative system. In Section 6.3, we provide several case studies that demonstrate how this can be done.

6.3 Case Studies From Example Systems

In this section, we provide a more in-depth description of several AR systems for remote guidance and in particular the methods used to evaluate the systems. The goal is not to showcase a particular interaction technique or interface design but to highlight more the various qualitative and quantitative measures that were used and hopefully learn how some of these measures can be applied to another setting. These evaluation summaries are taken from long papers that the reader can check for more details about the full system.

6.3.1 Impact of Gaze Tracking on Remote Guidance Collaboration

The first case study, drawn from [55], explores the impact of gaze tracking in an AR system for remote collaboration and guidance. It is becoming more common to have eye-tracking integrated into AR displays, so an interesting research question is how this might impact remote collaboration on a physical task. In face-to-face conversation, eye gaze provides an important nonverbal implicit communication cue, and so it could be expected that the same might occur in a remote collaborative system.

To explore this, we created a prototype wearable AR system that combined a head-mounted display, with a head-worn camera, and an eye-tracking system (see Figure 6.2). The HMD was the Brother AiRScouter [56] with an 800×600 pixel full-color monocular see-through display with a 22.5° field of view. To this display, we added a Microsoft Lifecam 500 pointing at the user's eye and used the Pupils Labs [57] open-source eye-tracking software, which is capable of tracking the eyes at 30 frames per second. Finally, we needed to add a second outward-facing camera, the Logitech C920 web camera to capture a view of the user's workspace.

This hardware was connected to a desktop computer, on which was shown a simple user interface showing the AR user's view and a red dot showing their current gaze point (see Figure 6.2b). On this interface, the remote collaborator could also point using a mouse pointer, and the pointer-enhanced video view was shown back to the AR user in their HMD. In this way, the person sitting at the computer could see a view of the AR user's workspace and what they are doing and provide feedback using voice and pointing gestures.

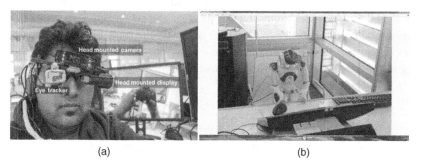

(a)   (b)

Figure 6.2   (a) Head-mounted prototype. (b) Remote user's desktop view.

(a)   (b)

Figure 6.3   (a) Remote expert view. (b) Local workers constructing models.

In order to evaluate this system, we needed a collaborative task and some task conditions to compare between. Using a process similar to that described in Section 6.2, we decided on a brick construction task. The AR user is seated at a table with real plastic bricks, and the remote user provides instructions on how to assemble the bricks. Figure 6.3 shows the setup with the remote expert watching the user's head-mounted camera view on a monitor and the local worker assembling the bricks. Although this is a simple task, it has the same characteristics of many other remote collaboration tasks, such as remote equipment repair, surgical assistance, or training.

The goal of the experiment was to explore the impact of remote pointing and gaze sharing on collaboration. The remote helper was given conditions with a virtual pointer for the local worker, either present or not, while the local worker had eye-tracking turned on or off. So, as Figure 6.4 shows, there were four conditions: NONE, E (eye-tracking on), P (pointer on), and BOTH (both eye-tracking and pointer available. As can be seen, this is a 2×2 experiment design with two factors (POINTER and EYETRACKER) and two states in each factor.

For each of the conditions, the remote helper was given an instruction manual with step-by-step instructions for assembling a simple block model. They would then assemble the model themselves to practice and when finished, begin the experimental task to help the local worker assemble the same model. Figure 6.5 shows one of the models assembled. There were four different models, and the order of the task conditions and models was counterbalanced to remove order effects in the experiment.

	EYETRACKER: No	EYETRACKER: Yes
POINTER: No	*NONE*	*E*
POINTER: Yes	*P*	*BOTH*

Figure 6.4   Experimental conditions.

Figure 6.5   Example model.

Following a Formative User-Centered Evaluation process, before conditioning the final experiment, we conducted pilot testing to get feedback from the users, and refine the test design. This included checking to make sure that the models chosen were about the same level of complexity and took the same amount of time to complete and making sure that the interface was easy to use.

After the pilot studies, the Summative Comparative Evaluation was conducted to compare between the four interface conditions. This was done as a within-subjects design where every subject experienced all four conditions in a counterbalanced order. The main hypotheses of this experiment were:

H1: Copresence – There will be a significant difference in feeling of copresence between the conditions that provided pointing and gaze cues, and those that do not.

H2: Performance – There will be a significant difference in performance time between the conditions that provided pointing and gaze cues, and those that did not.

They evaluated these hypotheses, and a range of different qualitative and quantitative measures were collected, including:

Quantitative cues:
- *Task performance time*: The time taken to complete each condition
- *Conversation analysis*: The number of phrases spoken

Qualitative cues:
- *Subjective copresence survey*: Eleven questions answered on a 7-point Likert scale
- *Subjective condition ranking*: Ranking of all of the conditions using different criteria

Table 6.3 Likert scale subjective survey questions.

Q#	Statement
Q1	I felt **connected** with my partner.
Q2	I felt I was **present** with my partner.
Q3	My partner was able to sense my presence.
Q4	My partner (*or for Remote Helper*: I) **could tell** when I (*or for Remote Helper*: my partner) **needed assistance**.
Q5	I **enjoyed** the experience.
Q6	I was able to **focus** on the task activity.
Q7	I am confident that we **completed** the task correctly.
Q8	My partner and I **worked together** well.
Q9	I was able to **express** myself **clearly**.
Q10	I was able to **understand partner**'s message.
Q11	**Information** from partner was **helpful**.

A recording was made of the subjects' conversation for each condition, and from this a transcript made of the spoken words. This could then be used to identify distinct phrases and communication patterns. After the subjects completed each condition, they answered a subjective survey with 11 questions about communication and copresence. For example, questions such as "I felt connected with my partner" and "I felt I was present with my partner" (see Table 6.3). These were answered on a Likert scale from one to seven, where one equals "Strongly Disagree" and seven equals "Strongly Agree." The Likert scale questions were based on validated questions used in an earlier video conferencing experiment [58]. Finally, after subject completed all of the conditions, they were asked to rank the four conditions in a number of categories, including which condition made them feel connected with their partner or helped them to enjoy the task (see Table 6.4). An open-ended interview was also conducted at the end of the experiment to capture people's free expression of feedback.

As can be seen, the most important thing was that a variety of quantitative and qualitative measures were used to help evaluate the key hypotheses. For hypothesis H1 about performance, the performance time and Likert questions about performance (Q7, Q8) would give an indication of the impact of the pointer and gaze cues. While for hypothesis H2 about copresence, the Likert questions about connection, ranking score, and conversation measures would be relevant.

Once the data was collected from the experiment measures, it was important to analyze it. Since we were conducting a summative evaluation, we could use

Table 6.4 Ranking criteria questions.

No.	Ranking criteria
	Which condition was best …
C1	at helping you to **enjoy** the task?
C2	at making you feel **connected** with your partner?
C3	at helping you stay **focused** on the task?
C4	at making you feel that you were **present** with your partner at the same workspace?
C5	for you (or the partner) to **know** that the partner (or you) **needed assistance**?
C6	at helping you **understand** the partner's message?

a variety of statistical techniques to see if there was a significant difference between the results for each condition. The performance time was analyzed using a two-way analysis of variance (ANOVA) test: the Likert survey was analyzed using an ANOVA test with the Wilcoxon Signed Rank test to compare between conditions, while the ranking results were compared with the Friedman test with Wilcoxon Signed Rank and Bonferroni correction to determine significance between conditions. Different statistical tests were needed because the average performance time is a continuous measure, while Likert scale and ranking questions are discrete.

The full experimental results are provided in [55], and will not be discussed in detail here because this section is about how to conduct the evaluation, not the research output itself. However, at a high level there were a number of key results: (1) the pointer and gaze cues significantly improved performance time, (2) the pointer cue significantly improved the perceived quality of communication, collaboration, and copresence for both the remote helper and local worker, and (3) the condition with both visual cues (BOTH) was ranked as the best in most criteria while the condition with no cues (NONE) was ranked as the worst. There were also interesting lessons learned from the conversational analysis, such as sharing the visual pointer cue significantly reduced the amount of words needed by the remote helper to communicate effectively and also changed how they identified objects.

Overall this paper illustrates how evaluation for AR remote guidance systems can be done in an effective way. The system was developed considering the user needs; pilot studies were done to inform the interface and task design, and a formal evaluation study was conducted to compare several different conditions. In the evaluation study, a variety of qualitative and quantitative measures were collected and aligned with the experimental hypotheses that were being examined and the appropriate statistical analysis techniques used. The value of following an evaluation methodology like this is that research outputs will be more likely to be useful for informing future interface designs. In this case, the research showed that sharing gestures and gaze pointing can indeed improve collaborative performance and copresence in AR remote guidance systems.

6.3.2 Sharing Awareness Cues in Collaborative Mixed Reality

The second study expands on the work of the first and explores the impact of sharing viewpoint and gaze awareness cues in a hybrid AR/VR interface for remote collaboration. This study is drawn from [59] and presents an evaluation of how sharing different types of awareness cues could provide users with an important indicator of what their partner was doing in a collaborative Mixed Reality (MR) application. The previous case study had a remote expert helping a local worker perform a block assembly task on a table, viewing a real workspace through AR glasses. In this case we wanted to explore how a person in AR could collaborate with a person in VR on a room sized task when both people had equal roles.

Figure 6.6 shows the system set up. On the left side, a person in the real world uses an AR display to see virtual objects superimposed over their view. On the

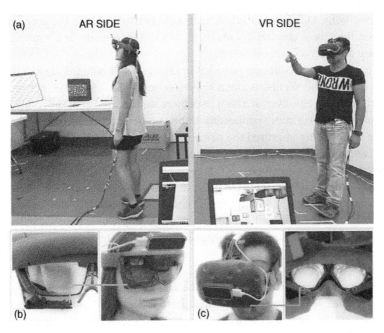

Figure 6.6 Hybrid AR/VR setup for collaboration. (a) Hololens AR display and HTC Vive VR display. (b) Hololens 2 with eye-tracking and gesture-tracking hardware. (c) HTC Vive with eye-tracking and gesture-tracking hardware.

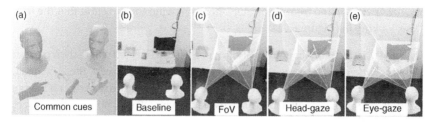

Figure 6.7 Communication cues shared. (a) Common cues. (b) Baseline. (c) FoV. (d) Head-gaze. (e) Eye-gaze.

right side a person in VR is immersed in a virtual copy of the AR user's room and can also see the same virtual content. In addition to the virtual objects shown in the environment, both users also see a representation of their collaborator, who is shown as a virtual head and pair of virtual hands (see Figure 6.7a). For the AR user they see a virtual head and hands from the VR user superimposed over the real world where the VR user is in the virtual environment. Similarly, the VR user sees the virtual head and hands of their partner appearing in their immersive virtual world. The AR and VR users also wear eye-tracking and hand-tracking hardware, so gaze and gesture cues can be transmitted between both locations.

The system was implemented using the Hololens AR display on the AR side, and the HTC Vive VR display on the VR side. Both of these displays were enhanced by adding the Pupil Labs AR/VR eye tracker[1] for eye-tracking. In addition, a Leap Motion[2] tracking module was added to the front of the display to enable hand tracking. Figure 6.6b shows the Hololens 2 with eye-tracking and gesture-tracking hardware, while Figure 6.6c shows the HTC Vive with the same hardware. The Leap Motion and Pupil Labs hardware needed to be connected to a PC to work, so both the Hololens and Vive displays were connected to PCs. These were networked together to exchange gaze and gesture cues.

The main goal of this experiment was to explore how different viewpoint cues would affect communication between AR and VR users. Figure 6.7a shows there were common cues that appeared in all experiment conditions, namely the virtual head and hands of the user's partner, showing in which direction they were facing and where they were pointing. Figure 6.7b shows the baseline condition with just the virtual avatar in it. In addition to this there were three other conditions that were tested:

Field of view (FoV): a virtual view frustum that appeared from the avatar's head in the direction that they were looking (Figure 6.7c).

Head-gaze: a virtual line that appeared from the center of the avatar's head, showing the direction that the head was pointing in (Figure 6.7d).

---

1 Pupil Labs AR/VR tracking website: https://pupil-labs.com/products/vr-ar/
2 Leap Motion website: https://www.ultraleap.com/product/leap-motion-controller/

Eye gaze: a virtual line from the avatar's head showing the direction that the user was looking in (Figure 6.7e).

Normally, in face-to-face conversation, a person can only tell where their partner is looking when they can see their face. However, as can be seen, the FoV, head, and eye-gaze conditions provide a virtual cue to show where their partner is paying attention, even when their virtual face is not in view. Based on this, there were several hypotheses:

H1: The baseline condition does not provide any additional cue, so we hypothesized that it would be the worst condition in terms of performance metrics and behavioral observation variables.

H2: The head-gaze and eye-gaze conditions provide a gaze pointer to identify the center of the FoV frustum and exact eye gaze location, respectively, which will enable users to perform better using these cues than the FoV-only condition.

H3: In terms of subjective opinions, the head gaze and eye gaze will be favored more than the baseline condition, as not having a cue will increase the collaborators' task load.

To study the effect of awareness cues on collaboration, we needed to design a collaborative task which encouraged each user to communicate with each other to complete the task. We created an experimental task which involved search and manipulation of virtual objects. The space about the user was filled with 25 virtual blocks placed randomly in the scene. On the AR side these blocks had numbers on them, while on the VR side the blocks had letters on them. The number or letter would only be revealed when the user gazed at the block.

The users were given a sequence of number/letter pairs (e.g. 1A, 5D, and 3C), and they had to search for the block with the correct combination. When the block was found, either user needed to pick it up with their virtual hands and place it in a target location on the side of the scene. The users could only either see numbers (AR user) or letters (VR user) and so were forced to communicate to find the target block. When both users were looking at the correct block it would change color, showing that it could be moved and the search task was complete. Figure 6.8 shows the view through the AR or VR display of user's performing the block searching and movement task.

In this experiment, there were four conditions (baseline, FoV, head gaze, and eye gaze) and a within-subjects experimental design was chosen where each pair of subjects experienced all four conditions in a counterbalanced order. For each condition, they needed to find and place eight target blocks. After each condition was completed, a variety of subjective and objective measures were collected, as shown in Table 6.5, along with the key. Subjective measures included a survey on social presence [60], the System Usability Scale [61], and some general usability

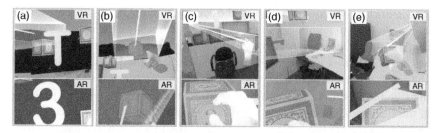

Figure 6.8 Snapshots from the actual footage captured during the collaboration, (top row) VR user's cropped screen captures, and (bottom row) AR user's full view captured by Hololens, (a) Both users needed to find a 3T block, (b) VR user pointed at the block, while both users cogazed at the incorrect block, 5T, (c) 3T was found and the AR user was grabbing it, (d) VR user found the placement target in front of the red book, and (e) AR user placed the block at the target and the trial was completed.

Table 6.5 The experiment measures used.

Measure type	Variable name
Performance metrics	– Rate of mutual gaze (objects identified/minute) – Task completion time (seconds)
Observed behaviors	– Number of hand gestures – Physical movement (meters) – Distance between collaborators (meters)
Subjective surveys	– Usability (System Usability Scale) – Social presence (Biocca Social Presence survey) – Semi-structured interview

questions. The objective measures included performance metrics around the task completion time and rate of mutual gaze and also observed behavior measures such as number of hand gestures, physical movement, and distance between collaborators. The different measures were chosen to explore the different research hypotheses. For example, for H1 (baseline condition has the worst performance), we needed to measure the task completion time and user behavior.

After conducting pilot tests to iteratively design the final user study, a total of 16 pairs took part in the study, with 9 women out of the 32 participants. In terms of previous experience, 6 participants had no previous VR experience, 10 participants had no previous AR experience, and 7 participants had no previous experience with any of the HMDs.

The full experimental results are provided in [59], but at a high level we found the following results. There was no significant difference in performance time between the conditions; however, there was a significantly lower rate of mutual gaze in the baseline condition than in the Eye gaze and the Head-gaze conditions, and the number of pointing gestures used was significantly higher in the baseline condition. Thus, hypothesis H1 was partially satisfied; no performance difference was found, but there was definitely a difference in behavioral cues. The Head-gaze and eye-gaze conditions did not produce better performance than the other conditions, and so hypothesis H2 was not satisfied. Finally, subjects felt that the usability of the Head-gaze condition was significantly better than the baseline condition, and it was easier to use, validating hypothesis H3. In addition, the baseline condition scored significantly lower in the copresence survey than the other three conditions.

One of the benefits of using a wide range of different evaluation measures was the discovery of unexpected results. In this case, we measured the movement of the AR and VR users, as shown in Figure 6.9. This shows a top-down view of the spaces with a heatmap of the user movement superimposed over it. As can be seen, the VR user moved more about the scene than the AR user. We also observed that the VR user used significantly more pointing gestures than the AR user in any of the conditions. One reason for this could be because the VR user had an HMD with a wider FoV than the AR display; 116° horizontal FOV for the HTC Vive

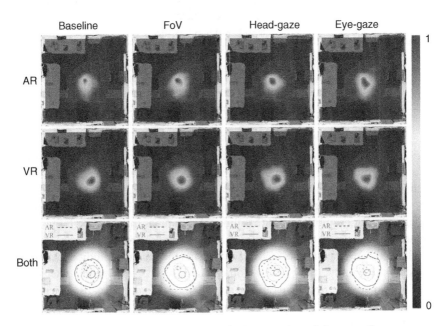

Figure 6.9 Heat map of physical movement in the scene by collaborators (Range: 0 – less time spent, 1 – more time spent): Movement of AR users (top), VR users (middle), and all users using contours with dotted line for AR users and solid line for VR users (bottom).

compared to 43° for the Hololens. This meant that the VR user could more easily see the virtual blocks, and so was often moving around the space and pointing them out to the AR user. This result was not anticipated before the start of the experiment.

This experiment showed that using additional awareness cues did not improve performance in the collaborative tasks, but did impact user behavior and perceived usability and copresence. For example, users pointed at least three times as much in the baseline condition than in the viewpoint awareness conditions (FoV and head or eye gaze), showing that providing an awareness cue reduced the need for pointing. In interviews after the experiment all participants reported difficulties of performing the task using the baseline condition and preferred some type of virtual awareness cues, with 10 pairs of the 16 favoring the Head-gaze cue. It also seems that Head-gaze cues were sufficient for the task, and the higher resolution awareness information available from the eye-tracking was not necessary.

In terms of experiment design, the overall lesson is that using a range of different measures enabled the impact of different interface designs to be found even when they did not provide a performance difference. It also showed the value of using different measures, such as user movement, to capture more subtle experimental results. Using the results from this case study and others, in the next section, we summarize some guidelines for evaluating collaborative AR systems for remote guidance.

6.4 Guidelines

From the case study examples presented above, and the larger literature reviews, there are a number of guidelines that can be applied for evaluating collaborative AR systems.

First, it is important to follow a formative evaluation process when designing evaluation studies. This includes the use of previous related work to design the AR interface for remote collaboration and knowledge of task scenarios to create the experimental task. Pilot studies should also be conducted to iterate on the interface and experiment design and ensure that the experiment can be run smoothly and that the users can interact with the collaborative interface in the way expected.

Second, it is important to use a variety of subjective and objective measures. As shown in the case studies above, different performance, behavioral, and survey measures were recorded to capture the full impact of the different collaborative interfaces.

Third, it is important to use validated measures where possible. In both the case studies well established surveys were used, such as the System Usability Survey. This ensures that the results had validity, and reviewers are not questioning if there are biases or other problems in the survey questions used. There are many previous papers in video conferencing or collaborative virtual environments that papers can be drawn from.

Fourth, the measure used should support the experimental hypothesis. For example, both of the case studies had a hypothesis about the impact on the performance of different interface conditions, and so a performance measure was chosen for the evaluation.

Fifth, it is important to use the correct statistical techniques when analyzing the experimental data capture. If the experiment captures a range of qualitative and quantitative data in continuous and discrete forms, then different statistical techniques will need to be used.

Finally, researchers should be willing to explore their own experimental measures. For example, in the second case study, the subjects' movement was recorded, and a heat map was made to measure the motion through the experiment, leading to a new finding.

6.5 Directions for Research

Although methods for evaluating AR remote guidance systems are becoming more sophisticated, there is still a need for more research in the area. First, there is a need for more publications on collaborative systems and for more evaluation studies to be conducted in general. Collaboration is an important topic for AR user studies, but few papers with a formal evaluation of collaborative AR have appeared. For example, a review of all the research published in the ISMAR conference [62], the main AR academic conference, found only nine papers on collaborative telepresence AR systems, less than 2% of all papers published. Clearly, there is a need for more evaluation studies of collaborative AR systems.

There is also an opportunity to explore how new technologies could be used to provide better evaluation measures. As shown in this chapter, early evaluation studies had people fill out subjective surveys after the user or capture performance measures. However, these are fairly simple measures and also are captured after the user completes a task. It may be difficult for the user to remember how they felt during the experiment. This limitation could be addressed by using new technology during the experiment. For example, EEG sensors could be used to continuously measure brain activity during a study. The EEG signal could then be analyzed to determine periods when the user was under heavy cognitive load, periods of stress, or emotional state. Other physiological sensors could also be used to provide a more reliable indication of user state, such as heart rate sensors, galvanic skin response (GSR) sensors, electro dermal activity sensors (EDA) and others.

In the past integrating a variety of physiological sensors into a user study meant that the subject would have to wear several devices. For example, Gupta et al. [63] report on a user study where the person wears an HMD, an EEG cap, a shimmer heart rate sensor, and GSR sensors. This can be difficult and uncomfortable for the user; with some EEG caps, it is impossible to wear them on the head, and then fit a HMD over the cap. It can also be very complicated to analyze the data from different sensors, all being captured at a different rate. Software libraries such as the Octopus Sensing Library [64] are designed to combine data from a variety of physiological sensors, but there is a need for more research in this area. Some HMDs are beginning to be developed with physiological sensors integrated in them, making it easier to capture physiological data during an experiment. For example, the Galea [65] VR HMD has 11 sensors in the headband and face plate, including EEG, EMG, EDA, PPG, EOG, and others.

Most of the collaborative AR guiding studies have been conducted between pairs of users, often one of whom is an expert and trying to guide the local worker to complete a physical task. However, there is a good research opportunity to conduct experiments with more than two participants. For example, two or three remote experts observe and give guided instruction to a local worker. Larger groups are interesting because the group members can begin to have side conversations that may impact overall performance. These studies are challenging because of the difficulty of finding multiple subjects to form groups, more complex data capture and analysis, and the increased chance of equipment failure.

Most of the published studies on using AR interfaces for remote guidance have relatively simple experimental measures, and few of them use conversational analysis. So, there is an opportunity to explore more the impact of collaborative AR technology on the speech between the participants. As described in the introduction, video conferencing studies often include a wide range of different conversational measures, often applied to a transcript of the users' speech. In contrast, few collaborative AR studies contain speech analysis, and if they do, most do not go beyond simply counting the number of phrases spoken. Research should be conducted on which conversational analysis tools are most appropriate for collaborative AR systems, and if there are new tools that could be developed.

Finally, there is an opportunity for more research on evaluating collaborative AR in natural settings, rather than in a laboratory. Laboratory studies provided a controlled environment to measure user response to the technology, but there is a risk that the user behavior in this environment will not match what happens in the real world. For example, users on a factory floor may find it difficult to hear the feedback from their remote partner over the noise of the machinery operating, and so other interface options are used.

6.6 Conclusion

This chapter has provided a summary of how to conduct evaluation studies of Augmented Reality remote guidance systems. As can be seen, there is a long history of conducting evaluation studies of video and audio conferencing systems, which can be drawn upon to inform the experiment design and measure that should be used.

Section two summarizes some of the previous collaborative AR studies and the features of those studies, such as the number of subjects in the study and the role of the participants. This collection of papers also highlighted some of the gaps in the previous research, such as the few studies that captured any communication measures.

The design methodology of Gabbard et al. [47] was also introduced. This provides a design methodology for combining together several different evaluation techniques for virtual environment system design. However the same approach can be used for AR remote guidance systems, as shown by the case studies reviewed in depth. Both of these followed a similar approach to Gabbard's methodology.

From these case studies and earlier related research, a number of design guidelines are presented that can be used to create robust evaluation studies. This list is not exhaustive, and it is expected that more suggested guidelines will be added as evaluation studies in collaborative AR become more common.

Finally, this chapter concludes with a list of possible research directions. Due to the relatively few studies conducted so far, there are many current topics that could be studied. Research in AR remote guidance systems is also advancing rapidly so other research directions will emerge over time.

It is hoped that this chapter will help the reader to become more proficient in their own evaluation studies, and create research outputs that will inspire others in the field and lead to the development of more effective systems for AR remote guidance.

References

1 Fletcher, H. and Galt, R.H. (1950). The perception of speech and its relation to telephony. *The Journal of the Acoustical Society of America* 22 (2): 89–151.

2 Draegert, G.L. (1951). Relationships between voice variables and speech intelligibility in high level noise. *Communication Monographs* 18 (4): 272–278.

3 Pickett, J.M. (1959). Low-frequency noise and methods for calculating speech intelligibility. *The Journal of the Acoustical Society of America* 31 (9): 1259–1263.

4 Riesz, R.R. and Klemmer, E.T. (1963). Subjective evaluation of delay and echo suppressors in telephone communications. *Bell System Technical Journal* 42 (6): 2919–2941.

5 Rothauser, E.H., Urbanek, G.E., and Pachl, W.P. (1971). A comparison of preference measurement methods. *The Journal of the Acoustical Society of America* 49 (4B): 1297–1308.

6 Chapanis, A. (1975). Interactive human communication. *Scientific American* 232 (3): 36–46.

7 Reid, A. (1977). Comparing the telephone with face to face interaction. *The Social Impact of the Telephone* 386–414.

8 Williams, E. (1977). Experimental comparisons of face-to-face and mediated communication: a review. *Psychological Bulletin* 84 (5): 963.

9 Short, J., Williams, E., and Christie, B. (1976). *The Social Psychology of Telecommunications*. Toronto/London/New York: Wiley.

10 O'Conaill, B., Whittaker, S., and Wilbur, S. (1993). Conversations over video conferences: an evaluation of the spoken aspects of video-mediated communication. *Human-Computer Interaction* 8 (4): 389–428.

11 Sellen, A.J. (1995). Remote conversations: the effects of mediating talk with technology. *Human-Computer Interaction* 10 (4): 401–444.

12 O'Malley, C., Langton, S., Anderson, A. et al. (1996). Comparison of face-to-face and video-mediated interaction. *Interacting with Computers* 8 (2): 177–192.

13 Heath, C., and Luff, P. (1991). Disembodied conduct: communication through video in a multi-media office environment. *Proceedings of the SIGCHI Conference on Human Factors in Computing Systems*, New Orleans Louisiana USA (27 April 1991–2 May 1991). ACM. pp. 99–103.

14 Whittaker, S. (1997). The role of vision in face-to-face and mediated communication. In: *Video-Mediated Communication* (ed. K.E. Finn, A.J. Sellen, and S.B. Wilbur), 23–49. Lawrence Erlbaum Associates Publishers.

15 Nyerges, T., Moore, T.J., Montejano, R., and Compton, M. (1998). Developing and using interaction coding systems for studying groupware use. *Human-Computer Interaction* 13 (2): 127–165.

16 O'Conaill, B. (1997). Characterizing, predicting and measuring video-mediated communication: a conversational approach. *Video Mediated Communication*.

17 Daly-Jones, O., Monk, A., and Watts, L. (1998). Some advantages of video conferencing over high-quality audio conferencing: fluency and awareness of attentional focus. *International Journal of Human-Computer Studies* 49 (1): 21–58.

18 McCarthy, J.C. and Monk, A.F. (1994). Measuring the quality of computer-mediated communication. *Behaviour & Information Technology* 13 (5): 311–319.

19 Boyle, E., Anderson, A., and Newlands, A. (1994). The effects of eye contact on dialogue and performance in a co-operative problem solving task. *Language and Speech* 37 (1): 1–20.

20 Tang, J.C. and Isaacs, E. (1992). Why do users like video? Studies of multimedia-supported collaboration. *Computer Supported Cooperative Work (CSCW)* 1: 163–196.

21 Anderson, A.H., Newlands, A., Mullin, J. et al. (1996). Impact of video-mediated communication on simulated service encounters. *Interacting with Computers* 8 (2): 193–206.

22 Bekker, M.M., Olson, J.S., and Olson, G.M. (1995). Analysis of gestures in face-to-face design teams provides guidance for how to use groupware in design. *Proceedings of the 1st Conference on Designing Interactive Systems: Processes, Practices, Methods, & Techniques*, Ann Arbor Michigan USA (23–25 August 1995). ACM. pp. 157–166.

23 Ekman, P. and Friesen, W.V. (1969). The repertoire of nonverbal behavior: Categories, origins, usage, and coding. *Semiotica* 1 (1): 49–98.

24 Schmalsteig, D., Fuhrmann, A., Szalavari, Z., and Gervautz, M. (1996). "Studierstube" – an environment for collaboration in augmented reality. *CVE '96 Workshop Proceedings*, Nottingham, Great Britain (19–20 September 1996).

25 Rekimoto, J. and Nagao, K. (1995). The world through the computer: computer augmented interaction with real world environments. *Proceedings of the 8th Annual ACM Symposium on User Interface and Software Technology*, Pittsburgh, PA, USA (14–17 November 1995). ACM. pp. 29–36.

26 Kuzuoka, H. (1992). Spatial workspace collaboration: a SharedView video support system for remote collaboration capability. *Proceedings of the SIGCHI Conference on Human Factors in Computing Systems*, Monterey California USA (3–7 May 1992). ACM. pp. 533–540.

27 Matthews, M.R., Cameron, K.H., Heatley, D., and Garner, P. (1993). Telepresence and the CamNet remote expert system. *Proceedings of Primary Health Care Specialist Group, British Computer Society*. pp. 12–15.

28 Armstrong, I.J. and Haston, W.S. (1997). Medical decision support for remote general practitioners using telemedicine. *Journal of Telemedicine and Telecare* 3 (1): 27–34.

29 Kraut, R.E., Miller, M.D., and Siegel, J. (1996). Collaboration in performance of physical tasks: effects on outcomes and communication. *Proceedings of the 1996 ACM Conference on Computer Supported Cooperative Work*, Boston Massachusetts USA (16–20 November 1996). ACM. pp. 57–66.

30 Fussell, S.R., Kraut, R.E., and Siegel, J. (2000). Coordination of communication: effects of shared visual context on collaborative work. *Proceedings of the 2000 ACM Conference on Computer Supported Cooperative Work*, Philadelphia, PA, USA (02–06 December 2000). ACM. pp. 21–30.

31 Billinghurst, M., Weghorst, S., and Furness, T. (1997). Wearable computers for three dimensional CSCW. *Digest of Papers. First International Symposium on Wearable Computers*, Cambridge, Massachusetts, USA (13–14 October 1997). IEEE. pp. 39–46.

32 Dünser, A., Grasset, R., and Billinghurst, M. (2008). *A Survey of Evaluation Techniques Used in Augmented Reality Studies*, 5–1. Christchurch, New Zealand: Human Interface Technology Laboratory New Zealand.

33 Dey, A., Billinghurst, M., Lindeman, R.W., and Swan, J.E. (2018). A systematic review of 10 years of augmented reality usability studies: 2005 to 2014. *Frontiers in Robotics and AI* 5: 37. https://doi.org/10.3389/frobt.2018.00037.

34 Swan, J.E. and Gabbard, J.L. (2005). Survey of user-based experimentation in augmented reality. *Proceedings of 1st International Conference on Virtual Reality*, vol. 22, Las Vegas, Nevada (22–27 July 2005). IEEE. pp. 1–9.

35 Billinghurst, M., Bee, S., Bowskill, J., and Kato, H. (1999). Asymmetries in collaborative wearable interfaces. *Digest of Papers. Third International Symposium on Wearable Computers*, San Francisco, California (18–19 October 1999). IEEE. pp. 133–140.

36 Bauer, M., Kortuem, G., and Segall, Z. (1999). "Where are you pointing at?" A study of remote collaboration in a wearable videoconference system. *Digest of Papers. Third International Symposium on Wearable Computers*, San Francisco, California (18–19 October 1999). IEEE. pp. 151–158.

37 Grasset, R., Lamb, P., and Billinghurst, M. (2005). Evaluation of mixed-space collaboration. *Fourth IEEE and ACM International Symposium on Mixed and Augmented Reality (ISMAR'05)*, Vienna, Austria (05–08 October 2005). IEEE. pp. 90–99.

38 de Souza Almeida, I., Oikawa, M.A., Carres, J.P., et al. (2012). AR-based video-mediated communication: a social presence enhancing experience. *2012 14th Symposium on Virtual and Augmented Reality*, Rio de Janiero, Brazil Brazil (28–31 May 2012). IEEE. pp. 125–130.

39 Chen, S., Chen, M., Kunz, A., et al. (2013). SEMarbeta: mobile sketch-gesture-video remote support for car drivers. *Proceedings of the 4th Augmented Human International Conference*, Stuttgart Germany (7–8 March 2013). ACM. pp. 69–76.

40 Gauglitz, S., Lee, C., Turk, M., and Höllerer, T. (2012). Integrating the physical environment into mobile remote collaboration. *Proceedings of the 14th International Conference on Human-Computer Interaction with Mobile Devices and Services*, San Francisco California USA (21–24 September 2012). ACM. pp. 241–250.

41 Gauglitz, S., Nuernberger, B., Turk, M., and Höllerer, T. (2014). In touch with the remote world: remote collaboration with augmented reality drawings and virtual navigation. *Proceedings of the 20th ACM Symposium on Virtual Reality*

Software and Technology, Edinburgh Scotland (11–13 November 2014). ACM. pp. 197–205.

42 Gauglitz, S., Nuernberger, B., Turk, M., and Höllerer, T. (2014). World-stabilized annotations and virtual scene navigation for remote collaboration. *Proceedings of the 27th Annual ACM Symposium on User Interface Software and Technology*, Honolulu Hawaii USA (05–08 October 2014). ACM. pp. 449–459.

43 Kasahara, S. and Rekimoto, J. (2014). JackIn: integrating first-person view with out-of-body vision generation for human-human augmentation. *Proceedings of the 5th Augmented Human International Conference*, Kobe, Japan (07–09 March 2014). ACM. pp. 1–8.

44 Poelman, R., Akman, O., Lukosch, S., and Jonker, P. (2012). As if being there: mediated reality for crime scene investigation. *Proceedings of the ACM 2012 Conference on Computer Supported Cooperative Work*, Seattle Washington USA (11–15 February 2012). ACM. pp. 1267–1276.

45 Sodhi, R. S., Jones, B. R., Forsyth, D., et al. (2013). BeThere: 3D mobile collaboration with spatial input. *Proceedings of the SIGCHI Conference on Human Factors in Computing Systems*, Paris France (27 April 2013–02 May 2013). ACM. pp. 179–188.

46 Dünser, A., Grasset, R., and Billinghurst, M. (2008). A survey of evaluation techniques used in augmented reality studies. *Technical Report TR-2008-02*. Christchurch, New Zealand: Human Interface Technology Laboratory.

47 Gabbard, J.L., Hix, D., and Swan, J.E. (1999). User-centered design and evaluation of virtual environments. *IEEE Computer Graphics and Applications* 19 (6): 51–59.

48 Norman, D.A. and Draper, S.W. (ed.) (1986). *User Centered System Design*. Hillsdale, NJ: Lawrence Erlbaum Associates.

49 Hix, D. and Hartson, H.R. (1993). *Developing User Interfaces: Ensuring Usability Through Product and Process*. New York: Wiley.

50 Shneiderman, B., Plaisant, C., Cohen, M.S. et al. (2016). *Designing the User Interface: Strategies for Effective Human-Computer Interaction*. Pearson.

51 Gabbard, J.L. (1997). A taxonomy of usability characteristics in virtual environments. Doctoral dissertation. Virginia Tech.

52 Nielsen, J. (1995). How to conduct a heuristic evaluation. *Nielsen Norman Group* 1 (1): 8.

53 Hix, D. and Hartson, H.R. (1993). *Developing User Interfaces: Ensuring Usability Through Product & Process*. Wiley.

54 Hix, D., Gabbard, J.L., Swan, J.E., et al. (2004). A cost-effective usability evaluation progression for novel interactive systems. *37th Proceedings of the Annual Hawaii International Conference on System Sciences*, Big Island, Hawaii (05–08 January 2004). IEEE. 10pp.

55 Gupta, K., Lee, G.A., and Billinghurst, M. (2016). Do you see what I see? The effect of gaze tracking on task space remote collaboration. *IEEE Transactions on Visualization and Computer Graphics* 22 (11): 2413–2422.

56 Brother AiRScouter Website. https://www.brother-usa.com/business/hmd%20 (03 March 2024).

57 Kassner, M., Patera, W., and Bulling, A. (2014). Pupil: an open source platform for pervasive eye tracking and mobile gaze-based interaction. *Proceedings of the 2014 ACM International Joint Conference on Pervasive and Ubiquitous Computing: Adjunct Publication*, Seattle Washington (13–17 September 2014). ACM. pp. 1151–1160.

58 Kim, S., Lee, G., Sakata, N., and Billinghurst, M. (2014). Improving co-presence with augmented visual communication cues for sharing experience through video conference. *2014 IEEE International Symposium on Mixed and Augmented Reality (ISMAR)*, Munich, Germany (10–12 September 2014). IEEE. pp. 83–92.

59 Piumsomboon, T., Dey, A., Ens, B. et al. (2019). The effects of sharing awareness cues in collaborative mixed reality. *Frontiers in Robotics and AI* 6: 5.

60 Harms, C., and Biocca, F. (2004). Internal consistency and reliability of the networked minds measure of social presence. *Seventh Annual International Workshop: Presence*, vol. 2004. Valencia, Spain: Universidad Politecnica de Valencia (13–15 October).

61 Brooke, J. (1996). SUS-A quick and dirty usability scale. *Usability Evaluation in Industry* 189 (194): 4–7.

62 Kim, K., Billinghurst, M., Bruder, G. et al. (2018). Revisiting trends in augmented reality research: a review of the 2nd decade of ISMAR (2008–2017). *IEEE Transactions on Visualization and Computer Graphics* 24 (11): 2947–2962.

63 Gupta, K., Hajika, R., Pai, Y.S., et al. (2020). Measuring human trust in a virtual assistant using physiological sensing in virtual reality. *2020 IEEE Conference on Virtual Reality and 3D User Interfaces (VR)*, Atlanta, GA, USA (22–26 March 2020). IEEE. pp. 756–765.

64 Saffaryazdi, N., Gharibnavaz, A., and Billinghurst, M. (2022). Octopus sensing: a Python library for human behavior studies. *Journal of Open Source Software* 7 (71): 4045.

65 Bernal, G., Hidalgo, N., Russomanno, C., and Maes, P. (2022). Galea: a physiological sensing system for behavioral research in Virtual Environments. *2022 IEEE Conference on Virtual Reality and 3D User Interfaces (VR)*, Christchurch, New Zealand (12–16 March 2022). IEEE. pp. 66–76.

7

Supporting Remote Hand Gestures over the Workspace Video

7.1 Introduction

Globalization is an inevitable trend in modern society, and the need for collaboration between remotely located individuals has increased substantially [1]. To meet this need, many systems have been proposed or developed in the literature (e.g. [2]). Technologies used to support remote collaboration include email exchanges, telephone calls, video conferencing, and video-mediated gesturing. However, most of these systems in use are typically developed for supporting group activities or performing tasks without having to referring to or operating on external physical objects [3].

There are a range of real-world situations in which remote expert guidance is required for a local novice to complete physical tasks. For example, in telemedicine a specialist doctor guiding remotely a nonspecialist doctor or nurse performing surgery for a patient [4]; in remote maintenance an expert guiding remotely a technician into repairing a piece of equipment [5]. Particularly in the field of the industrial and mineral extraction, complex technologies such as fully automated or semi-automated equipments, teleoperated machines, are being introduced to improve productivity. Consequently, the maintenance and operation of these complex machines is becoming an issue. Operators/technicians rely on assistance from an expert (or more) in order to keep their machinery functioning. Personnel with such expertise, however, are not always physically located with the machine. Instead, they are often in a major metropolitan city while the technicians maintaining equipment are in rural areas where industrial plants or mine sites may be located. Therefore, there is a growing interest in the use and development of technologies to support the collaboration between a maintenance worker and a remote expert.

It is often challenging to support interactions when collaborations take place over a distance, relying on computer-mediated communication and interactions, and even more so to support collaborations between a remote helper and a

Computer-Supported Collaboration: Theory and Practice, First Edition.
Weidong Huang, Mark Billinghurst, Leila Alem, Chun Xiao, and Troels Rasmussen.

local worker. For such remote collaborations, systems need to provide different interfaces and functions to support the specific actions taken by the worker and the helper and to facilitate communications and interactions between them. In general, one of the main issues when participants collaborate remotely is that there is no longer common ground for them to communicate in a way in which they do when they are colocated. Clark and Brennan [6] define common ground in communication as a state of mutual agreement among collaborative partners about what is referred to. In the scenario of an expert guiding a worker on physical tasks, the expert speaks to the worker by first bringing attention to the object that they are going to work on. To achieve this, the referential words such as "this,", "that,", along with gestures such as hand pointing, head nodding, eye contacts, and facial expressions, may be used [7]. Only when the mutual understanding is built can instructions on how to perform tasks be effectively communicated. As such, many attempts have been made to rebuild common ground. Among them, providing shared visual spaces is one that has been studied most. According to Tang et al. [8], "A shared visual workspace is one where participants can create, see, share and manipulate artifacts within a bounded space.". Real-world examples include whiteboards and tabletops.

When collaborating face-to-face on physical tasks, people use a range of visual cues available to them as a starting point and reference to communicate and interact with each other, therefore facilitating the negotiation of common ground during the process. The visual cues include facial expressions, body language, and actions of the partner, as well as the view of the task objects and environments [3]. A series of studies have discussed and demonstrated that providing remote collaborators with access to these visual cues is beneficial to the completion of collaborative tasks. The visual cues are often provided in the form of video view of the workspace [9]. Further, prior research has indicated that the reason why face-to-face communication is more efficient than computer-mediated communication is mainly because in the face-to-face condition, participants are able to perform gesturing more easily, and the gestures are visually available to all participants [7, 8, 10]. This suggests the importance of providing a shared visual space to the collaborators and supporting gesturing in the visual space.

A number of systems have been developed to support remote guidance, providing a shared visual space and using the space for gesturing. For example, Ou et al. [11] developed a DOVE (Drawing over Video Environment) system that integrates gestures of helper into the live video of the workspace to support collaboration on physical tasks. The main feature of the system is that the system allows a remote helper to draw on video streams of the workspace while providing task instructions. In their system, the gestures they support are mainly pointing and sketching. Kirk et al. [10, 12] presented a system MixedEcology. This system supports collaborative physical tasks through a mixed reality surface that aligns and integrates the ecologies of the local worker and the remote helper. In this system, the gestures of the helper's hands are captured by a video camera and projected onto the desk of the worker. MixedEcology aims to promote mutual awareness between participants by supporting the remote guiding through a mixed reality surface in which the remote helper's hands are overlaid on top of the local worker's hands. Kurata et al. [13, 14] developed a Wearable Active Camera with a Laser pointer system (WACL). In this system, the worker wears a steerable camera/laser head. The helper is allowed to control this steerable camera remotely, set his own viewpoint independently, and point to real objects in the task space with the laser spot. Kuzuoka et al. [15] developed GestureMan systems in which remote gestures are conveyed by a mobile robot through the use of a laser pointer.

Despite the progress made for supporting remote collaboration, current systems either assume that the workspace of the worker is limited to a fixed desktop or support only limited gestures such as pointing and sketching. In mining environments, traditional desktop workspaces are rare; the workspace conditions are usually dusty and unpredictable. Workers are often required to walk around to inspect the machine and fetch tools during maintenance. In addition, complex hand gestures are needed to facilitate the communication process. How to support the richness of hand gestures for an expert guiding a mobile worker located in a nontraditional desktop environment has not been fully understood.

In an attempt to fill this gap, we have developed a system called HandsOnVideo, following a participatory design approach. The key drivers for designing this system are:

1. Support the mobility of the worker
2. Support the richness of hand gestures beyond pointing and digital sketching
3. Easy to use with natural interaction
4. Can be used in a mining environment

To be more specific, HandsOnVideo uses a near-eye display to support mobility and uses unmediated representations of hands to support remote gestures of the helper over the video of the worker's workspace. A usability evaluation has been conducted, and the results confirm the usability and usefulness of this system and also indicate possible improvements for future work.

This chapter is an extension of its conference versions [16, 17]. The remainder is organized as follows. We first provide a review of the literature on systems for maintenance guidance. The technical specifications and features of HandsOn-Video are described, followed by a usability study. Design tradeoffs and system limitations we have encountered are discussed. Finally, we conclude the chapter with a short summary and future work.

7.2 Related Work

In this section, we selectively review related work. Systems for remote guidance are reviewed first. Then, a review of previous research on supporting remote gestures is presented.

7.2.1 Systems for Remote Guidance

There are many real-world collaborative scenarios in which the worker is engaged in a mobile task or performing tasks on objects that are consistently moving. The mobility of the worker presents unique challenges for system design, and a few attempts have been made by researchers to address the challenges.

Kuzuoka et al. [15] developed a system called GestureMan for supporting remote collaboration. GestureMan uses mobile robots as communication media. The instructor controls the robot remotely, and the operator receives instructions via the robot. In their system, the robot is mounted by a three-camera unit for the environment of the operator. It also has a laser pointer for hitting the intended position and a pointing stick for indicating the direction of the laser pointer. The movement of the robot is controlled by the instructor using a joystick.

Kurata et al. [13, 14] developed the WACL system that involves the worker wearing a steerable camera/laser head. WACL allows the remote instructor not only to independently look into the worker's task space, but also to point to real objects in the task space with the laser spot. In their system, the laser pointer is attached to the active camera head, and it can point to a laser spot. Therefore, the instructor can observe the worker independently of the worker's motion and can clearly and naturally instruct the worker in tasks.

Previous work in the area of remote guiding of mobile workers has mostly focused on supporting pointing to remote objects, using a projection-based approach such as the laser pointing system in WACL [13, 14], or using a see-through-based approach such as in REAL [18]. While pointing (with a laser or a mouse) is an important aspect of guiding, research has indicated that projecting the hands of the helper supports a much richer set of nonverbal communications and, hence, is more effective for remote guiding (e.g. [10, 12]). In Section 7.2.2, we review the work in this space.

7.2.2 Supporting Remote Gestures

Importance of gestures can be intuitively illustrated by hand movements that we use together with verbal and nonverbal communication in our everyday life. In fact, the use of hand gestures in support of verbal communication is so natural that they are even used in communication when people speak on the phone. Recent empirical studies have also shown that gestures play an important role in building common ground between participants in remote guiding [7].

Given that gesturing is of such importance to collaborative physical tasks, a variety of systems are being developed to facilitate remote gesturing (e.g. [3, 6, 7, 10]). Most of these systems are explicitly built with the intention of enabling a remote helper (expert) to guide the actions of a local worker, allowing them to collaborate over the completion of physical tasks. Results have so far suggested that such tools can increase performance speed and also improve the worker's learning of how to use the system and perform tasks (when compared to standard video-mediated communication methods).

More specifically, Fussell et al. [7] introduced a system in which the helper can perform gestures over the video streams. In their system, gestures were instantiated in a digital form. A user study conducted by the authors demonstrated the superiority of digital sketches over cursor pointers. More recently, Kirk et al. [10] explored the use of gestures in collaborative physical tasks using augmented reality. In particular, the guiding is supported through a mixed reality surface that aligns and integrates the ecologies of the local worker and the remote helper. The system allows the helper to see the objects in the worker's local ecology, the worker's actions on the objects in the task space, and their own gestures toward objects in the task space. The work of both Fussell et al. and Kirk et al. demonstrated the importance of supporting remote gestures. However, how gestures can be better supported with a mobile worker has not been fully understood.

7.3 HandsOnVideo

Our literature review suggests the following requirements for remote guiding systems in industry:

- The need to support the mobility aspect of the task performed by the worker using wearable computers and wearable cameras.
- The need to allow helpers to guide remotely using their hands in order to provide reference to remote objects and places but also support procedural instructions.

In this section, HandsOnVideo is introduced to address the above needs. In particular, HandsOnVideo captures the hand gestures of the helper and projects them onto a near-eye display worn by the worker. It is composed of (1) a helper user interface that is used to guide the worker remotely using a touchscreen device and an audio link and (2) a mobile worker interface that includes a wearable computer, a camera mounted on a helmet, and a near-eye display (a small device with two screens); see Figure 7.1. In the following subsections, the design of our remote guiding system and the technical platform are described in more detail.

Figure 7.1 Worker interface.

7.3.1 Worker Interface Design

When it comes to display information to a mobile worker, there are a range of types of displays that can be employed for this purpose [19]. For example, hand-held displays, wrist-worn displays, shoulder-worn displays, head-worn displays, displays that are embedded in the environment, displays that are specifically set up in the workspace, or devices that project images on the surfaces of arbitrary objects. According to Holler and Feiner [19], displays used for mobile AR can be generally classified into two categories: displays that make use of resources in the environments and displays that the worker carries on the body or by hand. The display used in the system of Fussell et al. [7] mentioned in Section 7.2 falls in the first category. Displays in this category require space, supporting facilities, and sophisticated infrastructure to be readily available. For remote guiding systems, one would need camera, projectors, monitors, desks, cables, power source, and proper lighting conditions. The list could go on. All these devices need space and require strict environmental conditions to work properly. Even setting them up and making them all work together requires extra expert instructions. Considering that our system is designed to be used in mining sites, the environment can be unpredictable. We therefore limit our options to the second category: displays that the worker carries on the body or by hand.

Since local workers need to move around and use their hands to operate on physical objects, it is not practical to use hand-held displays. In regard to head-worn displays, optical see-through and video see-through displays have been used for mobile AR guiding systems [19, 20]. Optical see-through displays are semi-transparent. They overlay computer-generated images on top of the worker's view of the real world. On the other hand, video see-through systems present an indirect and mediated view of the environment. They combine video feeds from cameras with computer-generated images and display the video representation of the real world in front of the user's eyes. Although both displays can be useful for general mobile AR systems, applications for mining sites have strict requirements in terms of human and environmental factors [21]. These factors include safety, ease of use, changing light and dust conditions of environments, and so on. The end users had indicated to us that mining sites are often very dusty. Dusts can easily spread on the surface of optical displays, blocking the view of the worker. This hardly makes optical see-through a practical option for consideration. In addition, both optical and video see-throughs offer a limited view of the worker's workspace. And not being able to see the surrounding environment fully and directly is risky for the workers in mining sites.

Workers usually wear a helmet while working in mining sites for safety reasons. We therefore make use of the helmet and attach a near-eye display under the helmet. As shown in Figure 7.1, the near-eye display is light, easy to put on, and comfortable to wear compared to other head-worn displays. The worker can easily look up and see video instructions shown on the two small screens, and at the same time, he/she can see the workspace in front of him/her with little constraint. We also tested the display with real users. The feedback from them during the design process was very positive with the near-eye display.

7.3.2 Helper Interface Design

We adopted a participatory approach for the design of the helper interface. Our aim was to come up with a design that fulfills the users' needs, and that is as intuitive to use as possible. Our initial step consisted of observing maintenance workers and developing a set of requirements for the helper user interface (UI) based on our understanding of their needs.

- The need for supporting complex hand movements such as "take this and put it here," "grab this object with this hand," and "do this specific rocking movement with a spanner in the other hand."
- Mobility of the worker during the task, as they move from being in front of the machine to a tool area where they access tools, to the back of the machine to check valves, etc.
- The helper may need to point/gesture in an area outside the field of view of the worker. Therefore, there is a need to provide the helper with a panoramic view of the remote workspace.

We then designed a first sketch of the interface consisting of a panoramic view of the workspace and a video of the worker's view. The video provides a shared visual space between the helper and the worker that is used by the helper for pointing and gesturing with their hands (using unmediated gesture). This shared visual space, augmented by the helper's gestures, is displayed real time on the near-eye display of the worker (image + gestures).

The helper UI consists of:

- A shared visual space that displays, by default, the video stream captured by the remote worker's camera. This space occupies the central area of the touch table.
- A panoramic view of the worker's workspace, which the helper can use for maintaining an overall awareness of the workspace. This view can also be used by the helper for bringing the worker to an area that is outside their current field of view. The panoramic view occupied the lower end of the touch table.
- Four storage areas, two on each side of the shared visual space, to allow the helper to save a copy of the shared visual space in case there is a need to reuse it.

We performed four design iterations of our UI, testing and validating each design with a set of representative end users on the following three maintenance/repair tasks (Figure 7.2):

- Repairing a photocopy machine
- Removing a card from a computer motherboard
- Assembling Lego toys

Over 12 people have used and trialed our system, providing valuable feedback on how to improve the helper UI and, more specifically, the interactive aspect of

Figure 7.2 Maintenance and assembly task.

the UI: the selection of a view, the changing of the view in the shared visual space, and the storage of a view. The aim was to perform these operations in a consistent and intuitive manner for ease of use. The overall response from our representative end-user pool is that our system is quite intuitive and easy to use. No discomfort has been reported to date with the near-eye display of the worker system.

7.3.3   The Platform and Technical Specifications

Our platform draws on previous experience in the making of the REAL system, a commercial, wearable, low-power augmented reality system. REAL employs an optical see-through visor (LiteEye 750) for remote maintenance in industrial scenarios. In particular, HandsOnVideo makes use of the XVR platform [18], a flexible, general-purpose framework for VR and AR development. The architecture of our system is organized around two main computing components: the worker wearable device and the helper station, as seen in Figure 7.3.

Wearable computers usually have lower computing capability with respect to desktop computers. To take into account the usual shortcomings of these platforms, all our software has been developed using an Intel Atom N450 as a target CPU (running Microsoft Windows XP). It presents a reasonable heat dissipation requirement and a peak power consumption below 12 W, easily allowing for battery operation. A Vuzix Wrap 920 HMD mounted on a safety helmet was used as the main display of the system. The arrangement of the display is such that the upper part of the worker's field of view is occupied by the HMD screen. As a result, the content of the screen can be seen by the worker just looking up, while the lower part remains nonoccluded. With such an arrangement, what is displayed on the HMD gets used as a reference, but then the worker performs all their actions by directly looking at the objects in front of him/her. CMOS USB camera (Microsoft

Figure 7.3   The helper control console (a) and the worker wearable unit (b).

Lifecam HD) is mounted on top of the worker's helmet (as seen in Figure 7.3). This allows the helper to see what the worker is doing in their workspace. A headset is used for the worker-helper audio communication.

The main function of the wearable computer is to capture the live audio and video streams, compress them in order to allow network streaming at a reasonably low bit rate, and finally deal with typical network-related issues like packet loss and jitter compensation. To minimize latency, we use a low-level communication protocol based on UDP packets, data redundancy, and forward error correction, giving us the ability to simulate arbitrary values of compression/decompression/network latency, with a minimum measured value around 100 ms. Google's VP8 video compressor [22] is used for video encoding/decoding, and the Open Source SPEEX library is used for audio, with a sampling rate of 8 KHz. Please note that at the same time, the wearable computer also acts as a video/audio decoder, as it receives live streams from the helper station and renders them to the local worker.

The main component of the helper station is a large (44 in.) touch-enabled display. The display is driven by a NVidia GeForce graphic card mounted on a Dual Core 2.0 GHz Intel workstation (Windows XP). The full surface of the screen is used as a touch-enabled interface, as depicted in Figure 7.4.

Occupying the central portion of the screen is an area that shows the video stream captured by the remote worker camera: it is on this area that the helper is using their hands to guide the worker. On the side of the live stream, there are four slots, initially empty, where, at any moment, it is possible to copy the current image of the stream. This can be useful to store images of particular importance

Figure 7.4 Layout of the helper screen.

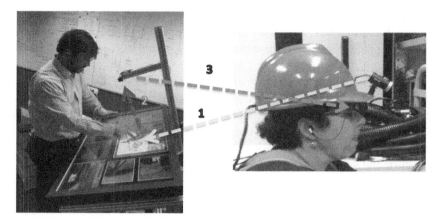

Figure 7.5 Data capture and display.

for the collaborative task or snapshots of locations/objects that are recurrent in the workspace. Another high-resolution webcam (Microsoft Lifecam HD) is mounted on a fixed support attached to the frame of the screen and positioned to capture the area on the screen where the video stream is displayed (see Figure 7.5): the camera captures what is shown on the touch screen (see arrow 1) and the hand performed by the helper over that area too (see arrow 2). The resulting composition (original image plus the hand gesture on top) is once again compressed and streamed to the remote worker to be displayed on the HMD (see arrow 3). The overall flow of information is represented in the diagram of Figure 7.5.

7.4 User Testing

In this section, we present a user study we conducted with representative end users. The main objectives were to collect feedback on the usability of HandsOn-Video and to identify possible directions for future work.

7.4.1 Design

The helper station of the system was located in a room, while the worker station was in a workshop room where the experimental environment was set similar to that of a mining site. Both rooms were about 20 m away from each other. The helper and worker could talk to each other through a headphone. Since the system was developed specifically for mining workers, end users who had experience with remote collaboration were asked to perform two different physical tasks. The users were randomly grouped in pairs, with one playing the role of helper and the other playing the role of the worker. If one participant played as a helper in the first task, then in the second task, he or she changed to play as a worker. For each session, the whole process was video recorded on both helper and worker sites for further analysis.

There were also a questionnaire session after each task and a discussion session in the end for each pair. There were two questionnaires with one for helper and the other for worker. The two questionnaires included the same Likert-style questions about ease of learning, ease of use, environment awareness of the workspace, sense of copresence, perceived task performance, and interaction. Open questions specific to the role played in the task and associated interfaces were also included.

7.4.2 Participants

Six staff members volunteered to participate in the study. Two of them are workshop people maintaining equipment on a daily basis. The other two were software engineers who have been working on remote collaboration projects for a number of years. The rest of the participants were managers supervising maintenance and collaboration projects.

7.4.3 Tasks

Two types of task were used. One is the assembly task using Lego toy blocks. This task has been used in previous research for similar purposes [7, 10]. This task is considered representative because it has a number of components that can be found in a range of physical tasks such as assemble, disassemble, select, move, and rotate. During the task, the worker was asked to assemble the Lego toys into a complex model under the instruction of the helper.

The other task is repair task. This is a real task that may occur in mining sites. Since we did not have access to mining equipment, we used the repair of a PC as our second task. During this task, the worker was asked to take the cover of the PC off, replace one part inside the PC with another, and put the cover back in place under the guidance of the remote helper.

At the start of each task, the manual on how to construct the Lego model or how to fix the PC was provided to the helper. The helper was instructed that he could provide verbal and gestural instructions to the worker at any time but was not allowed to show any part of the manual to the worker. The worker, on the other hand, had no idea about what steps were needed to complete the tasks.

During the experiment, the toy blocks and the PC parts were placed in different locations in the workspace; the worker had to move around the workspace to collect them and get the task done. Also, there were also obstacles being deliberately placed between the locations; the worker had to avoid them while moving around.

This was to test whether the worker was able to be aware of the environment while he walked with a near-eye display.

7.4.4 Procedure

The study was conducted in pairs. The first two participants were gathered in the meeting room of the helper station. They were informed about the procedure of the study. The helper interface and the worker interface were introduced. They were also given a chance to get familiar with the system and try out the equipment. During the introduction, the participants could ask questions and answers were provided by two experimenters.

When ready, the two participants were randomly assigned roles. Then they went to the corresponding rooms where the helper or worker station was located. On each site, there was also an experimenter providing further assistance to the participant, recording videos, observing, and taking notes of the communication behavior.

The participants performed the Lego task first. After the first task, each participant was asked to fill out the helper or worker questionnaire depending on his role. Then, the participants switched roles, went to the corresponding rooms, and proceeded to perform the second task: repair of a PC, followed by the questionnaires.

After finishing the two tasks and the two questionnaires, the participants went to the meeting room where they were debriefed about the purposes of the study first. Then, a semi-structured interview followed. They were encouraged to ask questions, propose ideas and further improvements, debate on the issues, and comment on the system. The whole session for the two tasks for each pair took about one hour.

7.4.5 Observations

All pairs of participants were able to complete their assigned tasks within reasonable periods of time. The main components of the helper interface: the shared visual space and the panoramic view were frequently used during the guidance. The helpers were able to perform a range of gestural actions over the shared space while giving verbal instructions. It was also seen that the helpers were able to identify the locations of PC parts and toy blocks and guide their collaboration partners to the specific locations using the panoramic view of the workspace. The worker was able to walk around the workspace without apparent difficulties. This demonstrates that the worker was able to be aware of the environment with the near-eye display. The communication between the pair seems smooth and effective. During the observation, we also identified some usability issues, which were detailed as follows.

Figure 7.6   Confusion with the view being displayed.

Figure 7.7   Limitations of the near-eye display.

Confusion with the view being displayed: The worker's view on the near-eye display does not correspond to what they see in their physical workspace (see Figure 7.6). This might be due to the network delay, as mentioned by participants: "video lag is annoying," "video delay makes it harder to use," and "video lag is not good."

During the study, we observed from three of the participants that there could be an issue of spatial awareness with the use of the near-eye display (see Figure 7.7):

- One of the participants seemed to have difficulties locating a computer that was next to him. This participant used the near-eye display as his main source of information. He hardly used the natural and unmediated view of his workspace. This participant did not feel confident moving around his workspace.
- Another participant adjusted the display frequently. It is likely that he did not notice that he can switch between the views of the help and workspace simply by looking up the near-eye display, and without having to adjust the display.
- One participant wore the near-eye display very low, and hence, the focus of his attention was more on the instruction than on the task place. This also resulted in limited spatial awareness.

This issue seems to indicate that the near-eye display, if not worn properly, may lead to a focus of the attention on the help provided rather than conducting the task while checking the help being displayed. A further exploration of how the near-eye display should be configured is needed in order to prevent such issues from happening again.

7.4.6 Questionnaire Results

Six participants filled two questionnaires: the helper questionnaire and worker questionnaire. We got 12 copies of questionnaires in total. The detailed responses from the participants were presented as follows.

First, both helper and worker questionnaires included six usability questions to be answered in a Likert scale fashion, from 1: strongly disagree to 7: strongly agree. The results are shown in Table 7.1:

As can be seen from Table 7.1, in general, the participants thought that the HandsOnVideo system is easy to learn and use. On average, copresence and environmental awareness were rated just above being neutral, while perceived task performance and interaction were rated relatively high. Specifically, the participants perceived the system more useful when they played helper than when they played worker in terms of usability measures except copresence. Although t-tests indicated that these differences were not statistically significant, the higher ratings with the helper role suggest that the participants were more comfortable with the helper interface and that on the other hand, they might need more time to get used to the worker interface.

In regard to copresence, generally speaking, copresence was rated relatively low compared to other measurements; it is reasonable since both the helper and worker knew they were located in different rooms, and we did not expect that the system would present a sense of "being together" as strong as virtual environments would do. On the other hand, we expected that the helper and

Table 7.1 Results of Likert score questions.

Question	Worker	Helper	Average
Ease of learning	5.00	6.00	5.50
Ease of use	4.83	5.66	5.25
Task satisfaction	5.50	5.66	5.58
Copresence	4.50	4.16	4.33
Awareness of environment	4.66	5.00	4.83
Perception of interaction	4.83	5.66	5.25

worker might have different levels of senses of "being together," as there were different interfaces on both sides. The helper used relatively large touch display that showed the view of the worker's workspace, while the worker used the small near-eye display that showed the workspace with the hands of the helper. As can be seen from Table 7.1, copresence on the worker side was rated as 4.50, which is relatively higher than what was rated on the helper side (4.16). This indicated that the worker had a greater sense of copresence. According to Li et al. [9], this difference might be the result of one of the key features that our HandsOnVideo system offered: the worker being able to see hands of the helper.

Second, the questionnaires also had open questions asking about their experience with the study and asking for their feelings about specific interface features. The comments indicated that the participants were generally positive about the system. They appreciated being able to perform hand gestures and see the helper's hands via the near-eye display. Examples of user comments include "the system should be useful in many situations"; "The near-eye display helped me to see what my partner could see.". The pros and cons mentioned by the participants were summarized as follows.

Pros:

- Helper UI is intuitive, enjoyable to use, and easy to learn.
- Guiding with hands is natural; being guided by seeing hands is easy to understand.

Cons:

- Latency is annoying: users get the audio instruction first and have to wait for the visual instructions
- Image quality is not very good
- Jerkiness of the images affects users

Specifically, in response to the question: *What is your view of seeing the hands of person helping you?* Participants playing the role of worker commented that it was useful. "It was useful. I could see what he was pointing to, even though he couldn't see the exact colors or shapes on his screen, I could tell which object he meant."

On the other hand, in response to the question: *Please explain your experience of guiding/instructing using your hands.* Participants playing the role of helper commented that it was helpful.

"I had to make sure I only had one hand on the screen. It was pretty easy to use. Might be useful to have some feedback of hand gestures thought, perhaps if we could see what the partner sees."

7.5 Discussion

Our usability evaluation confirmed the usability and usefulness of HandsOnVideo for supporting real-world scenarios in which a remote expert guides a mobile worker performing physical tasks in a nontraditional-desktop environment. The users were able to complete assigned tasks with quality and satisfaction in a reasonable time. The rating results of the usability measures indicated that users were generally positive about the system.

As far as we are aware, HandsOnVideo is the first system that uses the near-eye display to support mobile remote guiding. Although the display may partially block the local view, our usability study demonstrated that the use of it in such types of systems is promising. First, it is small and light and requires little hardware and environmental support. This is ideal for supporting the mobility of the worker in nontraditional-desktop environments. Second, on the worker side, both the near-eye display and the scene camera are attached to the peak of a helmet. Therefore, the two devices move with the worker at the same time. This ensures that the view of the camera and the view of worker are consistent. Although it is always desirable to make the two views the same, they are still different in the current setting. However, we were satisfied that none of our users had raised issues in relation to reference and orientation mapping, which is usually an issue when different viewpoints are used. This indicated that the view difference in the worker interface is small enough to avoid any noticeable negative consequences.

It is worth noting that although both HandsOnVideo and the mixed ecology system of Kirk and Fraser [10] use unmediated hand representations for remote gestures, different approaches are used to represent hands in these two systems. The former combined the hands with the live video of the workspace and shown on the near-eye display, while the latter directly projects the helper hands into the workspace. At first glance, presenting hand gestures on external monitors seems to require extra effort in shifting attention between workspace and monitor. This perception is also reflected in the comments of our users. However, prior empirical research has shown that the location of gesture output, no matter whether it is on an external monitor or it is on the surface of the workspace, does not make any significant differences in performance of collaborative physical tasks [10]. In addition, in our system, effort on attention shift has been reduced to minimum: the near-eye display is located above the eyes of the worker; seeing hand gestures is just a matter of an eyelid lift.

Designing a usable remote guiding system requires a close involvement of end users and the ability to capture and address the interaction issues they raise while using the system. This is not an easy task and during the system design and development process, we have encountered a number of challenges. These challenges include:

- The trade-off between the richness of the gesture supported by the system and the resulting latency it introduces. In our system, the two-way sequential process of capturing/encoding/streaming/decoding visual information introduces some intrinsic latency between the time the image is captured and it is displayed on the screen. We are currently exploring means by which we can extract the hands of the helper from the shared visual space and display the hands on the local video view of the worker.
- The trade-off between the quality of the image/video projected and network latency. We are currently exploring ways in which we can provide a high-resolution video of a subset of the shared visual space.
- The trade-off between supporting the mobility of the worker while maintaining spatial coherence. (1) Because of the worker's mobility, there are sometimes discrepancies between the view projected in the worker's display and the view the worker has of his physical workspace. This may disorient workers. There is a need for workers to maintain spatial coherence. We are currently exploring gesture-based interactions to allow the worker to change the view displayed on their view. (2) As the worker moves around, the shared visual view changes, and gesturing on a changing target could become challenging. To address this issue, gesturing (pointing to a location or an object and showing orientation and shape, etc.) was initially performed on a still image. Helpers were required to freeze the video view in order to gesture. Gesturing on a still image not only adds an extra workload for the helper but also results in the shared visual space not being synchronized with the view of the physical workspace. We are currently investigating how to allow the helper to gesture on the video view.

We believe that exploring these tradeoffs in series of laboratory experiments will provide a solid basis on how to design useful remote guiding systems.

7.6 Conclusion and Future work

In summary, in this chapter, we have reviewed the literature on remote guiding and put forward the case for supporting the richness of gesture and mobility of the worker. We described HandsOnVideo, a system for a remote helper guiding a mobile worker working in nontraditional desktop environments. The system was designed and developed using a participatory design approach. Our key research drive was to develop a remote guiding system that is truly useful, enjoyable, easy to use, reliable, effective, and comfortable for end users. The design approach we have taken has allowed us to test and trial a number of design ideas. It also enabled us to understand from a user's perspective some of the design tradeoffs. The usability study with end users indicated that the system is useful and effective. The users were also positive about using the near-eye display for mobility and instructions and using unmediated representations of hands for remote gestures.

Our future work is to investigate the expansion of the current system to a mobile helper station [23, 24]. In the remote guiding system currently developed, the gesture guidance is supported by a large touch table. A fully mobile remote guiding system using similar technologies for the two parts of the system, the expert station and the operator station, will be easily deployable and adaptable in the mining industry.

We are currently engineering a rugged version of the system for initial field deployment and field studies. Industry deployment and the study of system use in its real context are crucial in understanding the human factors issues prior to prototype development and commercialization of the system.

The deployment of a rugged HandsOnVideo system to a mine site would allow us to investigate the following questions:

- What is required for mining operators to use the system effectively?
- What measurable benefits can be achieved from the system used in a mine, such as productivity and safety?
- What ROI on maintenance cost could be obtained by means of a large deployment of several similar units?

Acknowledgment

We would like to thank the Future mine research theme of the CSIRO Minerals Down Under flagship for sponsoring and supporting this research effort. This chapter is a reprint of Alem et al. [1] with minor edits. Permission is granted from Bentham Open.

References

1 Alem, L., Huang, W., and Tecchia, F. (2011). Supporting the Changing Roles of Maintenance Operators in Mining: A Human Factors Perspective. *The Ergonomics Open Journal.* 4: 81–92. https://doi.org/10.2174/1875934301104010081.

2 Karsenty, L. (1999). Cooperative work and shared visual context: an empirical study of comprehension problems in side-by-side and remote help dialogues. *Human-Computer Interaction* 14 (3): 283–315.

3 Kraut, R.E., Fussell, S.R., and Siegel, J. (2003). Visual information as a conversational resource in collaborative physical tasks. *Human-Computer Interaction* 18: 13–49.

4 Palmer, D., Adcock, M., Smith, J., et al. (2007). Annotating with light for remote guidance. *Proceedings of the 19th Australasian Conference on Computer-Human Interaction: Entertaining User Interfaces*, Adelaide, Australia (28–30 November 2007). OZCHI '07, vol. 251. New York, NY: ACM. pp. 103–110.

5 Kraut, R.E., Miller, M.D., and Siegel, J. (1996). Collaboration in performance of physical tasks: effects on outcomes and communication. *CSCW '96: Proceedings of the 1996 ACM Conference on Computer Supported Cooperative Work*, Boston, Massachusetts, USA (16-20 November 1996). ACM. pp. 57–66.

6 Clark, H.H. and Brennan, S.A. (1991). Grounding in communication. In: *Perspectives on Socially Shared Cognition* (ed. L.B. Resnick, J.M. Levine, and S.D. Teasley), 127–149. American Psychological Association.

7 Fussell, S.R., Setlock, L.D., Yang, J. et al. (2004). Gestures over video streams to support remote collaboration on physical tasks. *Human-Computer Interaction* 19: 273–309.

8 Tang, A., Boyle, M., and Greenberg, S. (2004). Display and presence disparity in Mixed Presence Groupware. *AUIC '04: Proceedings of the Fifth Conference on Australasian User Interface*. Australian Computer Society, Inc, Dunedin, New Zealand (18-22 January 2004). pp. 73–82.

9 Li, J., Wessels, A., Alem, L., and Stitzlein, C. (2007). Exploring interface with representation of gesture for remote collaboration. *Proceedings of the 19th Australasian Conference on Computer-Human Interaction: Entertaining User Interfaces*, Adelaide, Australia (28–30 November 2007). OZCHI '07, vol. 251. New York, NY: ACM. pp. 179–182.

10 Kirk, D. and Stanton Fraser, D. (2006). Comparing remote gesture technologies for supporting collaborative physical tasks CHI '06. *Proceedings of the SIGCHI Conference on Human Factors in Computing Systems*, Montréal Québec Canada (22-27 April 2006). ACM. pp. 1191–1200.

11 Ou, J., Fussell, S.R., Chen, X., et al. (2003). Gestural communication over video stream: supporting multimodal interaction for remote collaborative physical tasks. *ICMI '03: Proceedings of the 5th International Conference on Multimodal Interfaces*, Vancouver B.C., Canada (05-07 November 2003). ACM. pp. 242–249.

12 Kirk, D., Crabtree, A., and Rodden, T. (2005). Ways of the hands. *ECSCW'05: Proceedings of the Ninth Conference on European Conference on Computer Supported Cooperative Work*. Springer-Verlag New York, Inc, Paris, France (18-22 September 2005). pp. 1–21.

13 Kurata, T., Sakata, N., Kourogi, M., et al. (2004). Remote collaboration using a shoulder-worn active camera/laser. *Eighth International Symposium on Wearable Computers*, Arlington, Virginia (31 October 2004–03 November 2004). ISWC 2004, vol. 1. IEEE. pp. 62–69.

14 Sakata, N., Kurata, T., Kato, T., et al. (2003). WACL: supporting telecommunications using – wearable active camera with laser pointer. *Proceedings Seventh*

IEEE International Symposium on Wearable Computers, White Plains, NY, USA (21–23 October 2003). pp. 53–56.

15 Kuzuoka, H., Kosaka, J., Yamazaki, K., et al. (2004). Mediating dual ecologies. *CSCW '04: Proceedings of the 2004 ACM Conference on Computer Supported Cooperative Work*, Chicago, Illinois, USA (06-10 November 2004). ACM. pp. 477–486.

16 Alem, L., Tecchia, F., and Huang, W. (2011). HandsOnVideo: towards a gesture based mobile AR system for remote collaboration. In: *Mobile Collaborative Augmented Reality: Recent Trends* (ed. L. Alem and W. Huang), 127–138. New York, NY: Springer.

17 Huang, W. and Alem, L. (2011). Supporting hand gestures in mobile remote collaboration: a usability evaluation. *Proceedings of the 25th BCS Conference on Human Computer Interaction*, BCS Learning & Development Ltd, Newcastle-upon-Tyne, UK (4–8 July 2011).

18 R.E.A.L. (REmote Assistance for Lines), (c)(TM) SIDEL S.p.a. and VRMedia S.r.l. http://www.vrmedia.it/Real.htm (accessed October 2010).

19 Hollerer, T. and Feiner, S. (2004). Chapter 9: Mobile augmented reality. In: *Telegeoinformatics: Location-based Computing and Services* (ed. H. Karimi and A. Hammad), 221–260. Taylor & Francis Books Ltd.

20 Rolland, J.P. and Fuchs, H. (2000). Optical versus video see-through head-mounted displays in medical visualization. *Presence Teleoperators and Virtual Environments* 9: 287–309.

21 Horberry, T.J., Burgess-Limerick, R., and Steiner, L.J. (2011). *Human Factors for the Design, Operation, and Maintenance of Mining Equipment.* CRC Press.

22 The Google WebM project. Google Inc. http://www.webmproject.org/ (accessed October 2010).

23 Alem, L., and Huang, W. (2011). Developing Mobile Remote Collaboration Systems for Industrial Use: Some Design Challenges. *Proceedings of the 13th IFIP TC13 Conference on HumanComputer Interaction (Interact11)*, Lisbon, Portugal (05–09 September 2011). Springer, Berlin, Heidelberg. https://doi.org/10.1007/978-3-642-23768-3_53.

24 Huang, W. and Alem, L. (2013). HandsInAir: A Wearable System for Remote Collaboration. CSCW '13: Proceedings of the 2013 conference on Computer supported cooperative work companion. San Antonio Texas USA (23–27 February 2013). pp. 153–156. https://doi.org/10.1145/2441955.2441994.

8

Gesturing in the Air in Supporting Full Mobility

8.1 Introduction

Rapid advancements in technology have made it possible for remotely located people to break barriers of distance in space and collaborate with each other in similar ways as they do when they are colocated [1]. Nowadays, collaboration between individuals across the globe and organizations has become an essential part of our daily activities. The past decades have seen a fast-growing interest among researchers and engineers in developing systems to support remote collaboration [2]. However, most of these systems aim to support collaborations in which individuals play similar roles. Relatively less attention has been given to collaborative activities in which collaboration partners have distinct roles, particularly with one partner playing the role of *helper* and the other playing the role of *worker*.

More specifically, as technologies become increasingly complex, our dependence on external expertise to understand and use the technology is growing rapidly. There is a range of real-world situations in which assistance from a remote helper is needed for a local novice worker to accomplish collaborative physical tasks [3]. Such tasks require remote collaborators to work together manipulating physical objects in the real world, involving complex coordination between verbal communication and physical actions. For example, an ultrasound examination is one of the medical checks that require specific expertise to conduct. However, such expertise is often limited in supply and not always available locally. In some cases, there is a need for a remote radiologist to guide a nonspecialist doctor or nurse operating an ultrasound machine to conduct a quality diagnostic ultrasound scan. Other examples include a remote expert providing technical support for an onsite technician to maintain or repair a piece of equipment and a remote instructor helping a disabled student at home to complete arts and crafts homework.

It has been widely agreed that one of the main issues with remote collaboration is that there is no longer a common ground for collaboration partners to communicate the same way as they do when they are colocated [4]. A series of

Computer-Supported Collaboration: Theory and Practice, First Edition.
Weidong Huang, Mark Billinghurst, Leila Alem, Chun Xiao, and Troels Rasmussen.
© 2024 The Institute of Electrical and Electronics Engineers, Inc. Published 2024 by John Wiley & Sons, Inc.

studies have been conducted demonstrating that providing remote collaborators with access to a shared visual space helps to achieve common ground and can be beneficial to the completion of collaborative tasks (e.g. [3]). According to Tang et al. [5], "a shared visual workspace is one where participants can create, see, share and manipulate artifacts within a bounded space." For remote collaboration, shared visual spaces are often provided in the form of video views of the workspace of the worker [6].

Further, prior research has indicated that the reason why face-to-face communication is more efficient than video-mediated communication is mainly because in the face-to-face condition, collaboration participants are able to perform gestures over the task objects, and these gestures are visually available to all participants [7]. This suggests that it is important to support gesturing in the shared visual space for effective remote collaboration.

A number of systems have been developed to support remote guidance by providing a shared visual space and using the space for gesturing (e.g. [8]). However, existing systems often confine collaborators to fixed desktop settings. The value of remote guiding technology in supporting mobility of the collaboration and providing easy access to remote expertise has not been fully explored:

Mobility: During the guiding process, workers may be required to walk around to fetch tools and inspect machines, while helpers may need to go to different locations to look for materials or information. For example, in a call center, a service provider often deals with customers who may use a range of devices or machines of different models. He/she often needs to walk around to look for the specific manual for that device, or have a closer look at the sample machines for various purposes such as identifying the right model.

Accessibility: When remote expertise is required, it is often urgent and helpers may be out of their office and on the move. For example, in a manufacturing factory, when a sophisticated machine suddenly breaks down, onsite maintenance technicians require urgent input from a remote expert, as the time lost in the machine not being running translates into a loss in productivity.

In an attempt to address the mobility and accessibility issues, we developed HandsInAir, a wearable system for mobile remote guidance, which was briefly reported in a poster [9]. In the remainder of this chapter, we first present a theoretical background for our research, with a focus on shared visual spaces and the role of remote gestures for collaborative physical tasks. Then, we briefly review approaches of supporting remote gestures that have been used in previous research, followed by the presentation of the "hands-in-the-air" approach. Next, we introduce our HandsInAir system with detailed user interfaces and system specifications, followed by a usability study. Finally we conclude the chapter with a brief discussion and a short summary.

8.2   Background

8.2.1   Shared Visual Space

In performing collaborative physical tasks, people interact with each other via various communication channels. The interpersonal communication can be more effective when collaborators share a greater amount of common ground, which includes mutual knowledge, beliefs, attitudes, and expectations. Previous research has demonstrated the value of shared visual space in achieving common ground (e.g. [10]). In particular, according to Kraut et al. [11], shared views of a workspace play at least three interrelated roles:

- Maintain situational awareness
- Aid conversational grounding
- Promote a sense of copresence

First, to have a successful collaboration, collaborators need to have ongoing awareness of the task and their partner. This awareness can be used to plan what to say and what to do next, serving as a mechanism to coordinate between their verbal utterances and physical actions. Such awareness can be obtained through shared visual views of the workspace because collaborators can see what is happening. Second, effective communication largely depends on how much mutual knowledge they have about the task and their partner. More specifically, as stated by Gergle et al. [6], "speakers form utterances based on an expectation of what a listener is likely to know and then monitor whether the utterance was understood. In return, listeners have a responsibility to demonstrate their level of understanding." Information needed for building such mutual knowledge can be obtained from shared views of workspace. Third, when collaboration takes place among individuals who are physically distributed, it is important to help collaborators to feel connected. Enhancing the sense of copresence has proved to be beneficial to the success of the collaboration, and shared views of workspace help to promote such a sense of "being together" [12, 13].

8.2.2   Remote Gestures

Although a shared visual space is helpful for grounding or establishing common ground between collaboration partners, it is not feasible, if not impossible, to provide all visual information that is available to colocated collaborators to remote collaborators due to bandwidth limitations and limited cognitive capacity of humans [3, 11]. Therefore, in developing systems for remote collaboration, it is important to determine what visual information is the most important and make sure this information is provided in an appropriate way.

Observational studies of remote collaboration on physical tasks have revealed that collaborators speak and act in relation to the position and status of objects in the workspace and ongoing activities of each other in the environment Their speeches and actions are intricately dependent on each other while speaking, they constantly use hand gestures to clarify and enhance their messages. Fussell and her colleagues conducted a series of studies on collaborative physical studies and found that not only speech but also gestures and actions were used for grounding and that the use of gestures improved task performance (e.g. [10, 14]). With access to the shared visual space, helpers allocate most of their attention to workers' hands and task objects [6, 15, 16]. All these findings indicate that it is important to support remote gestures for developing tools for remote collaboration on physical tasks.

Further, Fussell et al. [14] classified hand gestures into two groups: pointing gestures and representational gestures. The former indicates the direction of movement or the locations of task objects, while the latter represents the form and nature of task objects or actions to be taken with the objects. The authors conducted two studies to investigate the role of these two types of gestures and how the gestures could be effectively conveyed to the remote site. The first study used a system that was mouse-based and supported remote pointing only, while the second study used a system that used pen-based drawings to represent hand gestures. The results indicated that only a simple cursor pointing was not enough for effective collaboration, while pen-based drawings of remote gestures resulted in communication and performance being as good as that in colocated collaboration.

More comprehensively, Kirk and Fraser [7, 17] conducted a series of studies that compared different ways of conveying remote gestures, including projected hands, video-presented hands, and sketches. These studies investigated the effects of gesture formats on both immediate task performance and longer-term knowledge development (learning). They found that gesturing with an unmediated representation of hands led to significantly better performance of collaborative physical tasks.

8.2.3 Supporting Remote Gestures

A number of systems have been proposed or developed in the literature to support remote gestures using various technologies for remote guidance. In this subsection, we briefly review prior approaches with a focus on how remote gestures are performed by the helper.

8.2.3.1 Agent-Based Remote Gestures

In this approach, helper gestures are delivered by an agent located at the workspace of the worker such as a laser pointer or a stick. For example, in the

WACL system of Sakata et al. [18], the worker wears a steerable camera attached to a laser head. The helper can independently control the camera to see the workspace and point to the real object via the laser pointer. In this setting, the helper is sitting at a desk operating the laser pointer. The GestureMan systems of Kuzuoka et al. [19] also employed the agent approach. In their systems, the helper uses joystick to control a mobile robot that is located on the worker site. The helper points to the object on the screen and this gesture is conveyed by the mobile robot through the use of a pointing stick and a laser pointer.

8.2.3.2 Digital Annotations/Gestures
In this approach, digital sketching is used to represent gestures. For example, Ou et al. [20] developed a DOVE (Drawing over Video Environment) system that integrates gestures of helper into the live video of the worker's workspace. In this system, the helper uses a digital pen and performs gestures by drawing on the video streams of the work environment while providing verbal instructions. Palmer et al. [21] and Gauglitz et al. [2] also used this approach in developing their remote guidance systems.

8.2.3.3 Projected Hands
In this approach, the helper hands are directly projected into the worker's workspace and aligned with the associated objects. For example, Kirk and Fraser [7, 17] presented a mixed ecology system. This system requires the helper to gesture at the desk while looking at the monitor in the front. His hands are captured by a video camera, and the captured hands are directly projected onto the desk of the worker. Yamashita et al. [22] developed an immersive system called T-room. This system also uses projected hands to support remote gestures but with additional images of helpers shown on the vertical walls of the T-room.

8.2.3.4 Hands over Workspace Videos
In this approach, the helper performs gestures over the workspace videos shown on a computer display. The hand movements of the helper are captured by a camera and displayed to the worker. For example, in the SharedView system of Kuzuoka [23], the helper is required to stand at the side of a display that shows the video of the worker's workspace. He uses his hands to gesture on the objects over the display. The combined video of the helper's hands and the worker's workspace is captured by a camera and then sent to the worker's side and displayed on the head-mounted display worn by the worker. This approach was also used in the HandsOnVideo system of Alem et al. [8].

8.3   System Overview

Our HandsInAir system includes two parts: a helper station and a worker station. The two stations are connected through a wireless network. In this section, we first explain how remote gestures are supported in HandsInAir. Then, the system's hardware and software implementation is presented, which is followed by information on how the system works.

8.3.1   Hands in the Air

In order to support the mobility of collaborators, we used wearable computers to run the system software and near-eye devices to display visual information. In previous approaches to supporting remote gestures, helpers were often required to work within a fixed desktop setting and need to use, touch, or control physical objects to perform gestures. This setting is no longer feasible to set up when they are mobile. As a result, a major challenge we faced, among others (e.g. [24, 25]), was how to support the richness of remote gestures when the support of a physical display/screen or other operational objects for the helper was no longer available.

In developing HandsInAir, we implemented an approach that was to meet this challenge [26]. As shown in Figure 8.1, the helper wears a helmet mounted with a camera and a near-eye display, and he is able to perform gestures in the air for guiding purposes. The near-eye display we used in HandsInAir is a device with two small screens. The user can look into the screens and see a virtual display that is equivalent to a 67-in. screen as viewed from 10 ft away. More details are introduced in the following subsection.

8.3.2   Hardware and Software Implementation

Both the helper and worker stations have the same hardware configuration. As shown in Figure 8.2, the hardware used to implement each wearable unit consists of a helmet mounted with a Microsoft Lifecam Webcam on top and a Vuzix 920 wrap near-eye display beneath the brim. Both the camera and the near-eye display

(a)   (b)

Figure 8.1   Helper performs gestures in the air (a) and the near-eye display (b).

Figure 8.2   User interface.

are connected to a wearable PC worn by the user. The camera of the worker station is used to capture the workspace of the worker, while the camera of the helper station is used to capture the hands of the helper. The near-eye display is used to display the combined video of the worker's workspace and the helper's hands. There is also an audio connection between the two sides to support verbal communication.

A software application is running on each wearable PC to provide the necessary functions of video processing/transmission and network communication between the two stations. The software is developed in C++ on Windows XP machines utilizing a number of open-source libraries. Both the worker and helper stations simultaneously act as a video server and a video client. The worker station acts as a server, sending local camera feeds of the workspace and, as a client to the helper station, receiving video feeds of the helper's hands. Likewise, the helper station acts as a video client receiving workspace feeds and sending video feeds containing the helper's hands. The Intel OpenCV open-source computer vision library is used to implement an Adaptive Skin Detector, which extracts the helper's hands from video feeds of the helper camera and combines them with corresponding video feeds of the workspace (see Figure 8.3). This detector is also used to display the combined videos on the near-eye displays of the helper and the worker.

Network connections are realized at the low level by opening up streaming connections as the wearable PCs on both sides simultaneously send and receive a sequence of images. The images are compressed with JPEG compression prior to sending, and decompressed upon receipt using the open source IJG (Independent JPEG Group) LibJPEG library to avoid sending costly raw image data and to maintain a real-time frame rate on both sides.

8.3.3   How the System Works

How the system works is illustrated in Figure 8.4. Once a connection is established, the system initializes two video streams between the stations. First, the scene video from the worker camera is fed to the helper station and displayed on the near-eye display. The helper examines the video, talks to the worker, and performs gestures

1. A work scene captured by the worker camera

Figure 8.3 Illustration of combining a hand gesture and a workspace scene.

2. A hand gesture captured by the helper camera

3. Combination of the gesture and the scene

Figure 8.4 Illustration of camera captures and the content of near-eye displays. Source: Alem et al. [26]/with permission from Springer Nature.

which are captured by the helper camera. The hands are extracted without the background and combined with the scene video. What is shown on the helper's near-eye display is continuously updated with the combination. In other words, the helper is able to see his hands performing gestures at the task artifacts on the display. The extracted hand images are also sent to the worker side, combined with the scene video, and displayed on the near-eye display. This allows the worker to see the unmediated hand gestures. The worker hears the audio instructions, sees the visual aids by looking up in the near-eye display, when necessary, and performs operations as instructed by the helper.

8.4 Usability Study

As described in the last section, HandsInAir not only allows collaborators to be mobile but also enables helpers to perform pointing and more complex hand gestures over the remote objects shown on the virtual display. We followed a participatory design approach during the design and development of this system. In this section, we present a user study that we conducted to validate its usability.

8.4.1 Design

There were two separate rooms, with one hosting the helper station and the other hosting the worker station. Users were recruited to complete representative physical tasks collaboratively. The participants were randomly grouped in pairs, one playing the role of helper and the other playing worker. The participants of each pair were each located in one of the two rooms according to the role being assigned. The helper and worker could talk to each other through speaker/microphone headsets. The whole task process was video recorded on both helper and worker sides for further analysis.

There was also a questionnaire session at the end of the task: one for the helper and one for the worker. Both included questions asking participants to rate the system usability. In this particular study, the usability was evaluated from the following perspectives:

- Ease of learning
- Ease of use
- Usefulness
- Task satisfaction
- Mobility
- Perception of interaction
- Environment awareness
- Copresence
- Perception of hand gestures

Questionnaires also included open questions. The open questions asked participants' experiences of using the system and their opinions on possible further improvements.

8.4.2 Participants

Twenty people volunteered to participate in the study. Fifteen of them were male and the rest were female. They were aged between 20 and 40. By the time they participated in the study, none of them had experience of using HandsInAir, or any other systems of the same kind.

8.4.3 Task

In this study, the participants were asked to assemble a set of loose Lego toy pieces. Previous user studies in the literature have also used the assembly of toy blocks for similar purposes (e.g. [27]). Since toy assembly has components that can be found in a range of real-world physical tasks such as assemble, disassemble, select, and rotate, this task is considered representative for remote guidance on physical objects [3]. During the task, the worker was asked to assemble the toy pieces into a prespecified complex model under the instruction of the helper.

There was an instruction manual for the helper. The guiding manual was divided into three parts, and the parts were placed separately in different locations in the room. The helper needed to go to the first place, pick up the manual do the guiding, and then go to the next until the end of the task. The helper was instructed that he could provide verbal and gestural instructions to the worker at any time but was not allowed to show any part of the manual to the worker. The worker, on the other hand, had no idea about what steps were needed to complete the task.

To mimic the general workplace settings of workers, we used a workshop room for the location of the workers (see Figure 8.5). The workshop was full of equipment and tools and was composed of a number of work areas. The toy pieces were placed in different locations. The worker had to move around the workspace to collect them and get the task done. To test whether the worker was aware of the environment while he walked with a near-eye display, small obstacles were

Figure 8.5 The workshop room setup.

deliberately placed in the trajectory of the worker. The worker had to avoid them while moving around. To prevent workers from tripping over, only light empty boxes were used as obstacles. The helper room had tables and chairs, and it was about 20 m away from the worker room.

8.4.4 Procedure

Before the experiment, an introduction session was given to the two participants of each pair. They were gathered in a meeting room. First, they were given a short tutorial and a brief demonstration on how the system worked. The helper interface and the worker interface were introduced. Then, the task and the procedure of the study were explained. The two participants were also given a chance to get familiar with the system and try out the equipment. During this session, the participants could ask questions, and answers were provided by two experimenters.

When they were ready, the two participants were randomly assigned roles, one as a helper and the other as a worker. Then, each of them was led to the corresponding room where the helper or worker station was located. On each site, there was also an experimenter providing further assistance to the participant, putting the wearable backpack on, recording videos, observing, and taking notes of the collaboration behavior.

Once the connection was established on both sides, the participants started performing the guiding task on the Lego toys provided. After the task was completed, each participant was asked to fill the helper or worker questionnaire, depending on his role. After finishing their questionnaires, the participants went back to the meeting room for a semi-structured interview. They were encouraged to ask questions, propose ideas and further improvements, debate on the issues and comment on the system. The whole experiment for each pair took about 40 minutes.

8.4.5 Results

8.4.5.1 Observations

All participants were able to perform and complete the assembly task without obvious delays. It seemed that the participants were comfortable with the system. Helpers were able to gesture in the air while looking at the video in the near-eye display and giving verbal instructions to workers. Workers were able to assemble the toy pieces with their hands while receiving verbal instructions from helpers and looking at the visual aids shown in the near-eye display. It appeared that the participants were able to communicate with each other smoothly and effectively via both the visual and verbal channels provided by HandsInAir.

There were no apparent difficulties observed for the helpers to collect instruction materials and gesture in the air. While walking, they tended to slow down or stop to perform gestures. It seemed natural for them to perform pointing gestures using one hand, or perform representational gestures using two hands.

For workers, it was observed that they were able to avoid obstacles on the way. They generally became more careful when they were close to an obstacle. The workers could also easily locate and fetch the toy pieces required by following the instructions given by the helper.

8.4.5.2 Usability Ratings

The usability of HandsInAir was tested based on a mix of positive and negative statements. Each statement was to evaluate one specific aspect of the usability. These statements are listed in Table 8.1.

The participants rated the extent to which they agreed with the statements, based on a scale of 1–7, with 1 being "strongly disagree," 7 being "strongly agree," and 4 being "neutral." For the purpose of analysis, user ratings were first transferred so that higher ratings meant better usability and then averaged across the participants. The average of the obtained ratings was computed for the overall usability. The results are shown in Table 8.2.

Table 8.1 Statements for usability ratings.

Usability	Statement
Ease of learning	I found that the system was easy to learn
Ease of use	I found that the system was difficult to use
Usefulness	I found that the system was useful for remote guiding tasks
Task satisfaction	I was disappointed with my task performance
Mobility	I felt that I was free to move around
Environment awareness	I felt that I was unaware of my physical surroundings
Copresence	I felt that my partner and I were at the same location
Perception of interaction	I felt that I was engaged with my partner during the task
Perception of gestures (helper)	I found it difficult to point to objects
Perception of gestures (helper)	I found it easy to demonstrate assembly of objects
Perception of gestures (worker)	I found it difficult to understand which objects my partner was pointing to
Perception of gestures (worker)	I found it easy to follow my partner's hand gestures to assemble objects

Table 8.2 Average values and standard deviations of user ratings for individual and overall usability measures.

	Average	StDev
Ease of learning	6.05	1.10
Ease of use	5.50	0.89
Usefulness	6.15	0.67
Task satisfaction	5.30	0.86
Mobility	5.70	0.92
Environment awareness	5.55	0.94
Copresence	4.35	0.88
Perception of interaction	4.90	0.79
Perception of pointing gestures	5.75	0.97
Perception of representational gestures	5.25	0.85
Overall usability	**5.45**	**0.89**

As can be seen from Table 8.2, overall, the participants were positive about the usability of HandsInAir with a rating of 5.45 on a scale of 1–7. More specifically, the participants rated the usefulness of the system at the highest value of 6.15, indicating that HandsInAir was considered useful. The participants were also positive about ease of learning, ease of use, task satisfaction, mobility, environment awareness, perception of interaction, and hand gestures, while copresence was rated just above being neutral (4.35).

To explore possible differences between helpers and workers, we looked at their user ratings separately, and the results are shown in Figure 8.6. It can be seen

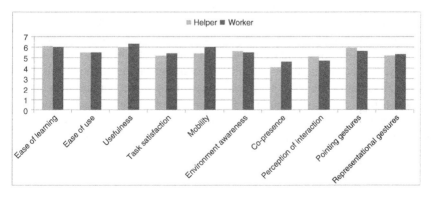

Figure 8.6 Average usability ratings from helpers and workers.

that all usability measures were rated positive (above 4). Ease of use was rated the same. Helpers gave higher usability ratings for ease of learning, environment awareness, perception of interaction, and pointing gestures, while workers rated higher for usefulness, task satisfaction, mobility, copresence, and representational gestures. However, statistical tests revealed that these differences in the ratings between helpers and workers were not statistically significant at the significance level of 0.05.

8.4.5.3 User Comments

In regard to user responses to the open questions, the participants were generally positive about the system. User comments included "(the system is) very useful for remote guiding." "It is perfect when I do not know how to do and want to be guided." Helpers appreciated being able to perform hand gestures without any physical constraints and commented that gesturing with hands in air is a "cool factor." For example: "I can explain my intention with the help of my hands easily." "It is very helpful, especially for the complex tasks which is hard to instruct by speaking only." "Using my hands for this task is the most appropriate and the best way of guiding the remote worker." "I have a much better idea of what workers can see, much better remote awareness of working environment than other methods. Feel like you can nearly directly interact with the remote environment." "This is very useful and an intuitive way of guiding." "I can do anything I like with my hands (showing shapes and orientations)." However, it was also mentioned that "pointing in the air can be tiring after a while."

Workers also found helpful being able to see the helper's hands via the near-eye display. For example: "It is helpful to be able to see gestures." "It is good that both parties can see and hear the same 'scene' during the task." "It is easy to use." "Seeing the helper's hands is very helpful and instructive." "Hand gestures are useful for knowing which object they want you to interact with. Otherwise they would have to spend much time explaining what object they want you to use."

Further, the qualitative user feedback indicated that the system provided both easy access to the helper and a good user experience with mobility. For example, "I enjoyed being able to move around in between giving instructions, this gives me the freedom to attend to other tasks if need be. I would not have this option if I was using a desktop computer." "Sure! I can have the helper pack in the boot of my car. Then I can use it at any time/anywhere." "Being able to move around is great." "I can sit or stand so that I can be in a comfortable position." "I don't feel restricted in anyway. I can be standing or sitting anywhere I want." "It was a great experience. I felt relaxed." "To me the value of this system is that you can access the expert/helper wherever they are as long as they are wearing the gear with them; you do not need for your expert to be in a specific room. Experts are highly mobile workers. Accessing them should be as easy as calling them on their mobile phone."

Two participants mentioned that they tried to perform hand gestures while walking and pointed out that this might not be a feasible thing to do. One commented that "I wanted to walk and guide at the same time, but I find that I can't do it. As soon as I try to gesture, I find it hard to keep walking because my attention is on guiding and subconsciously feel dangerous to walk at the same time, without looking at the floor/surrounding. As soon as I stopped guiding, I can walk again." The other participant commented that "I can monitor what my partner is doing while I am on the move. I can provide audio instructions while on the move, but the moment I need to give instructions requiring pointing and showing with my hand how to perform a task, I need to be static."

On the other hand, user comments also suggested some areas for further improvements. For example, some participants mentioned that during the task, they sometimes had to keep switching between the near-eye display and the workspace. This could make their eyes tired if they use the system for a long time. We are planning to use a see-through device to replace the near-eye display to avoid such frequent switches between the video and the real world. Other suggestions made by participants were related to the limitations of network bandwidth, system process capacity and the hardware that we currently could provide. We believe that these limitations could be removed when more powerful technologies and devices become available to us. For example, the quality of videos and images of hands could be further improved by using higher resolution displays and cameras; 3D cameras could be used so that users could have a better understanding of the spatial relationships between objects (e.g. [28]).

8.5 Discussion

8.5.1 Usability

The study results confirmed the usability and usefulness of HandsInAir in supporting a remote mobile helper who guides a mobile worker in performing physical tasks. Our observations revealed that participants were able to complete the tasks comfortably without apparent difficulties. The user feedback and usability ratings also indicated that HandsInAir can be useful and usable for real-world use.

In particular, the study participants were positive about the mobility support provided by the system to the collaborators. According to their feedback, the mobility support allows a worker to access a remote helper more easily. Also, helpers are enabled to continuously engage with the system and their partner when they move around during the guiding process. Participants who played the role of helper also considered gesturing in the air as being intuitive and effective.

8.5.2 Gesturing in the Air

To meet our goal of freeing helpers from a fixed position, we implemented an approach that enables helpers to perform hand gestures in the air. This is achieved by combining the videos of helper's hands and worker's workspace. The combined video is displayed on the near-eye display. Therefore, all the helper needs to do is to look at the objects shown in the video and gesture in the air, thus removing the reliance on a desktop display/screen.

The current approach also allows the system to convey unmediated hand gestures to the worker. It has been demonstrated that this type of gesture is associated with better task performance and copresence (e.g. [13, 17]). Further, the use of a wearable computer and a near-eye display effectively frees helpers from traditional fixed desktop settings. However, it is important to note that the mobility support in this system is to ensure that the collaborators can continuously engage with each other and the system while moving around. It is not our purpose and expectation that the helper is required to *gesture* while walking and that the worker is required to *manipulate objects* while walking.

8.5.3 Limitation of the Studies

Due to our limitations in accessing real-world resources, we used the assembly of Lego toy pieces as the experimental task, recruited volunteers as the targeted users, and conducted the studies in simulated workshop settings. However, these were done at the cost of realism and generalizability. As demonstrated in prior research (e.g. [23]), testing the system with real users in real-world workplaces for an extended period of time would allow us to systematically examine usability issues and provide us with unbiased insights into the usability of the system, thus being more desirable.

8.6 Concluding Remarks and Future Work

We have argued that confining collaborators to a fixed position is a limiting factor. This is because, in many real-world situations, collaborators need to be away from a fixed position for various purposes. In this chapter, we have presented HandsInAir, a wearable system for mobile remote collaboration that can be used in these situations. The system has been tested for its overall usability and its ability to support the mobility of collaborators, and the test results are positive.

We are currently planning two field trials, one in an aircraft manufacturing factory and the other in a mining site. Our field partners are currently recording performance data for the current practices of remote guidance. They will also record data for the time period for which HandsInAir is used. This will allow us to conduct comparative studies to understand the benefits of the system. We will also conduct onsite observations to understand user behavior changes before and after the use of the system and to investigate research questions such as how users interact with each other and with the system and how their hands, visual focus, body, and verbal communication coordinate together when mobility is an essential part of their collaboration.

Acknowledgment

This chapter is a reprint of Huang and Alem [1] with minor edits. Permission is granted from J.UCS.

References

1 Huang, W. and Alem, L. (2013). Gesturing in the Air: Supporting Full Mobility in Remote Collaboration on Physical Tasks. *Journal of Universal Computer Science.* 19: 1158–1174. https://doi.org/10.3217/jucs-019-08-1158. Reprinted with minor edits.

2 Gauglitz, S., Lee, C., Turk, M. and Höllerer, T. (2012). Integrating the physical environment into mobile remote collaboration. *Proceedings of the 14th International Conference on Human-Computer Interaction with Mobile Devices and Services (MobileHCI'12)*, San Francisco California USA (21–24 September 2012). ACM. pp. 241–250.

3 Fussell, S.R., Setlock, L.D., Yang, J. et al. (2004). Gestures over video streams to support remote collaboration on physical tasks. *Human-Computer Interaction* 19: 273–309.

4 Clark, H.H. and Brennan, S.E. (1991). Grounding in communication. In: *Perspectives on Socially Shared Cognition* (ed. L.B. Resnick, J.M. Levine, and S.D. Teasley), 259–292. Washington, DC: American Psychological Association.

5 Tang, A., Boyle, M., and Greenberg, S. (2004). Display and presence disparity in Mixed Presence Groupware. *Proceedings of the Fifth Conference on Australian User Interface*, Dunedin, New Zealand (18–22 January 2004). ACM. pp. 73–82.

6 Gergle, D., Kraut, R. E. and Fussell, S. R. (2006). The impact of delayed visual feedback on collaborative performance. *Proceedings of the SIGCHI Conference on Human Factors in Computing Systems*, Montréal Québec Canada (22–27 April 2006). ACM. pp. 1303–1312.

7 Kirk, D.S. and Fraser, D.S. (2005). The impact of remote gesturing on distance instruction. *Proceedings of the Conference on Computer Supported Collaborative*

Learning, Taipei Taiwan (30 May 2005–04 June 2005). International Society of the Learning Sciences. pp. 301–310.

8 Alem, L., Tecchia, F., and Huang, W. (2011). HandsOnVideo: towards a gesture based mobile AR system for remote collaboration. In: *Recent Trends of Mobile Collaborative Augmented Reality Systems* (ed. L. Alem and W. Huang), 127–138. New York, NY: Springer.

9 Huang, W. and Alem, L. (2013). HandsinAir: a wearable system for remote collaboration on physical tasks. *Proceedings of the 2013 Conference on Computer Supported Cooperative Work Companion (CSCW '13)*, San Antonio, TX, USA (23–27 February 2013). ACM. pp. 153–156.

10 Fussell, S. R., Kraut, R. E., and Siegel, J. (2000). Coordination of communication: effects of shared visual context on collaborative work. *Proceedings of the 2000 ACM Conference on Computer Supported Cooperative Work*, Philadelphia, PA, USA (2–6 December 2000). ACM. pp. 21–30.

11 Kraut, R.E., Fussell, S.R., and Siegel, J. (2003). Visual information as a conversational resource in collaborative physical tasks. *Human-Computer Interaction* 18: 13–49.

12 Alem, L. and Li, J. (2011). A study of gestures in a video-mediated collaborative assembly task. *Advances in Human-Computer Interaction* 2011: 987830: 7 pages.

13 Kraut, R. E., Gergle, D. and Fussell, S.R. (2002). The use of visual information in shared visual spaces: informing the development of virtual co-presence. *Proceedings of the 2002 ACM Conference on Computer Supported Cooperative Work*, New Orleans Louisiana USA (16–20 November 2002). ACM. pp. 31–40.

14 Fussell, S.R., Setlock, L.D., and Parker, E.M. (2003). Where do helpers look? Gaze targets during collaborative physical tasks. *Proceedings of Extended Abstracts on Human Factors in Computing Systems*, Fort Lauderdale, Florida (5–10 April 2003). ACM. pp. 768–769.

15 Kirk, D. S., Crabtree, A., and Rodden, T. (2005). Ways of the hand. *Proceedings of the European Conference on Computer Supported Cooperative Work*, Paris, France (18–22 September 2005). Springer. pp. 1–21.

16 Kirk, D.S., Rodden, T. and Fraser, D.S. (2007). Turn it this way: grounding collaborative action with remote gestures. *Proceedings of the SIGCHI conference on Human Factors in Computing Systems*, San Jose California USA (28 April 2007–03 May 2007). ACM. pp. 1039–1048.

17 Kirk, D.S., and Fraser, D.S. (2006). Comparing remote gesture technologies for supporting collaborative physical tasks. *Proceedings of the SIGCHI Conference on Human Factors in Computing Systems*, Montréal Québec Canada (22–27 April 2006). ACM. pp. 1191–1200.

18 Sakata, N., Kurata, T., Kato, T., et al. (2003). WACL: supporting telecommunications using – wearable active camera with laser pointer. *Proceedings of the*

Seventh International Symposium on Wearable Computers, White Plains, NY, USA (21–23 October 2003). IEEE. pp. 53–56.

19 Kuzuoka, H., Kosaka, J., Yamazaki, K., et al. (2004). Mediating dual ecologies. *Proceedings of the 2004 ACM Conference on Computer Supported Cooperative Work*, Chicago Illinois USA (6–10 November 2004). ACM. pp. 477–486.

20 Ou, J., Fussell, S.R., Chen, X., Setlock, L.D., and Yang, J. (2003). Gestural communication over video stream: supporting multimodal interaction for remote collaborative physical tasks. *Proceedings of the 5th Conference on Multimodal Interfaces*, Vancouver British Columbia Canada (5–7 November 2003). ACM. pp. 242–249.

21 Palmer, D., Adcock, M., Smith, J., et al. (2007). Annotating with light for remote guidance. *Proceedings of the 19th Australasian conference on Computer-Human Interaction: Entertaining User Interfaces (OZCHI'07)*, Adelaide, Australia (28–30 November 2007). ACM. pp. 103–110.

22 Yamashita, N., Kaji, K., Kuzuoka, H., and Hirata, K. (2011). Improving visibility of remote gestures in distributed tabletop collaboration. *Proceedings of the ACM 2011 Conference on Computer Supported Cooperative Work*, Hangzhou China (19–23 March 2011). ACM. pp. 95–104.

23 Kuzuoka, H. (1992). Spatial workspace collaboration: a SharedView video support system for remote collaboration capability. *Proceedings of the SIGCHI Conference on Human Factors in Computing Systems*, Monterey, CA, USA (3–7 May 1992). ACM. pp. 533–540.

24 Ghiani, G. and Paternò, F. (2010). Supporting mobile users in selecting target devices. *Journal of Universal Computer Science* 16 (15): 2019–2037.

25 Herskovic, V., Ochoa, S.F., Pino, J.A., and Neyem, A. (2011). The iceberg effect: behind the user interface of mobile collaborative systems. *Journal of Universal Computer Science* 17 (2): 183–202.

26 Alem, L. and Huang, W. (2011). Developing mobile remote collaboration systems for industrial use: some design challenges. In: *Human-Computer Interaction – INTERACT 2011. INTERACT 2011. Proceedings of the 13th IFIP TC13 Conference on Human-Computer Interaction (INTERACT'11)* (ed. P. Campos, N. Graham, J. Jorge, et al.), 442–445. Berlin, Heidelberg: Springer.

27 Huang, W. and Alem, L. (2011). Supporting hand gestures in mobile remote collaboration: a usability evaluation. *Proceedings of the 25th BCS Conference on Human Computer Interaction*. pp. 211-216.

28 Tecchia, F., Alem, L., and Huang, W. (2012). 3D helping hands: a gesture based MR system for remote collaboration. *Proceedings of the 11th ACM SIGGRAPH International Conference on Virtual-Reality Continuum and Its Applications in Industry (VRCAI '12)*, Singapore Singapore (02–04 December 2012). ACM. pp. 323–328.

9

Sharing Hand Gesture and Sketch Cues with a Touch User Interface

9.1 Introduction

In many real-world scenarios, people need help from a remote expert [1]. For example, when a broken machine needs to be fixed by an expert who is not present on location, when a doctor needs to conduct an operation on a patient with the essential help of a distant surgeon, or when a crime scene investigator needs help from a forensics expert.

To support these scenarios, researchers have developed a variety of technologies where multimodal communication cues (verbal and visual cues) can be shared between a local worker and a remote expert helper. Typically, these technologies add Augmented Reality (AR) visual communication cues, such as pointing, sketching, or hand gestures, on top of a video conferencing system where audio communication is available in order to improve the collaboration experience and task performance [2, 3]. Through the video conferencing system, a view of local workspace can be shared over live video, and the visual cues overlaid on this to represent spatial information from the remote helper such as position and orientation of an object.

For example, ReMoTe [4], WACL [5], GestureCam [6], and TeleAdvisor [7] are the systems that add visual communication cues on top of the videoconferencing for remote collaboration. More specifically, in the ReMoTe system [4], a remote expert can share his/her hand gesture on the shared live video. Generally, the goals of these systems are to reduce the cost and delay of the expert traveling to the onsite work location, as well as to up-skill the local taskforce.

In some of these interfaces, the shared live video view was from a camera attached to the local worker's head, which allows the remote helper to have the same first-person view as the local worker [4, 8]. While sharing the first-person view, the remote helper can perform hand gestures, sketching (a.k.a drawing annotations), or move a pointer, and the system superimposes and synchronizes

Computer-Supported Collaboration: Theory and Practice, First Edition.
Weidong Huang, Mark Billinghurst, Leila Alem, Chun Xiao, and Troels Rasmussen.
© 2024 The Institute of Electrical and Electronics Engineers, Inc. Published 2024 by John Wiley & Sons, Inc.

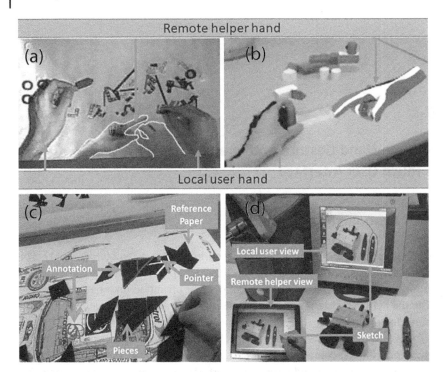

Figure 9.1 The use of AR gesture communication cues in the top two pictures (a,b) [2, 3] and AR sketches and pointer cues in the bottom two pictures (c,d) [8, 9]. Source: (a) Huang et al. [1]/with permission from Elsevier; (b) Sodhi et al. [3]/with permission from Association for Computing Machinery; Kim et al. [8]/with permission from IEEE; (d) Fussell et al./with permission from Taylor and Francis.

these cues in the shared live video. Figure 9.1 shows several examples of using AR technology to show visual cues in a remote collaborative interface.

In this chapter, we explore the use of the combination of raw hand gestures and sketch cues in an AR interface for remote collaboration. In the previous studies, Fussell et al. [9] and Kim et al. [10] explored the use of a pointer and sketch cues and compared them. Higuchi et al. [11] compared hand gestures only and hand gestures with a remote user's gaze pointer. However, researchers have not yet clearly investigated the use of the combination of raw hand gestures and sketch cues.

In our previous remote collaboration studies [12], using raw hand gesture cues had two main issues: (1) a finger can be too big to point at small objects (e.g. pointing at a nail, bolt, or nut from a group of them) and (2) it can be difficult to point to an object when the viewpoint of the live video is moving due to the local worker's head movement (as they are sharing the live video of local worker's first-person view). To solve these issues, we developed a system called HandsInTouch, which supports a unique remote collaboration touch user interface by including both raw hand gestures and sketch cues on a live video or still images.

This chapter makes the following contributions:

- It is the first formal study with a remote collaboration gesture interface sharing both raw hand gestures and sketch cues either on a live video or still images.
- It describes the setup of a user study to evaluate shared gestures and the effects of sketches in remote collaboration.
- Instead of using a virtual representation of the hand gestures, our system segments the raw hands with a single camera and shares them.
- It makes recommendations for the interface design of remote guidance systems and for future research directions.

In the following sections, we first review related works, then describe the HandsInTouch prototype and present a user study design with the prototype. Next, we present and discuss the study results, and finally, end with a conclusion section and directions for future work.

9.2 Related Work

There has been a significant amount of earlier research on systems for collaboration between local workers and remote experts. Our research is most related to earlier systems that share a first-person view of the local worker. For example, Fussell et al. [13] compared audio-only and audio–video links for remote collaboration on a physical task and found that the audio–video link did not have a significant benefit over the audio-only link. However, when researchers added visual communication cues on top of the audio–video link, they did significantly improve performance and created better remote collaboration. The most studied visual communication cues so far are pointers [5, 9, 10, 14], sketching [8, 15–17], and hand gestures [2, 3, 11, 12]. In the rest of this section, we describe previous research using each of these communication cues and discuss their results.

9.2.1 Pointer Cues

GestureCam [6] is one of the first systems to use a pointer for remote collaboration. In this case, a remote user could control a local camera and a laser pointer, enabling them to point at objects in the local workers workspace. This enabled the local worker to easily know which object the remote expert was talking about. Later, Sakata et al. [5] put a camera and laser pointer on a local worker's shoulder as a wearable system with increased portability. The camera and laser pointer were on a pan/tilt system under control of the remote expert, enabling them to freely look around the local worker's environment. Fussell et al. [9] integrated a virtual pointer into a shared live video, and Kim et al. [10] did the same with a handheld device, and in both cases, the remote expert could draft sketches on the live video. They compared pointer and sketch cues and found that the sketch cue was preferred and provided better performance in a remote collaboration task.

Recently, pointer cues have been used for sharing the user's gaze. Gupta et al. [18] developed a system for sharing the local worker's gaze and expert's mouse pointer. They found that both gaze and mouse pointers had a main effect on the task performance and users' feeling of copresence compared to conditions without gaze or pointer cues. Lee et al. [14] studied sharing both the remote expert and local worker's gaze using pointer cues and found a similar result: the remote and local users felt better connected with their partner. Higuchi et al. [19] compared hand gesture only and hand gesture with the remote expert's gaze pointer, and found that adding the gaze pointer helped participants to better understand each other.

9.2.2 Sketch Cues

There were a number of early collaborative systems developed that enabled sharing of sketching cues. For example, Tang and Minneman [20] developed "Videodraw," which took a live video of a task space and projected it into the other users space. Either of the users could sketch on paper and share views of the sketches using the system. Similarly, Ishii and Kobayashi [21] introduced "ClearBoard" in which users could sketch on a screen which was immediately synchronized with the screen of a collaborating partner, and found that collaboration is improved when participants were able to spend more time on collaboration domain tasks, instead of spending time on tasks to maintain the collaboration. These Early systems supported symmetric sketch cues where users at either end could draw on the live video.

Later, researchers introduced asymmetric and more lightweight systems. Fussell et al. [9] introduced "Dove," where a user could sketch and control a pointer on a shared live video of the third-person view from a camera attached to a tripod. Chen et al. [22] developed SEMarbeta, where a remote user helped a local user to fix a car using sketch cues and hand gestures. They compared a video-only condition to a condition with video, hand gestures, and sketch cues and found that users could complete the car fixing task quicker with the two additional cues. Gurevich et al. [7] used a laser projection to allow the remote user to overlay sketch cues directly on the local task space while supporting remote expert control of the local cameras.

One of the challenges with these systems was to have the virtual sketch annotations remaining fixed to the objects of interest when the local workers camera view changed. Kato and Billinghurst [23] employed AR techniques for stabilizing the user sketches with a marker-based tracking system that determined the real-world position of the remotely added sketches fixed in space. Instead of using a physical marker, Kim et al. [8] and Gauglitz et al. [16] used SLAM (Simultaneous Localization and Mapping) tracking to stabilize the user sketches in the real world and found that users preferred the stabilized sketches compared to nonstabilized sketches. One of the interesting features of their systems was being able to manually pause the live video to sketch on a frozen still image. Since the live video was from a camera attached to the local worker's head, it could be unsteady and moving according to the local user movement. To solve this issue, they provided stable still images for users to comfortably and easily sketch on it. Our system also includes the ability to capture and share still images, and sketch on the still images.

9.2.3   Hand Gesture Cues

Researchers have also explored how natural hand gesture cues can be shared between local workers and remote experts. While the system does not need computational support to show and share the local worker's hand gesture as it can be included in the shared live video, the system does to extract and display remote helper's raw hand gesture to the local user in the live video. Most previous studies included one video link for sharing a local workspace, but Kirk et al. [2] added a second video link that captured the remote user's hand gestures and projected them into the local user's space. Alem et al. [4, 24] implemented a similar system but increased the portability and designed a wearable system for the local user with a head-worn camera (HWC) and a head-mounted display (HMD). Huang and Alem [25], Kunz et al. [26], and Tang et al. [27] used computer vision techniques to extract the remote expert's hands by eliminating the background and overlaying the extracted hands onto the live video. Sodhi et al. [3] explored 3D hand gestures and spatial inputs by tracking the user's fingers with a depth camera. Huang et al. [28] also introduced a 3D hand gesture with a depth camera and additionally supported 3D stereo views.

9.2.4   Discussion

We have shown there has been a number of previous works on sharing pointers, sketches, and raw hand gestures in remote collaboration. Each of these visual cues has different benefits. A sketch interface can be more descriptive in sharing information compared to a pointer [9] while a pointer can help a remote user to quickly participate in collaboration [8]. In contrast, raw hand gestures can express diverse information such as pointing, shaping, and appreciating (i.e. thumbs up) ([29]). Moreover, the sketches typically remain visible for a longer time than pointer or hand gesture cues. As they have different benefits and characteristics, different types of tasks may require different visual communication cues [30]. It could be better to integrate them together and let the user switch between them according to their needs and personal preferences. However, there have been few studies exploring the benefit of combining hand gesture, pointer, and sketch cues in remote collaboration systems.

The most relevant work is Chen's study [22], which explored an interface with hand gesture and sketch cues. However, there are several significant differences between this work and our study. First, in exploring the interface with hand gesture and sketch cues, we compared it to an interface with only hand gestures, while they compared their system to a video-only condition with no additional visual cues. So, Chen's study [22] could not separate the effect of each of the two cues (gesture and sketch) in the experimental results. Second, we conducted a user study with two different task types while they tested their system only with one task. Third, the local user wears an HMD in our study, thus enabling the use of two hands, but their system uses a hand-held display (HHD) for a local user that requires one hand to hold it. Fourth, we measured the user's task load but they did not. Fifth, their system separates the hand gesture space from the display space at the remote helper end, while our system supports performing hand gestures on top of the remote helper's display space, so reducing artificial seam between the gesture space and physical workspace. Finally, they only collected qualitative data from interviews, but we collected both qualitative (from interviews) and quantitative data (from questionnaires and task completion time).

In summary, there have been many previous studies that have explored the user of gesture, pointing, and sketching cues in remote collaboration, but almost none of these have explored the combination of them. The most similar work to ours compares using hand gestures and sketching to a video-only condition, and has a number of other significant differences to our system. We are presenting the first research to compare the combination of gesture and sketching cues in remote collaboration with both qualitative and quantitative measures. In Section 9.3, we describe the prototype system we created for sharing gesture and sketching cues between remote collaborators.

9.3 Methods and Materials

In this section, we explain our prototype system and its user interface.

9.3.1 Prototype System

In this section, we describe the prototype HandsInTouch system we developed. We begin with a description of the hardware, then the user interface, and finally, the support for hand gestures and sketching.

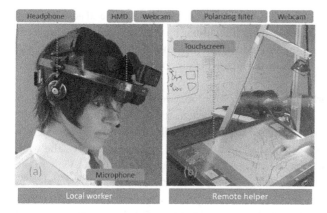

Figure 9.2 Local worker HMD with a camera attached (a) and the remote expert helper unit with a touch screen and a camera (b).

9.3.1.1 Hardware

HandsInTouch has two units (Figure 9.2): a local worker unit and a remote expert helper unit, connected via Ethernet port or Wifi. The local worker unit includes a custom-built headset consisting of an HMD (Vuzix Wrap 1200, a video see-through near-eye display with 852×480 pixels resolution, 35° field of view [FOV]), a webcam, a Windows tablet, and an audio headphone with a microphone (see Figure 9.2a). A webcam (Logitech C920) is attached to the front of the headset. The HMD and the webcam are connected to the Windows tablet for computational power support, and the video from the webcam is displayed on the HMD and shared with the remote helper. The headphone and microphone (Logitech H230) are for verbal communication with the remote helper.

In designing the headset, we tried to keep a space between the local worker's eyes and HMD so that the local worker could switch between the real-world view and the HMD view (the near-eye view). The local worker can see the workspace more clearly when directly watching it with their bare eyes and can switch to the HMD view to see the remote helper inputs (hand gestures or sketches). The real-world view was also to reduce the issues of small FOV (35°) and occlusion by the frame of the HMD display, and it would help understand the surroundings which is important for safety issues. Since the application domain could be the collaboration between a factory local worker and a remote expert, our local worker headset should allow the local workers to keep their peripheral vision open for safety reasons. Furthermore, the headset was designed to be easy to take off, and it also frees both hands of the local workers, enabling them to freely manipulate objects.

The remote helper unit has a touchscreen monitor that allows natural and intuitive touch interactions and is connected to a laptop which processes computing tasks (see Figure 9.2b). A webcam is attached on the top using a support arm and is looking down at the screen for extracting the remote helper's hand gestures over the live video image with a polarizing filter, then sharing them back with the local worker. An audio headset that includes a microphone (Logitech H230) is plugged into a USB port of the monitor or the laptop for supporting verbal communication.

9.3.1.2 The User Interfaces and How HandsInTouch Works

With the hardware devices, our prototype system starts by connecting two units via Ethernet port or Wifi. Figure 9.3 shows its overall software architecture. Once the connection has been established and the whole system is running, the webcam on the local worker's headset captures and sends the workspace view to the remote helper's screen as a live video. For the live video, the image from the webcam is compressed with JPEG compression prior to sending and decompressed upon receipt by the IJG (Independent JPEG Group) LibJPEG C++ library to avoid sending raw image data and to maintain a real-time live video.

The screen of the helper monitor is divided into three sections: left, large middle, and right (see Figure 9.2b). The large middle section displays the live video feed from the local worker showing the workspace, and a still image can be taken from the live video by the remote helper pressing a button at the bottom of the section. The captured snapshot image is immediately displayed in the large middle section and the remote helper can then return back to the live video by pressing the same button used for taking the still image.

In the left section, there is a list of still images that are taken from the live video. After using a snapshot still image, the remote helper can simply leave it and return it to the live video view or can save it by pulling the image to the list area with two-finger touching and dragging interactions. The remote helper can navigate

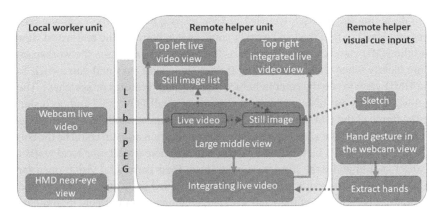

Figure 9.3 Overall software architecture of our system. The lines with arrows show the information flow of system operation, and the dotted lines show the remote helper's inputs.

the list by swiping up and down on it and can retrieve a still image from the list into the large middle section by selecting it. Above the list, a live video of the local worker's view is playing regardless of the remote helper's display mode. This video is not overlaid with any sketches or extracted hand gestures, which gives the remote helper a constant view of the workspace even when a still image is shown in the middle section.

In the right section, there is an identical view of local worker's HMD near-eye view, which helps the remote helper to know that the system is working well and what the local worker is seeing.

All interactions for sketch and hand gesture made by the remote helper are captured in the large middle section (described more in Section 9.3.1.3). The remote helper sees the live video or still image and performs hand gestures while speaking to the local worker. The hands of the remote helper will be captured without the background information due to the filter mounted in front of the helper webcam. The captured helper hands are then integrated into the live video or still image and returned back to the local worker unit to be displayed on the HMD near-eye view (see Figure 9.4). The remote helper could sketch on a still image by touch screen interaction, and the still image view including sketches are returned back to the local worker unit to be displayed on the HMD near-eye view. The sketches could be erased by pressing a "Erase" button bottom of the screen.

Our system shares the remote helper's hand gesture or/and sketches as a live video after integrating them into the large middle view, and the live video is returned back to local worker's HMD near-eye view. For the hand gesture or/and sketches integrated live video, we used the same mechanism of sending pure live video from a local end to the remote helper with the LibJPEG library. The integrated sketches are synchronized between the large middle section (remote

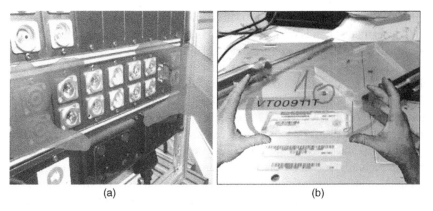

(a)   (b)

Figure 9.4 Workspace video with hands (a) and a snapshot with sketches and hands (b) shown on the HMD display of the local worker.

end) and the HMD near-eye view (local end), but the extracted remote helper hands are only displayed on the local worker view while the remote helper can see their real hands. Thus, the view on the large middle section at remote end is almost identical with the local user's HMD near-eye view. We did not collect the frame rate for sharing the live video with hand gestures, but there was no visible delay. We did not add software support for an audio channel for verbal communication, but participants in the user study could verbally communicate as they were in the same room (but could not see each other or their partner's workspace).

To sum up, with our system, the remote helper can see live video of the worker's workspace in the large middle section, see what the worker sees in their display in the right section, take a picture of the workspace for future references or for sketching, store pictures, perform hand gestures, draw sketches, erase sketches, and switch between the video mode and the image mode by pressing a button when needed. This remote helper switching from the video mode to the image mode for using the sketch cue was especially useful when an object was too small to point to with hand gestures, or when there was a need to give a more detailed description on a fixed view of the objects for a relatively long time.

9.3.1.3 Hand Gestures and Sketches

The hand gesture cue is available in both the live video and still image modes. We used polarizing filters to extract the remote helper's hands from the video feed of the webcam above the touch-screen monitor (see Figure 9.2b). With the C++ OpenCV library,[1] the background is filtered out by measuring color of each pixel of the image from the webcam and extracting the user's hands. The extracted hands are integrated into the shared still image or the pure live video, the hand gesture integrated view is compressed with LibJPEG library, sent to local worker's system, and then displayed on the HMD near-eye display in real time. At the remote end, the extracted hands were not overlaid on the shared still image or live video because the remote user could see their real hands.

Sketch cues are available in the still image modes, and the remote helper can sketch with touch-screen interaction by drawing directly on the screen. Live sketches can be shown on the local worker's near-eye display in real time, and automatically fade away after 2.5 seconds or permanently remain on the still image, depending on a configuration chosen by the remote helper during the system setup stage. When the permanent sketch option is chosen, the sketches can be manually erased by pressing a button at the bottom of the screen. The automatic fade-away function was inspired by Fussell's sketch study [9], and it has the benefit of reducing the level of visual clutter. The permanent sketch option

---

1 https://opencv.org/

was adopted from Kim's sketch study [8] with still images, and the accumulated sketches have the benefit of being more descriptive with the combination of several sketches. Sketches drawn on an image can also be saved when the image is stored into a slot of the left section of the interface. Unlike hand gestures, the live sketch is displayed not only on the local worker's HMD near-eye view but also on the remote helper's large middle view.

9.3.2 User Study Design

Using the prototype, we conducted a user study to compare the gesture and sketching interface with the gesture-only interface. In the user study, we collected the task completion time, interview results, and user rating scales from usability and NASA task load index questionnaires. We prepared two user interfaces to compare and explore the effect of combining hand gesture and sketch cues.

(1) *HandsOnly*: sharing the raw hand gestures from a remote helper to a local worker.

(2) *Hands + Sketches*: sharing the raw gestures and sketches (the HandsInTouch interface) from a remote helper to a local worker. In this condition, the remote participant can selectively use the sketch cue when they like.

9.3.2.1 Tasks

Since Kim et al. [8] reported that the use of visual communication cues in remote collaboration depends on the task type, we prepared two types of physical tasks for each interface condition: (1) a Lego assembling task (Lego task) and (2) a real-world laptop repair task (Repair task) (see Figure 9.5). The main difference between the two task types is the size of objects being manipulated. Lego blocks

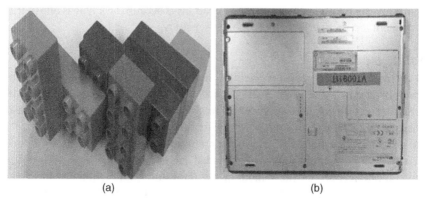

(a)   (b)

Figure 9.5 Examples of the target shape of Lego block task (a) and the component part used in the laptop repair task (b).

are large while objects in the Repair task, such as screws, can be small. Using these two task types in one study can help to compare and test the usefulness of sketching in different task conditions.

Lego assembly tasks are often used in the evaluation of collaborative systems because they have a number of procedures that can be found in real-world tasks, such as assembly, disassembly, selection, moving, attaching, and object rotation [7–9, 28]. In this task, the worker was asked to make a predefined model with blocks under the instruction of the remote helper. The remote helper had step-by-step instructions for assembling the Lego model. In the Repair task, the worker was asked to repair a computer with steps such as opening the device, taking a component out, replacing the component, and then putting all the components back in place under the guidance of the remote helper. Similar repair tasks have also been used in previous studies [22, 24].

We employed a within-subject design, so each participant performed tasks in all conditions. Participants were randomly grouped in pairs. In each pair, one subject played the role of the remote expert helper while the other was the local worker. To avoid learning effects, the order of tasks and interface conditions were randomized. After a pair of participants completed the four experimental tasks with the two interfaces and two task types, we switched their roles and asked them to do the tasks again, with another four experimental tasks.

To prevent a pair from solving the same task, we prepared eight experimental tasks (four Lego assembly tasks and four Laptop Repair tasks). In preparing the tasks, we tried to make them have the same difficulty level by using the same number of objects. For the four Lego tasks, we chose eight Lego blocks and prepared four different target models. For the four Laptop Repair tasks, we prepared two laptops and chose four screws, two inner computer components, and one cover panel to manipulate. However, we did not seriously consider the difference in difficulty level between Lego and Laptop Repair tasks because task type is one of the dependent factors that we explore in the user study.

9.3.2.2 Measurement

There was a research assistant helping the user experiment and acting as an observer quietly taking notes during all task sessions. Task performance for each condition was timed and cognitive load was also measured using the NASA Task Load Index (NASA-TLX) survey [31]. After all task sessions, participants filled out a usability questionnaire asking their agreement level on a scale from 1 to 7 (see Table 9.1). The questionnaire included ten common questions (Q1–Q10), three questions only for the remote helper (Q11 and Q13, highlighted in gray), and four questions only for the local worker (Q11–Q14, highlighted dark grey).

After the subjects completed all of the experimental tasks, there was a final interview, where subjects were individually asked questions about the system usability,

Table 9.1 Questions in the usability questionnaire.

#	Questions in usability questionnaires
Q1	The system was easy to learn
Q2	The system was easy to use
Q3	The system was useful for remote guiding tasks
Q4	I was satisfied with my own task performance
Q5	I was satisfied with the task performance with my partner as a group
Q6	I was engaged with my partner during the task
Q7	I felt as if I was colocated with my partner
Q8	It was easy to point to objects (It was easy to understand which objects my partner was pointing to)
Q9	It was easy to demonstrate object assembly (It was easy to follow my partner's demonstration)
Q10	The sketches were useful
Q11	Using hand-touch interaction to sketch was natural and helpful
Q12	I found that the interfaces intuitive
Q13	It was easy to switch the view mode between a live video and a still image
Q11	Watching sketches on still images was useful
Q12	The new-eye display (on HMD) was useful
Q13	I found being able to adjust the position of near-eye display and the viewing angle of camera useful
Q14	I found the helmet is comfortable to wear

The questions with light gray background were only for the remote helper, and the ones in dark gray were for the local worker.

their performance, and other topics. We separately interviewed each participant from a pair one by one to reduce the effect of hearing the other person's answers.

9.3.2.3 Participants

Sixteen third- or fourth-year university students were recruited as participants on a volunteer basis, and the participants in a pair knew each other as friends. Ten of them were male, while six were female. They were aged between 20 and 22 (*Mean* = 21.3; *SD* = 0.87), and none of them had prior experience of using remote guidance systems similar to HandsInTouch. Three of them had experience with HMD and all of them had experience of using video conferencing system.

9.3.2.4 Procedure

The user study started by welcoming the pair of participants into the laboratory room. After distributing the information statement and participants signing on the consent form, participants were informed about the procedure of the study and how the system worked. They were also given a chance to try out the equipment, get familiar with the system operations, and ask questions. They were told that there was no time limit but that they should complete their task as soon as possible. Before trying each experimental condition (interface), participants with the remote expert helper role had time to learn and understand the solution of the assigned tasks and how to give instructions to the local worker.

Then, the participants started to perform a given task with a given interface. After completing each task, they were asked to give ratings for items specified in the NASA-TLX form. After all tasks, each participant was given the usability questionnaire to fill out. This was followed by a short discussion between the research assistant and the participants for further feedback. The whole session took about one hour.

9.4 Results

Given the small sample size and data of task completion time was not normally distributed (Shapiro–Wilk test: $p = .0316$), we used the Wilcoxon Signed Rank test, a nonparametric statistical hypothesis test that is used to compare two related samples, for our statistical analysis. The significance level of all tests was set at 0.05.

9.4.1 Task Performance and Task Load

Figure 9.6 shows the average task completion time. For the Lego task, participants took 65 seconds on average with HandsOnly condition, 3 seconds shorter than with the Hands + Sketches condition ($Mean = 68$ seconds). However, the Wilcoxon test showed that this difference was not statistically significant ($Z = -0.362; p = 0.718$), so there was no main effect of interface condition on task performance time for the Lego task.

For the Repair task, participants took 8 seconds longer on average with the HandsOnly condition ($Mean = 77$ seconds) to complete the task than they did with Hands + Sketches ($Mean = 69$ seconds). The Wilcoxon test showed that performance was significantly improved when using sketches and hand gestures together compared to when using only hand gestures for the remote helpers to guide the Repair task ($Z = -2.145; p = 0.031$).

The average ratings of the NASA-TLX score are shown in Figure 9.7. The task load was rated on a scale from 1 to 7. For the Lego task, the average perceived task load was 4.65 with the HandsOnly condition, while the load increased to 4.97

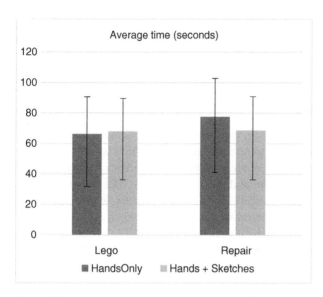

Figure 9.6 Average time spent for Lego task and Repair task.

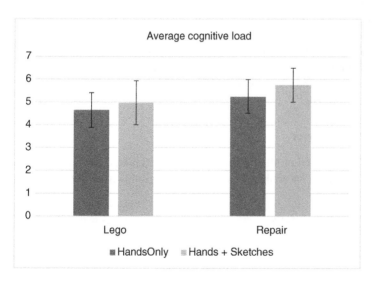

Figure 9.7 Average cognitive load for Lego task and Repair task.

with the Hands + Sketches condition. However, the Wilcoxon test showed that this difference was not statistically significant ($Z = -1.035$; $p = 0.298$).

For the Laptop Repair task, the average perceived task load was 5.23 with the HandsOnly condition. However, the average task load increased to 5.75 with the Hands + Sketches condition. The Wilcoxon test showed that this difference was statistically significant ($Z = -2.171$; $p = 0.03$) and indicates that using both hand gestures and sketches created significantly higher workload for the Laptop Repair task.

9.4.2 Overall Usability

After completing all four trials (with two experimental conditions and two types of tasks), we asked participants to rate the usability of the HandsInTouch (gesture + speech) interface. We first analyzed internal consistency for data reliability with all the rating items of the usability questionnaire. The rating questions answered by the remote helpers had a high level of consistency, with Cronbach's α value of 0.667. The rating questions answered by local participants also had a high level of consistency with Cronbach's α value of 0.873.

The results of overall usability survey are shown in Figure 9.8. The perceived overall usability was 5.58 on average for the remote helpers while it was 5.86 for the local workers. A one sample Wilcoxon Signed Rank test ($\alpha = 0.05$) showed that the remote helpers felt the HandsInTouch system had a high level of usability by rating it significantly higher than the neutral value of 4 ($Z = -3.065$, $p = 0.002$). The local workers also felt that the HandsInTouch system had a high level of usability by rating significantly higher than the neutral value of 4 ($Z = -3.068$, $p = 0.002$). In comparison between the remote helpers' and local workers' ratings, the Wilcoxon test revealed that this difference was approaching significance ($Z = -1.921$; $p = 0.054$), indicating that workers considered the system more usable than the helpers thought.

9.4.3 Sketch Usefulness

Our study mainly focused on the benefit of having sketching cues in addition to the hand gesture cue, so we analyzed participant ratings only from questions Q10

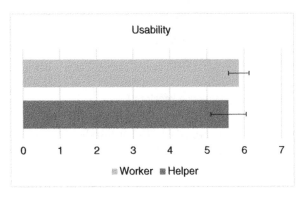

Figure 9.8 Average ratings of overall usability.

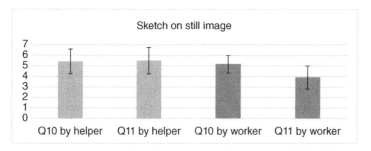

Figure 9.9 The average ratings from Q10 and Q11 by the remote helpers and local workers.

("The sketches were useful") and Q11 ("Using hand-touch-interaction to sketch was natural and helpful" for the remote helper and "Watching sketches on still images was useful" for the local worker) relating to the use of sketch cues from the usability questionnaire (see Figure 9.9). To have a better understanding, we separately analyzed each question.

The average ratings of remote helpers were 5.41 and 5.5 for Q10 and Q11, and those of local workers were 5.17 and 3.92 for Q10 and Q11. A one sample Wilcoxon Signed Rank test showed that the remote helpers felt the sketch cues were useful (Q10, $Z = -2.709$, $p = 0.007$) and the touch interaction for sketching was natural (Q11, $Z = -2.705$, $p = 0.007$) as their ratings were significantly higher than the neutral value 4. The local workers also felt the sketch cues were useful (Q10, $Z = -2.913$, $p = 0.004$). However, interestingly, they did not agree that watching sketches on still images was useful, as their ratings did not show a significant main effect compared to the neutral value 4 ($Z = -0.277$, $p = 0.782$). These results indicate that although the local workers considered that sketches were useful in general, delivering sketches via still images was perceived as being relatively less useful.

9.4.4 Observations and Log Data Results

In general, participants were able to quickly learn how to use the system. They did not need much time to get familiar with the system as there was no complex interactive procedures for handling the interfaces. On the worker side, some participants took some time to adjust the camera and near-eye display into comfortable positions. But once set, they had no problems in coordinating the HMD near-eye display view and the bare eye real-world view for task performance when needed. They were all able to use natural hand gestures, sketching, and verbal cues to complete the assigned tasks.

We describe our observation of how participants used the HandsInTouch as listed below and compared the frequency of using hand gesture and sketch cues from the log data to support our observation:

1. Participants had better collaboration with easy step-by-step instructions rather than complex but comprehensive instructions. Complex hand gestures were frequently used by the remote helpers at the beginning. This caused many workers to be confused, and the helpers quickly stopped using complex patterns. For example, they tried to use two hands to show the whole shape of the Lego, which was confusing to the worker. They then learned they needed to explain the task simply step by step.

2. Sketches were generally used as a last resort when the worker was struggling to complete the task. Participants used the hand gesture in majority of collaboration, but they used the sketch cue when it was difficult to explain with hand gesture. From the log data, we found that every pair of participants used hand gesture cue more than sketch cue within the HandsInTouch condition for both Lego and repair tasks (by using paired t-test; Lego: $t(15) = 10.559$, $p < 0.001$; repair task: $t(15) = 12.223$, $p < 0.001$).

3. Sketching was more frequently used in the complex task (the Laptop Repair task in this study) while the remote helper kept using the permanent sketches option (default value was for the sketches to permanently remain). From the log data, we found that participant used the sketch cue significantly more with the repair task than with the Lego task (by using paired t-test; $t(15) = 2.892$, $p = 0.011$).

4. There was a notable difference in the communication styles of the remote expert helpers. For example, some predominantly used verbal communication cues; some used gestures accompanied by words and short sentences. Some used pointing gestures only, while others tried to use combinations of various types of gestures.

5. There was an improvement in performance when remote helpers coordinated their speech in conjunction with their hand gestures.

6. The best approach was selecting a desired object, and then showing the relative position and orientation with another object by using two fingers. This was commonly achieved by pointing toward a piece(s) then using hand gestures to show how they should be aligned. This was done either using two fingers side by side, to show blocks being parallel, or one on top of each other to show the orientation being perpendicular.

7. While the local workers changed their viewpoint between the HMD near-eye view and the bare eye real-world view, they seemed uncomfortable and confused about sudden change from a live video to a still image view when this occurred in the HMD near-eye view.

9.4.5 Qualitative Results

We also collected qualitative data in addition to the quantitative data. When asked about the usefulness of sketches, participants were generally positive. For example, sketching was "brilliant for arrows and drawing when anything more complex would be an issue (with hand gesture)." "I think this worked well and was a good way to specify exactly what he/she wanted done without head movement getting in the way." "Useful for pointing small objects." "Audio instructions plus live footage allows me to follow instructions and correct my mistakes easily." "Good. It only makes sense to use still image as (the viewpoint of) the real-time video was changed constantly." Some were positive but with reservations. For example, "This is very helpful, but better image resolution is needed." "The annotation tool may be more helpful than speech if more complex tasks were used."

However, on the other hand, our current way of sharing still images with sketches needs further improvement. In our current implementation, after a snapshot is taken, it is displayed on the worker's HMD near-eye display and helper's large middle section, so as a result, the live video view of the workspace was fully blocked on both sides (although remote helpers can still have a live video of the workspace in the top left corner of the screen). This could potentially cause confusion for the local worker if the still image was suddenly shared without any notification from the remote helper. One participant commented that snapshot could give them an impression that the system was malfunctioning, and "It is hard to tell that is a snapshot." We also have observed that sharing still images on the HMD near-eye view forced local workers to change their gaze point to see the workspace (with their bare eyes), even though it was not desirable. For example, "I felt that I had to look up at the small screen (near-eye display view) to get the information from remote partner, but I did not watch it as I automatically changed my viewpoint to see the workspace directly." "I felt cumbersome to look at the display for instructions and the real world to perform the task."

In terms of usability, both the remote helpers and local workers were very positive. However, workers rated the usability higher than the remote helpers, indicating that the system was considered more useful by workers than by helpers. This may come from the user's feeling that encoding (sending) the information on the remote helper end was more difficult than decoding (receiving) it on the local worker end. For example, one user said "it was not as easy to give directions as it was to receive them."

Participants also made comments on other features of the system, especially the responsiveness of the sketching and shared video. For example, "showing live footages (sketches) from the same viewing of the worker is the best way to deal with a remote problem." "Being able to see what the partner was seeing (at the right section of remote helper's touch screen) helped me respond quickly." "Live videos helped my partner make good suggestions so I did not need to explain my situation (the remote helper had better understand of workspace)." "Workspace live video is extremely useful. I can use point of reference for description (with hand gesture)." "The current setup makes it easy for me to respond to what my partner is doing."

Some participants made some suggestions for future improvements. For example, "slow down the camera speed/response time (slow down the view change when the viewpoint is moving by the local worker head movement) so it is easier for the helper to see what they are doing." "It may help if the worker can make annotations (sketches) as well." "I felt that resolution of the camera and images could have been better." "The top right display on the screen (where the worker's near-eye display view is shown) should be enlarged or made clearer."

9.5 Discussion

In this section, we discuss our results first, then the implications of the interface design of remote guidance systems and their applications in real life. Finally, we discuss the limitations of the user study and how these could be overcome in the future.

9.5.1 Discussion of Our Results

Our data analysis showed that both response time and task load were not significantly different in the HandsOnly and Hands + Sketches conditions for the Lego task. In other words, the additional sketch cue did not make any difference for collaboration on the Lego task. This is not surprising because the Lego blocks are sufficiently large to point with fingers, and assembling them can be achieved with simple spatial information such as position and orientation of a block rather than complex object manipulation such as screwing in a hole and inserting an object at a specific orientation.

However, for the Laptop Repair task, there was an effect of using sketching cues on both performance and task load. Participants took a significantly shorter time to complete the task when sketches were available to them. This is consistent with our expectation that sketching is helpful when the target object is too small to be handled with hand gestures, and when the task includes complex object manipulation such as screwing in a hole and inserting an object at a specific orientation. More specifically, in this task, screws and a memory card slot were very small or thin compared to the user's fingers in the hand gesture cue, so the sketch cue with thin lines could be more useful. Moreover, the Laptop Repair task might require complex remote helper demonstration, which a sketch cue might be suitable for it. The sketch cue on the still image can remain permanently, so the accumulation of several sketches with hand gestures could help to properly explain more complex object manipulations.

Surprisingly, the better task performance for the complex task (the Laptop Repair task) in the Hands + Sketches condition was achieved at the cost of significantly higher task effort. There are five possible reasons for this. First, the sketch cue in the still image required more interaction compared to the hand gesture cue as it needed switch interaction from the live video to the still image view, and return back to the live video after sketching. Moreover, when sketching, the remote helpers might switch their viewpoint from the large middle section (where a still image was displayed to sketch on it) to the left top live video section to see the local worker's reaction to their sketches. Second, as previously reported, the sketch cue was used for complex task when the hand gesture was not enough, so the remote helpers might have felt a higher task load with the task complexity. Third, they might feel time pressure when sketching on the still image because they want to quickly return back to the live video view to see what is going on in the large middle section. Fourth, we had observed that workers seemed uncomfortable when the live video was suddenly changed to a still image view (for remote helper to sketch on the steady and stable view), and some local workers thought the system was malfunctioning, which could induce anxiety and the increased anxiety and stress could contribute to a feeling of higher task load [32, 33]. Fifth, since the still image did not show the current state of workspace, the local worker and remote helper should consider the difference between the current task space and the (past) task space shown on the still image while reading or drawing the sketches. As the difference between the current and past task spaces was greater, the local worker might have an increased cognitive load while viewing sketches on the still images. All of these factors could add up, resulting in increased cognitive load in the Hands + Sketches condition.

However, even with these five possible reasons of disadvantage of the sketches, there were occasions needing sketch cues as the hand gestures were not enough to describe complex tasks, especially with small objects. Moreover, it should be noted that the five possible reasons for the increased task load were from using still images together with the sketching cue rather than an attribute of the sketch cue itself: firstly, switching to still image view from a live video required additional user interaction. Secondly, suddenly switching was not comfortable for the local worker. And thirdly, replacing a live view of the task space by the still image view also interrupted smooth collaboration. An alternative option would be using the automatic freezing and unfreezing methods of Kim et al. [17], which pauses a live video when the user starts sketching and restarts it when the sketching finishes. This reduces the number of user interactions and the amount of time needed for still image view.

9.5.2 Implications for the Design of Remote Guidance Systems

Our discussions and findings above have the following implications for the design of remote guidance systems:

1. Sketches are more useful when the manipulation of objects cannot be easily described by words or simple pointing hand gestures.
2. When fingers are too big to point, making an object bigger with a zooming interaction, or using a smaller replacement for hand fingers such as a mouse pointer or a pointing arrow could be useful options.
3. A system with sketch cues should minimize the time needed for sharing the still image, and should keeping a live view of what is happening in the local task space. This could include implementing sketch cues on the live video without worrying about losing reference with the help of vision tracking system [8, 16] and sharing the still image only when the remote helper sketches on it, using Kim's [17] auto-freeze interface method.
4. There should be a notification that helps the local worker to know when the remote helper changes the view mode between the still image and the live video, so the local user can properly react according to the remote helpers' view change.
5. To maximize the benefit of supporting richness of hand gestures, make the remote helpers hands half transparent with a different color, so that the local worker can still see objects under the remote helper's hands and distinguish between their hands, and the helpers hands.

9.5.3 Implication for the Application of the System

The system has many potential applications in real life for remote collaboration on physical tasks, particularly for scenarios where the local worker has to be mobile and complex operation is required, especially when the objects are too small to be pointed at by a finger. For example, when a machine breaks down in a mining site, urgent help is needed to fix the machine. However, it is often that the expert who can fix it is not on-site, and it is time-consuming and costly to have the expert come to the mining site. In this case, HandsInTouch could be useful as it enables the expert to remotely guide the local technician to fix the machine. Our system is particularly useful as the design of the worker helmet allows the local worker to not only see the remote helper's instruction in the HMD near-eye display but also directly see the real-world task space to ensure safety and better understanding. Using a bare-eye view supports a better FOV and creates less motion sickness compared to a live camera view [34].

9.6 Limitations

Our experiment design has some limitations, which could have affected our data analysis results and findings. First, the results and findings may only apply in the context of this study, and caution needs to be taken before generalizing them. The sample size was relatively small for data analysis, and participants were all from a student population. As the system was developed to address industry needs, more users recruited from the target industries will help increase the reliability of data analysis and external validity.

Secondly, during the experiment, we had a research assistant acting as an observer taking notes. However, to fully understand how our system affects collaboration, analysis of verbal communications and collaborating behaviors (such as the number of times they used sketching or gestures and how long they spent on this) will be needed to help gain more insights into possible effects. Moreover, our results was mostly depend on the subjective user data rather than the objective data of interface use. This would be solved in the future work through using electroencephalography (EEG) system for measuring cognitive load, and through measuring the user's microsaccades eye movement that tells mental fatigue [35] and feeling of task difficulty [36] for investigating mental effort during the collaboration.

Third, these days, techniques for action recognition [37–39] from sensor-generated data are mature, and they can be used to have better remote collaboration by recognizing remote helper's activities then notifying it to the local helper to have aligned activities. For example, after recognizing remote helper's sketch activities, system can notify it to the local worker, so the local worker can focus on the HMD near-eye view to receive the sketch information rather than seeing workspace with the bare-eye view. This recognition and notification are especially true with the repeated hand gestures and sketch activities patterns in temporal sequence [39]. For this, the system needs to extract activity feature, learning process, and recognition step. We will extend our system by adding these functions for activity recognition and better collaboration.

Fourth, the tasks were prepared for an hour-long user study, so the task completion time was shorter than most real-life remote collaboration tasks. For example, on average, the time taken for Repair task was little more than 60 seconds. More realistic tasks that take a longer time will help us test the usability of the system and the usefulness of hand gestures and sketches in a more thorough manner.

In terms of technical implementation, we did not implement technology advances such as using a vision tracking system to stabilize sketches in the real world [8, 16] or implementing an independent view for the remote helper [30, 40]. However, we developed this system for practical use, and it does not have issues that the previous studies with these advanced technologies had: (1) the vision tracking techniques are not reliable when the local environment is too bright or too dark, (2) the tracking can be lost when there are too many objects. Support for an independent view could also require attaching a 360° camera onto the HMD, which could be bulky and cause safety issues as it can disturb quick local worker's reaction (such as taking off it before running away from danger). Moreover, sharing 360° camera views requires a high-speed internet connection.

9.7 Conclusion

In this chapter, we have briefly reviewed related work on a remote guidance system and support of hand gestures and sketching cues. Previous studies did not explore the effect of using both hand gestures and sketching cues together, so we introduced and investigated the use of our HandsInTouch system. A user study was presented by comparing the HandsInTouch (hand gesture + sketch cues) system to a condition using only hand gestures, and the results showed that users were positive about the usability of HandsInTouch for remote guidance and found that adding sketching cues was useful for complex collaborative tasks. One issue with the HandsInTouch interface was using still images for the user to sketch on, so losing the live video view of the task space. This may have a negative impact on the user experience and task load. With these results, we discussed six suggestions for designing a remote guidance system and the limitations of our work.

In the future, we will improve our system based on our design recommendations and further investigate the effects of different gesture combinations. We will especially add an auto-freeze function for better use of sketch cues with the hand gesture cues and compared three conditions: (1) only hand gesture, (2) hand gesture + sketch, (3) hand gesture + sketch + a pointer. We will also explore how to integrate a user's gaze pointer to have a better awareness of where the other user is looking at. Another line of future work would be to compare our HandsInTouch interface with the SEMarbeta interface of Chen et al. [22] to verify the performance differences in time and cognitive load according to the device for local workers: a HMD and a tablet. Especially, in measuring cognitive load, we plan to use the EEG system which tells brain activities.

Acknowledgment

This chapter is a reprint of Huang et al. [1] with minor edits. Permission is granted from Elsevier.

References

1 Huang, W., Kim, S., Billinghurst, M., and Alem, L. (2019). Sharing Hand Gesture and Sketch Cues in Remote Collaboration. *Journal of Visual Communication and Image Representation.* 58: 428–438. https://doi.org/10.1016/j.jvcir.2018.12.010.

2 Kirk, D., Rodden, T., and Fraser, D.S. (2007). Turn it this way: grounding collaborative action with remote gestures. *Proceedings of the SIGCHI Conference on Human Factors in Computing Systems*, San Jose California USA (28 April – 03 May 2007). ACM. pp. 1039–1048. https://doi.org/10.1145/1240624.1240782

3 Sodhi, R. S., Jones, B. R., Forsyth, D., et al. (2013). BeThere: 3D mobile collaboration with spatial input. *Proceedings of the SIGCHI Conference on Human Factors in Computing Systems*, Paris France (27 April – 02 May 2013). ACM. pp. 179–188. https://doi.org/10.1145/2470654.2470679

4 Alem, L., Tecchia, F., and Huang, W. (2011a). Remote tele-assistance system for maintenance operators in mines. *11th Underground Coal Operators' Conference.* University of Wollongong and the Australasian Institute of Mining and Metallurgy. pp. 171–177.

5 Sakata, N., Kurata, T., Kato, T., et al. (2003). WACL: supporting telecommunications using wearable active camera with laser pointer. *Seventh IEEE International Symposium on Wearable Computers, 2003. Proceedings*, White Plains, NY, USA (21–23 October 2003). *IEEE.* pp. 53–56. http://doi.ieeecomputersociety.org/10.1109/ISWC.2003.1241393

6 Kuzuoka, H., Kosuge, T., and Tanaka, M. (1994). GestureCam: a video communication system for sympathetic remote collaboration. *Proceedings of the 1994 ACM Conference on Computer Supported Cooperative Work*, Chapel Hill, NC (22–26 October 1994). ACM. pp. 35–43. https://doi.org/10.1145/192844.192866

7 Gurevich, P., Lanir, J., Cohen, B., and Stone, R. (2012). TeleAdvisor: a versatile augmented reality tool for remote assistance. *Proceedings of the SIGCHI Conference on Human Factors in Computing Systems*, Austin, USA (05–10 May 2012). ACM. pp. 619–622. https://doi.org/10.1145/2207676.2207763

8 Kim, S., Lee, G., Sakata, N., and Billinghurst, M. (2014). Improving co-presence with augmented visual communication cues for sharing experience through video conference. *2014 IEEE International Symposium on Mixed and Augmented Reality (ISMAR)*, Munich, Germany (10–12 September 2014). *IEEE.* pp. 83–92. https://doi.org/10.1109/ISMAR.2014.6948412

9 Fussell, S., Setlock, L., Yang, J. et al. (2004). Gestures over video streams to support remote collaboration on physical tasks. *Human-Computer Interaction* 19 (3): 273–309. https://doi.org/10.1207/s15327051hci1903_3.

10 Kim, S., Lee, G.A., and Sakata, N. (2013). Comparing pointing and drawing for remote collaboration. *2013 IEEE International Symposium on Mixed and Augmented Reality (ISMAR)*, Adelaide, Australia (01–04 October 2013). *IEEE.* pp. 1–6. https://doi.org/10.1109/ISMAR.2013.6671833

11 Higuchi, K., Chen, Y., Chou, P. A., et al. (2015). ImmerseBoard: Immersive telepresence experience using a digital whiteboard. *Proceedings of the 33rd Annual ACM Conference on Human Factors in Computing Systems*, Seoul, Republic of Korea (18–23 April 2015). ACM. pp. 2383–2392. https://doi.org/10.1145/2702123.2702160

12 Huang, W. and Alem, L. (2013a). Gesturing in the air: supporting full mobility in remote collaboration on physical tasks. *Journal of Universal Computer Science* 19 (8): 1158–1174.

13 Fussell, S.R., Kraut, R.E., and Siegel, J. (2000). Coordination of communication: effects of shared visual context on collaborative work. *Proceedings of the 2000 ACM Conference on Computer Supported Cooperative Work*, Philadelphia, PA, USA (02–06 December 2000). ACM. pp. 21–30. https://doi.org/10.1145/358916.358947

14 Lee, G. A., Kim, S., Lee, Y., et al. (2017a). Improving collaboration in augmented video conference using mutually shared gaze. *International Conference on Artificial Reality and Telexistence & Eurographics Symposium on Virtual Environments*, Adelaide Australia (22–24 November 2017). pp. 197–204. http://dx.doi.org/10.2312/egve.20171359

15 Fakourfar, O., Ta, K., Tang, R., et al. (2016). Stabilized annotations for mobile remote assistance. *Proceedings of the 2016 CHI Conference on Human Factors in Computing Systems – CHI '16*, San Jose, USA (07–12 May 2016). New York, NY: ACM Press. pp. 1548–1560. https://doi.org/10.1145/2858036.2858171

16 Gauglitz, S., Nuernberger, B., Turk, M., and Höllerer, T. (2014). In touch with the remote world: remote collaboration with augmented reality drawings and virtual navigation. *Proceedings of the 20th ACM Symposium on Virtual Reality Software and Technology*, Edinburgh, United Kingdom (11–13 November 2014). ACM. pp. 197–205. https://doi.org/10.1145/2671015.2671016

17 Kim, S., Lee, G.A., Ha, S., et al. (2015). Automatically freezing live video for annotation during remote collaboration. *Proceedings of the 33rd Annual ACM Conference Extended Abstracts on Human Factors in Computing Systems*, Seoul, Republic of Korea (18–23 April 2015). ACM. pp. 1669–1674. https://doi.org/10.1145/2702613.2732838

18 Gupta, K., Lee, G.A., and Billinghurst, M. (2016). Do you see what I see? The effect of gaze tracking on task space remote collaboration. *IEEE Transactions on Visualization and Computer Graphics* 22 (11): 2413–2422. https://doi.org/10.1109/TVCG.2016.2593778.

19 Higuchi, K., Yonetani, R., and Sato, Y. (2016). Can eye help you?: effects of visualizing eye fixations on remote collaboration scenarios for physical tasks.

Proceedings of the 2016 CHI Conference on Human Factors in Computing Systems, San Jose, USA (07–12 May 2016). ACM. pp. 5180–5190. https://doi.org/10.1145/2858036.2858438

20 Tang, J.C. and Minneman, S.L. (1991). VideoDraw: a video interface for collaborative drawing. *ACM Transactions on Information Systems (TOIS)* 9 (2): 170–184. https://doi.org/10.1145/123078.128729.

21 Ishii, H. and Kobayashi, M. (1993). Integration of interpersonal space and shared workspace: ClearBoard design and experiments. *ACM Transactions on Information Systems* 11 (4): 525–532. https://doi.org/10.1145/159764.159762.

22 Chen, S., Chen, M., Kunz, A. et al. (2013). SEMarbeta: mobile sketch-gesture-video remote support for car drivers. *Proceedings of the 4th Augmented Human International Conference*, Stuttgart Germany (7–8 March 2013). ACM. pp. 69–76. https://doi.org/10.1145/2459236.2459249

23 Kato, H. and Billinghurst, M. (1999). Marker tracking and HMD calibration for a video-based augmented reality conferencing system. *Proceedings 2nd IEEE and ACM International Workshop on Augmented Reality (IWAR'99)*, San Francisco, California (20–21 October 1999). IEEE. pp. 85–94. https://doi.org/10.1109/IWAR.1999.803809

24 Alem, L., Tecchia, F., and Huang, W. (2011b). HandsOnVideo: towards a gesture based mobile AR system for remote collaboration. In: *Recent Trends of Mobile Collaborative Augmented Reality* (ed. L. Alem and W. Huang), 127–138. New York: Springer https://doi.org/10.1007/978-1-4419-9845-3_11.

25 Huang, W. and Alem, L. (2013b). HandsinAir: a wearable system for remote collaboration on physical tasks. *Proceedings of the 2013 Conference on Computer Supported Cooperative Work Companion (CSCW '13)*, San Antonio, TX, USA (23-27 February 2013). USA: ACM. pp. 153–156. https://doi.org/10.1145/2441955.2441994

26 Kunz, A., Nescher, T., and Kuchler, M. (2010). Collaboard: a novel interactive electronic whiteboard for remote collaboration with people on content. *2010 International Conference on Cyberworlds (CW)*, Singapore (20–22 October 2010). IEEE. pp. 430–437. https://doi.org/10.1109/CW.2010.17

27 Tang, A., Neustaedter, C., and Greenberg, S. (2007). VideoArms: embodiments for mixed presence groupware. In: *People and Computers XX—Engage* (ed. N. Bryan-Kinns, A. Blanford, P. Curzon, and L. Nigay), 85–102. Springer https://doi.org/10.1007/978-1-84628-664-3_8.

28 Huang, W., Alem, L., and Tecchia, F. (2013c). HandsIn3D: supporting remote guidance with immersive virtual environments. *Human-Computer Interaction – INTERACT 2013*, Cape Town, South Africa (2–6 September 2013). *Lecture Notes in Computer Science*, vol. 8117. pp. 70–77. https://doi.org/10.1007/978-3-642-40483-2_5

29 Goldin-Meadow, S. (1999). The role of gesture in communication and thinking. *Trends in Cognitive Sciences* 3 (11): 419–414. https://doi.org/10.1016/S1364-6613(99)01397-2.

30 Kim, S., Billinghurst, M., and Lee, G. (2018). The effect of collaboration styles and view independence on video-mediated remote collaboration. *Computer Supported Cooperative Work (CSCW)* 27: 1–39. https://doi.org/10.1007/s10606-018-9324-2.

31 Hart, S.G. and Staveland, L.E. (1998). Development of NASA-TLX (Task Load Index): results of empirical and theoretical research. *Advances in Psychology* 52: 139–183. https://doi.org/10.1016/S0166-4115(08)62386-9.

32 Chen, I.J. and Chang, C.C. (2009). Cognitive load theory: an empirical study of anxiety and task performance in language learning. *Electronic Journal of Research in Educational Psychology* 7 (2): 729–745: 17pp. 3 Charts.

33 Conway, D., Dick, I., Li, Z., et al. (2013). The effect of stress on cognitive load measurement. *14th IFIP TC 13 International Conference on Human-Computer Interaction – INTERACT*, Cape Town, South Africa (2–6 September 2013). New York, USA: Springer-Verlag New York, Inc. pp. 659–666. https://doi.org/10.1007/978-3-642-40498-6_58

34 Lee, G.A., Teo, T., Kim, S., and Billinghurst, M (2018). A user study on MR remote collaboration using live 360 video. *2018 IEEE International Symposium on Mixed and Augmented Reality (ISMAR)*, Munich, German (16–20 October 2018). IEEE. pp. 153–164. https://doi.org/10.1109/ISMAR.2018.00051

35 Di Stasi, L.L., McCamy, M.B., Catena, A. et al. (2013). Microsaccade and drift dynamics reflect mental fatigue. *The European Journal of Neuroscience* 38 (3): 2389–2398. https://doi.org/10.1111/ejn.12248.

36 Siegenthaler, E., Costela, F.M., McCamy, M.B. et al. (2014). Task difficulty in mental arithmetic affects microsaccadic rates and magnitudes. *The European Journal of Neuroscience* 39 (1): 1–8.

37 Liu, Y., Nie, L., Han, L., et al. (2015) Action2Activity: recognizing complex activities from sensor data. *24th International Joint Conference on Artificial Intelligence (IJCAI 2015)*, Buenos Aires Argentina (25–31 July 2015). AAAI Press

38 Liu, Y., Zhang, L., Nie, L., et al. (2016a) Fortune teller: predicting your career path. *Proceedings of the Thirtieth AAAI Conference on Artificial Intelligence (AAAI'16)*, Phoenix, Arizona, USA (12–17 February 2016). AAAI Press. pp. 201–207.

39 Liu, Y., Nie, L., Liu, L., and Rosenblum, D.S. (2016b). From action to activity: sensor-based activity recognition. *Neurocomputing* 181: 108–115. https://doi.org/10.1016/j.neucom.2015.08.096.

40 Lee, G.A., Teo, T., Kim, S., and Billinghurst, M. (2017b). Mixed reality collaboration through sharing a live panorama. *SIGGRAPH Asia 2017 Mobile Graphics & Interactive Applications*, Bangkok, Thailand (27–30 November 2017). ACM. p. 14. https://doi.org/10.1145/3132787.3139203

10

Augmenting Hand Gestures in 3D Mixed Reality

10.1 Introduction

A remote helper guiding a local worker in performing physical tasks is one type of remote collaboration that is often seen in industrial settings [1]. Much research has been conducted investigating how to support such collaborations and significant progress has been made both in developing theoretic understanding and in developing system solutions. It has been widely agreed that one of the main issues associated with remote collaboration is the loss of common ground [2]. When colocated, collaborators share common ground and are able to constantly use hand gestures to clarify and ground their messages while communicating with each other verbally [3]. However, when collaborators are geographically distributed, such common ground no longer exists, resulting in them not being able to communicate the same way as they do when colocated. Prior research has shown that providing remote collaborators with access to a shared visual space helps to rebuild common ground, thus improving the performance of collaborative tasks (e.g. [4]). A shared visual space is one where all collaborators can see the task-related objects and monitor the status of the task at hand. Further, when colocated, collaborators talk to each other and act on physical artifacts in the workspace accordingly. During this process, hand gestures are also frequently used to complement their verbal communication. Prior research has found that support of hand gestures in addition to speech leads to significantly better performance of remote collaboration (e.g. [5]). All these findings indicate that in order to support remote collaboration on physical tasks, it is important to not only provide a shared visual space but also to support remote gestures.

A number of remote guiding systems have been reported in the literature. These systems allow a remote helper to guide a field worker to perform a physical task using visual aids in addition to audio communication and differ in how remote gesture is supported. Some systems use mediated-hand approaches. For example, in the DOVE of Ou et al. [6], helpers are able to use a pen to draw digital

Computer-Supported Collaboration: Theory and Practice, First Edition.
Weidong Huang, Mark Billinghurst, Leila Alem, Chun Xiao, and Troels Rasmussen.
© 2024 The Institute of Electrical and Electronics Engineers, Inc. Published 2024 by John Wiley & Sons, Inc.

sketches to indicate the intended action to the worker. In the GestureMan of Kuzuoka et al. [7], remote gestures are delivered by a mobile robot located on the worker's side through the use of a pointing stick and a laser pointer. Other systems use unmediated-hand approaches. For example, in the SharedView of Kuzuoka [8], the helper performs gestures over the video feeds of the local workspace. The hands combined with the background video are recorded by a camera and displayed to the worker. In the T-room of Yamashita et al. [9], the helper's hands are recorded and directly projected onto the surface of the worker's workspace. A recent work by Alem et al. [10] uses a slightly different method. In their HandsOnVideo system, the hands of the helper are not projected onto the physical space of the worker; rather, they are fused with the video of the workspace and displayed on a near-eye display worn by the worker. During the guidance process, the worker gets to see a video of their workspace augmented with the helping hands, in addition to seeing their physical workspace.

Despite their differences in supporting remote gestures, the shared visual space is typically presented in the form of 2D video feeds by either displaying the video feeds on a monitor or projecting the video into a physical space. A recent user study by Huang and Alem [11] indicates that 2D remote guiding systems have limitations: users playing the role of the helper expressed the need to have more control over the view of the remote workspace. Helpers wanted to be able to change their point of view independently from their partner's point of view; they wanted to be able to zoom in and out without having to ask their partner to move closer or further to the object of interest. They complained that showing the workspace on a 2D display made it difficult for them to understand the spatial relationships between the objects. They also commented that there were simple hand gestures that they were unable to perform over the 2D video, such as pointing in between two aligned objects. In summary, helpers wanted a better perception of the spatial relationship between objects in the remote workspace, more control over the view of the remote workspace as well as the ability to gesture in 3D. Further, the same study also revealed that helpers had a relatively lower sense of copresence than their partners. Copresence has been shown to be associated with improved user experience and task performance [12]. Spatial understanding also has a role in affecting task performance [13]; in our collaborative setting, helpers need a good spatial understanding in order to make the correct judgments on the remote objects and guide their partners accordingly.

Research on 3D virtual environments has shown that these environments help improve spatial understanding [14]. Virtual environments also bring other benefits such as higher sense of copresence, improved spatial awareness, more accurate cognitive transfer between simulation and reality, and better task performance [13, 15]. These virtual environments have been shown to be useful in supporting general telecollaboration in which all collaborators work within the same virtual environment. In our collaboration scenario, one partner (e.g. the worker) is work ing in a real and physical environment while the other partner (e.g. the helper) gets a real-time 3D capture and reconstruction of his partner's workspace. While we acknowledge that the space in which the collaborative work is conducted is not virtual, we hypothesize that the 3D shared space will bring the same benefits reported in the collaborative virtual environments research.

A prototype system called HandsIn3D has been developed to explore this hypothesis. This system uses 3D real-time capturing and rendering of both the remote workspace and the helper's hands and creates a 3D shared visual space as a result of colocating the remote workspace with the helper's hands. The 3D shared space is displayed on a head-tracked stereoscopic head-mounted display (HMD) that allows the helper to perceive the remote space in 3D as well as guide in 3D.

This chapter is an extension of its conference version [16]. In the remainder of this chapter, we introduce HandsIn3D and present a user study we have conducted to demonstrate the usability of the system and verify the feasibility of our proposed approach. More specifically, Section 10.2 presents related work. Section 10.3 presents the system we have developed. Section 10.4 presents the user study and Section 10.5 presents a follow-up cross-study comparison. Finally the chapter concludes in Section 10.6.

10.2 Related Work

10.2.1 Remote Guidance Systems

In both the DOVE [6] and TeleAdvisor [17] systems, the helper was presented with an interface including a live video stream of the worker's workspace. The helper was able to interact with the interface and draw annotations on the screen of a tablet PC in relation to an object of interest. In regard to communication of hand gestures, the former system normalized hand sketches into prestored standard drawings and showed them in a monitor at the desk of the worker, while the latter projected hand sketches directly on top of the objects at the worker's location.

Instead of representing hand gestures as digital annotations, SharedView [8] allowed helpers to perform natural hand gestures over the video stream. In this system, the scene of the hands, together with the video, was captured by a camera and shown in a head-mounted display to the worker. User studies have been conducted to compare different ways of conveying hand gestures including digital representations and unmediated hands, and the results demonstrated that unmediated hand gestures had some advantages [5]. These advantages include preserving the richness of expression of remote gestures and reducing the costs of interpretation, resulting in better task performance, improved experience of interaction, and a higher sense of copresence. Based on this line of research, researchers have introduced additional devices, including projectors and/or video cameras, to set up an environment in which the helper can perform natural hand gestures on the surface of a desk rather than using a digital pen over a live video (e.g. [18]). In these systems, helpers looked at the video shown on display in front of them so that their hands could be accurately aligned with the object of interest, and the movement of their hands was captured by the camera located above the desk and projected by a projector or shown on a monitor to the worker.

However, projecting hands on a surface requires the surface to be flat so that the projected hands and their movement can be recognized by the worker. Further, showing hands on an external display on a desk confines helpers to a fixed desktop environment and requires extra effort in matching objects in the display and those in the workspace to follow guiding instructions. These constraints have largely prevented existing systems from being more widely useful, particularly in industrial fields such as mining sites and manufacturing factories where workspaces are usually not standard desktop settings, and users often need to be mobile. Recent research and development efforts in CSIRO on remote mobile tele-assistance technology were intended to address these issues. This effort has resulted in a series of remote guidance systems that were developed to support the real-time delivery of expert services to local fields, particularly for machine repair and maintenance (e.g. [19]). Although each of these systems was developed for a specific user case and aimed to address a different set of research questions, the same approach was used to convey remote gestures. That is to extract hand images of the helper and combine them with the video of the local workspace. In HandsOnVideo [10], a camera was mounted above the display together with an optical filter placed between the display and the camera. While helpers performed gestures over the live workspace video, the filter filtered out the background information, resulting in only hand images being captured by the camera, which were then sent to the worker side to be combined with the workspace video. In HandsInAir [20], helpers performed gestures in the air by aligning their hands with the object in consideration shown in the video. At the same time, a skin-color-based algorithm was used to detect hand images, which were then sent to the worker side to be processed.

Instead of using digital gesture representations or unmediated hands, some systems used remotely controlled pointing devices located in the workspace to support the pointing gestures of the helper. For example, WACL [21] is a wearable system which included a shoulder-mounted interface device on the worker side. A camera and a laser pointer were attached to the device, and they were independently controlled by the remote helper. Camera gave the helper a live view of the task space while the laser pointer could be used for pointing and directing purposes. Although a pointer does not convey gestural information as rich as unmediated hands do, it can point directly to real objects, thus allowing workers to remain focused on the task at hand without having to split their attention between a display and the workspace.

10.2.2 Collaborative Virtual Environments

Various technologies in Virtual Reality (VR) have been used to build 3D collaborative environments for distributed participants to interact with each other and work on various collaborative activities [13, 22]. A large body of research has been conducted to investigate the benefits of 3D collaborative environments for various collaboration tasks. For example, Brown et al. [23] used recent virtual world technology to build an environment for the collaborative design of business processes. Monahan et al. [24] developed a desktop and web-based multi-user system. This system used Virtual Reality and multimedia technologies to mimic a real university and provided a collaborative e-learning environment.

Existing virtual collaborative systems have been demonstrated in various studies for their usefulness in the specific context in which they were developed. More specifically, 3D virtual environments can offer collaborators better visual and spatial cues for a better understanding of the objects in the environments [14], provide a higher sense of being together with another person [13], and induce a more accurate cognitive transfer of spatial awareness from virtual environment to reality [15], resulting in better performance of collaborative tasks. Although these studies were conducted using tasks that were specific to a range of collaboration scenarios, 3D collaborative environments have not been used in the context of remote guidance on physical tasks.

In our HandsIn3D system, we use Mixed Reality (MR) for remote collaboration. We keep the focus of our research on physical tasks but provide helpers with a better perception of spatial relationships between objects and give them more freedom to perform more natural hand gesturing, exploiting a full 3D reconstruction of the collaboration task that even takes into consideration the effect of inter-object occlusions.

The fundamental contribution of our research is therefore to demonstrate the value of a 3D paradigm in a Mixed Reality system to help real-time collaboration between the helper and the worker. We will describe in Section 10.3 our proof of concept system exploiting real-time capturing and rendering of 3D data as well as Mixed Reality techniques to enhance the workspace of the worker and display it to the helper. Some advanced features such as stereoscopic rendering, real-time head tracking, and virtual shadowing of objects are also implemented to study how to provide the helper with a closer-to-real feeling of immersion in the remote environment.

10.2.3 Implicit Visual Communication Cues and Cultural Factors

Implicit cues, such as eye gaze and body language, are used in face-to-face collaboration. But their usefulness and potential impact in enhancing user experience and improving task performance remains largely unexplored in the context of remote collaboration on physical tasks. However, this situation is changing. In addition to supporting explicit communication cues (e.g. voice, gesture) and creating collaborative AR/VR/MR environments, recent work has started to explore the effects of providing additional implicit cues for remote collaboration. For example, Higuch et al. [25] investigated in a study the effects of eye fixations of the helper on collaboration behaviors between the helper and the worker. In their study, helpers' eye gazes were tracked on a monitor screen that shows a physical workspace and then visualized onto the space by a projector or through an optical see-through HMD on the worker side. It was found that eye gazes served as an effective pointer to objects of interest and could help the worker predict the intentions of the helper. Gupta et al. [26] conducted a study that compared different visual cues including gaze tracking and pointing. It was found that the use of both eye tracking and a pointer significantly improved task performance and co-presence. Otsuki et al. [27] presented a simple add-on display called ThirdEye that represented a remote participant's gaze direction in local environments. The authors conducted an experiment in which the eye display and a face display were compared and it was found that an observer could perceive a remote looker's gaze direction more precisely with ThirdEye.

Human factors, social and cultural factors also have a role in remote collaboration [28]. Gao et al. [29] conducted a study examining the effects of showing how automated transcripts and bilingual dictionaries were used in a laboratory experiment. In this experiment, native speakers and non-native speakers collaborated via audio conferencing on a map navigation task and it was found that participants collaborated most successfully when the guiders were made aware of the detailed information on usage of transcripts and dictionaries by the followers. Nguyen and Fussell [30] conducted a study that explored how people in same-culture and cross-culture pairs used verbal cues to express involvement in computer-mediated communication situations. Their study found that people of different cultures used verbal involvement cues differently.

10.3 System Overview

10.3.1 Technical Challenges

Our HandsIn3D is a prototype system that was intended to demonstrate the proof of concept of using a 3D hand gesture paradigm for remote collaboration. Therefore, while the focus of our research is on geographically remote collaboration, we do not have to develop a system that fully works in real world situations. This would make our prototype system simpler and the implementation of it easier as we no longer needed to implement two separate systems on both sides as long as the interfaces of the worker and the helper meet the needs of real-world collaboration scenarios. For example, in our current system, the helper and the worker could sit close to each other with a board in the middle separating them giving the same sense that they were physically distributed. This made it possible for us to handle all the computation tasks by means of a single PC.

To fulfill our objectives, there were a few technical challenges that we needed to address. The first was that we needed to capture what was happening in the workspace of the worker. This was done by using a Kinect-like camera looking over the workspace from the behind of the worker. We also used a second Kinect-like camera looking over the helper workspace, capturing the helper gestures. Further, since the captured samples were too sparse to constitute a good visualization of the scene, an additional task of mesh triangulation was implemented so that those two clouds could be used in the rendering loop. The second challenge in our system was represented by rendering speed. A good frame rate is important as we wanted to use an HMD that was tracked both in position and in orientation to provide the helper with a strong feeling of immersion in the remote workspace. To maintain the frame rate at a reasonable level, the high-speed rendering was achieved in our system by leveraging GPU capabilities with a technique inspired by Maimone's work [31]. The third challenge was to provide the helper with a strong feeling of immersion in the remote collaboration space. This was done by casting virtual shadows from one 3D cloud to the other. Shadows provide a visual illusion of co-location, reinforcing the impression of the helper and the worker to be collaborating on the same physical space although the captured workspaces on both sides are physically disjoint. Further, in our system, we also implemented stereoscopic rendering and head tracking that was expected to give a stronger feeling of immersion to the users collaborating in the remote collaboration space.

Proper calibration is important when we need to combine video streams from different sources which are part of a single coordinated system. And this is the case for HandsIn3D. In our platform there are three different coordinate systems: the first one is related to the worker space (and in particular to the placement of the worker camera with respect to the captured space), the second one related to the helper space, and the third related to the placement of the optical tracker with respect to the helper HMD. Before the data can be fused, these three reference systems need to be calibrated. Overlapping the worker and the helper space requires quite a bit of precision: we have used a marker-based tracking toolkit (Aruco) in order to get the pose matrix of the cameras with respect to the marker. At setup time we placed two copies of the same marker at the center of the two workspaces and computed the camera pose of the two cameras. We used the computed camera pose matrix in the OpenGL rendering code, and in this way, we were able to overlap graphically the two workspaces with a good level of precision.

Calibration of the optical tracker data with respect to the fused workspace is not performed through the use of any AR marker, as the marker would be invisible to the IR cameras. Manual calibration is performed instead, measuring the displacement between the reference marker and the optical tracker frame or reference origin. It must be noted that while it is important to overlap precisely the worker and helper workspace (and this is obtained by means of the common AR-based marker), the accuracy of head tracking calibration has a lesser impact on the final experience, and small errors in position and orientation can still be tolerated before the misalignment becoming apparent to the helper wearing the HMD.

10.3.2 Hardware, Software and Implementation

Our platform includes the following hardware items:

- One PC equipped with a 4-core Intel Core i7-960 CPU, 8GB of RAM and Nvidia GeForce GTX 560 graphics board.
- Two Kinect-like sensors (Asus XTion LivePro).
- One 24 inches LCD monitor (non-stereoscopic).
- One VR1280 stereoscopic HMD produced by Virtual Research.
- One Optitrak TRIO 120 optical tracker.

HandsIn3D runs on 64-bit Windows 7, and uses the OpenNI 1.5.2.23 API along with a Kinect driver to communicate with the sensors. OpenGL is used for rendering, the OpenGL 4.2 shader language (GLSL) is used for programmable GPU operations, and FreeGLUT is used for windowing and user input. The OpenCV 2.4 computer vision library was utilized for accessing the Kinect data as well as to perform basic image processing operations.

The system starts the data processing with the 3D scene acquisition by the Kinect-like cameras installed on both the helper and the worker stations. OpenCV 2.4 libraries are used to access the RGB and Depth image array generated by the cameras. Both streams are acquired at 30Hz and at the resolution of 640×480 pixels. As widely known, the Kinect camera returns both the RGB colors of the points in front of the camera as well as the corresponding distance from the sensor, measured in mms, and we rely on the RGB and Depth image alignment built-in OpenNI. Real-time meshing of the Kinect data is then performed on both the data acquired by the worker and the helper camera. The two resulting meshes are then co-located in the same 3D space. In this way, the hands of the helper appear colocated with the worker space, and vice-versa. A 3D rendering of such shared space is then displayed on the LCD placed in the worker space (the LCD resolution is 1280×1024). For the helper, an immersive, stereoscopic view of the

3D shared space is displayed by means of the VR1280 HMD (resolution for each eye is 1280 × 1024). To increase the feeling of remote-presence of the helper in the worker space, real-time head tracking is performed of the helper HMD by means of an optical tracker. To provide to both users an additional depth cue and enrich the feeling of co-presence in the shared virtual space, artificial fake shadows are added to the scene: the 3D mesh of the captured hands of the helper is used to cast a virtual shadow on the 3D mesh of the captured worker space and vice-versa. More details of technical implementation of the system are reported in [32].

10.3.3 Overall Architecture

HandsIn3D is currently running on a single PC. It has two logical sections: the worker space and the helper space (see Figures 10.1 and 10.2). The worker performs a physical task at the worker space, while the helper provides guidance to the worker at the helper space. Figure 10.1 shows the layout of the worker space. A user sits at the desk performing a task on physical objects (for example, assembly of toy blocks). A 3D camera is mounted overhead to capture the workspace in front of the user including the hands of the user and objects. An LCD nonstereoscopic monitor is placed on the desk to display the 3D view of the workspace augmented by the guiding information.

Figure 10.2 shows the layout of the helper space. In this space, there is a 3D camera that captures the hands of the helper. The helper wears a stereoscopic HMD and sits in front of an optical head tracker. The HMD allows a realistic virtual

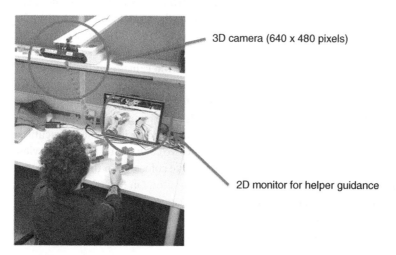

3D camera (640 x 480 pixels)

2D monitor for helper guidance

Figure 10.1   The *worker space* of our system. A 3D camera is used to capture what is in front of the user, and a monitor shows a 3D view of the scene augmented by guiding information.

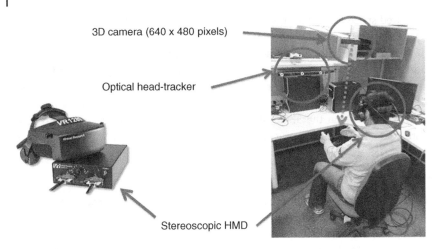

3D camera (640 x 480 pixels)

Optical head-tracker

Stereoscopic HMD

Figure 10.2 The *helper space* of our system. A 3D camera is used to capture the hand gestures performed by a remote helper. By using an HMD, the helper is immersed in a 3D rendering of the worker's workspace. The HMD position and orientation are tracked by means of optical markers.

immersion in the 3D space captured by the camera placed in the worker space, while the tracker tracks the HMD position and orientation.

The system functions as follows. On the worker side, the worker talks to the helper, looks at the visual aids on the screen, picks up and performs actions on the objects. On the helper side, the helper wears the HMD, looks into the virtual space (and looks around the space if necessary), talks to the worker, and guides him by performing hand gestures such as pointing to an object and forming a shape with two hands. During the process of interaction, the camera on the worker side captures the workspace in front of the worker, while the camera on the helper side captures the hands of the helper. The acquired 3D scene data from both sides are fused in real time to form a single common workspace, which we call *shared virtual interaction space*. This space is displayed in HMD. The image in Figure 10.3 is provided to help understand what is presented to the helper during the task execution: by fusing together the 3D meshes acquired by the two cameras, an augmented view of the workspace where the hands of the helper are colocated in the worker space is synthetically created, as shown in the LCD monitor. On the other hand, being presented with this augmented view, the worker can easily mimic the movements of the helper's hands and perform the Lego assembly task accordingly.

The main features of the system, therefore, include the following:

- Users can speak to each other.
- The helper can see the workspace of the worker via the shared virtual interaction space.

Figure 10.3 The *shared virtual interaction space* is visible in this picture of the worker space: in the LCD monitor, the 3D meshes captured from the worker space and the helper space are colocated and fused together. Four hands can be spotted in the virtual scene: two from the worker and two from the helper.

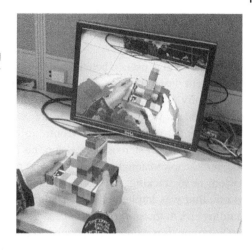

- The helper can perform hand gestures.
- The worker can see the hand gestures of the helper on the screen.
- The two hands of the worker are freed for manipulation of physical objects.

In addition, the shared virtual interaction space also implements additional features to improve the sense of 3D immersion for the helper. These features include (1) the objects and hands cast shadows in the space; (2) the HMD is tracked, which allows the helper to see the space from different angles.

10.4 Evaluation

To evaluate our 3D gesture-based interaction paradigm, we conducted a user study. The main purposes of the study were to test the usability of the system and to understand how users, particularly helpers, felt about the 3D user interface.

10.4.1 Method

We recruited 14 participants who had normal or corrected-to-normal vision on a volunteering basis to participate in the study. Upon their agreement to participate, the volunteers were randomly grouped as pairs to perform a collaborative task. In this study, we used the assembly of Lego toy blocks as our experimental task. Toy assembly is not a real-world task, but it has a number of components which can be found in real-word tasks, thus being considered representative. This task has also been previously used for similar evaluation purposes. During the task, the worker was asked to assemble the Lego toys into a reasonably complex model under the instruction of the helper. The helper was instructed that he could provide verbal and gestural instructions to the worker at any time. The worker, on the other hand, was not informed of and did not know what steps were needed to complete the task.

In order to give users a better appreciation of our new 3D interface in relation to different design options, following the assembly task, the pair were given the opportunity to explore and experience different levels of immersion. The levels of immersions were made based on the availability of several 3D features, and they were: (1) no stereoscopic vision, no head tracking and no hands shadow (2D interface), (2) stereoscopic vision, no head tracking and no hands shadow, (3) stereoscopic vision, head tracking, and no hands shadow, and (4) stereoscopic vision, head tracking and hands shadow (full 3D interface). This last feature is one that was implemented in HandsIn3D and that participants used in the guiding task at the start of the trial.

Questionnaires were administrated to the participants to seek their subjective feedback. There were two versions: worker questionnaire and helper questionnaire and they were distributed immediately after each pair of the participants completed their assembly tasks. These questionnaires asked participants to rate a set of usability metrics, answer some open questions and share their experience of using the system. The usability measures include both commonly used ones and those specific to the system. For more details, see Section 10.4.3.

10.4.2 Procedure

The study was conducted in a meeting room where each pair was observed by an experimenter. The helper space and worker space were separated by a dividing board so that the two participants could not see each other. Upon arrival, they were randomly assigned helper and worker roles and informed about the procedure of the study. The helper interface and the worker interface were introduced. They were also given the chance to get familiar with the system and try out the equipment. Then, the helper was taken to an office room where he/she was shown a model that needed to be constructed. The helper was given time to think about and plan how to do it and remember the steps.

Then, the helper went back to the experimental room and put the HMD on and the experiment started. After the assembly task, the pair of participants were asked to experience the different interface features listed in the last subsection in an informal style. The switch between the interface features was controlled by the experimenter. During the process, the participants were told which feature the system was using. They could play with the toy blocks and talk to each other about the assembly steps. But they were not allowed to comment and share how they felt about the system and its features. This was to ensure that their later responses to the questionnaires were not affected by each other's opinion. After going through all four features, each participant was asked to fill out the helper or worker questionnaire for the role played. Then, the participants switched roles and the above process was repeated again. Note that this time, the model to be constructed was different but with a similar level of complexity.

After finishing the assembly tasks and questionnaires, participants were debriefed about the purposes of the study, followed by a semi-structured interview. They were encouraged to share their experiences, comments on the system, ask questions and suggest improvements. The whole session took about one hour on average.

10.4.3 Results and Discussion

10.4.3.1 Observations
Since participants had never experienced this type of systems before, some of them were very shy about wearing an HMD, resulting in very few head movements. Participants needed prompting and encouragement in order to start moving their head around and change their field of view. This indicates that users may need to take some time to get used to system, as one user commented: "It took me about 10 seconds to adapt to the 3D viewpoints. But after that everything is fine."

Despite this, all participants were able to complete their assigned tasks without apparent difficulties. Their communications seemed smooth. Both helpers and workers looked comfortable performing tasks with the system. More specifically, workers were able to follow the verbal instructions from helpers and understand what they were asked to do by looking at the visual aids shown on the screen. Helpers were able to guide workers through the task process with the HMD worn on their head and using hand gestures.

10.4.3.2 Usability Ratings
The usability of HandsIn3D was tested based on a set of positive and negative statements. Each statement was to evaluate one specific aspect of the usability. The usability metrics used for evaluation included: ease of learning, ease of use, usefulness, individual task satisfaction, group task satisfaction, perception of interaction, perception of pointing and representational gestures, sense of copresence, sense of immersion, and perception of spatial relation.

Fourteen participants filled out two questionnaires each: the helper questionnaire and the worker questionnaire. This resulted in 28 responses in total. The participants rated the extent to which they agreed with the statements, based on a scale of 1–7, with 1 being "strongly disagree," 7 being "strongly agree," and 4 being "neutral." Note that helpers had two extra items to rate: perception of spatial relationship between objects and sense of immersion. For the purpose of analysis, user ratings were first transferred so that higher ratings meant better usability

Table 10.1 Average values and standard deviations of user ratings for individual and overall usability measures.

	Average	StDev
Ease of learning	5.96	0.92
Ease of use	5.96	0.74
Usefulness	6.10	0.87
Individual task satisfaction	5.89	1.13
Group task satisfaction	5.85	1.17
Perception of interaction	6.39	0.73
Perception of pointing gestures	6.00	1.12
Perception of representational gestures	5.67	1.09
Copresence	5.46	0.92
Spatial relation	5.64	0.84
Immersion	5.92	0.99
Overall usability	**5.89**	**0.96**

and then averaged across the participants. The average of the obtained ratings was computed for the overall usability. The results are shown in Table 10.1.

As can be seen from Table 10.1, overall, the participants were positive about the usability of HandsIn3D with a rating of 5.89 on a scale of 1–7. More specifically, the participants rated "perception of interaction" at the highest value of 6.39, indicating that HandsIn3D was considered highly interactive. The participants were also positive about all other usability measurement items.

To explore possible differences between helpers and workers, we looked at their user ratings separately, and the results are shown in Figure 10.4.

As can be seen from Figure 10.4, all items were rated greater than 4, which meant that both helpers and workers were generally positive about the system. Further, helpers rated the system relatively low for ease of learning and ease of use compared to workers. While the system made workers more satisfied with their own individual performance, helpers were more satisfied with the overall group performance. In addition, while helpers gave the same rating for being able to perform both pointing and representational gestures, workers seemed to perceive pointing gestures more easily than representational gestures. In regard to copresence, helpers reported a relatively higher sense of copresence compared to workers. This was what we expected since helpers were immersed in a virtual environment of the worker's space while workers saw what helpers saw on a 2D display. Given all these differences between helpers and workers, paired-sample

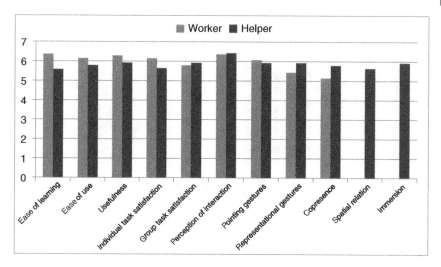

Figure 10.4   Average user ratings of the full 3D system (worker vs. helper).

Table 10.2   Statistical results of ratings between helpers and workers.

	t value	*p* value
Ease of learning	2.349	0.035
Ease of use	1.235	0.239
Usefulness	1.046	0.315
Individual task satisfaction	1.165	0.265
Group task satisfaction	−0.322	0.752
Perception of interaction	−0.322	0.752
Perception of pointing gestures	0.279	0.785
Perception of representational gestures	−1.242	0.236
Copresence	−2.386	0.033

t-tests were conducted to determine whether they were statistically significant at the level of 0.05. The results are shown in Table 10.2 and it was found that the ratings between workers and helpers were significantly different only in ease of learning ($t = 2.349$, $p = 0.035$) and in sense of copresence ($t = −2.386$, $p = 0.033$). Further, helpers reported a positive rating for their ability in perceiving spatial relation of objects at 5.64 and for sense of immersion at 5.92.

Participants were also asked to rate the 2D interface. For the purpose of comparison, the ratings for the 2D and the full 3D interfaces are shown in Figure 10.5.

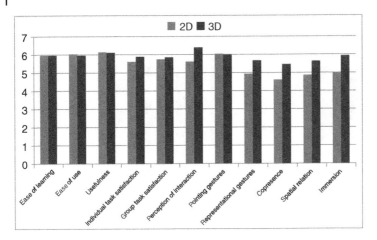

Figure 10.5 Average user ratings (2D interface vs. 3D interface).

As can be seen from Figure 10.5, participants were generally in favor of the 3D interface in terms of immersion, spatial relation, copresence, and representational gestures. They also had a stronger perception of interaction with the 3D system. For other usability measures, user ratings were different, but all were very close. Again, to test statistical significance of these differences, paired-sample t-tests were conducted. The results of these tests are shown in Table 10.3. It was revealed that the difference was significant for immersion ($t = 2.738$, $p = 0.017$), for spatial relation ($t = 3.015$, $p = 0.010$), for copresence ($t = 4.500$, $p < 0.001$),

Table 10.3 Statistical results of ratings between 2D and 3D.

	t value	*p* value
Ease of learning	0.000	1.000
Ease of use	0.402	0.691
Usefulness	0.328	0.745
Individual task satisfaction	−1.395	0.174
Group task satisfaction	−0.648	0.523
Perception of interaction	−4.032	0.000
Perception of pointing gestures	0.176	0.861
Perception of representational gestures	−3.689	0.001
Copresence	−4.500	0.000
Spatial relation	−3.015	0.010
Immersion	−2.738	0.017

for representational gestures ($t = 3.689$, $p = 0.001$), and for the perception of interaction ($t = 4.032$, $p < 0.001$). All this indicated that the introduction of our 3D immersive interface was beneficial not only in improving perception of spatial relation and ability of performing rich hand gestures, but also in terms of other usability measures such as sense of copresence, perception of interaction, and user task satisfaction.

10.4.3.3 User Experiences

Based on user responses to the open questions and user interviews, participants were generally positive about the system, as one participant stated that "it is very impressive and a great experience to use this system."

More specifically, participants appreciated the feature that workers are able to see, and helpers are able to perform hand gestures. A helper commented that "he (the worker) knew exactly what I meant by 'here,' 'this one,' and 'that one'." A number of workers simply commented that "(hand gestures were) easy to understand and follow."

Consistent with the usability ratings, the 3D interface has boosted a strong sense of copresence and immersion for helpers. It was commented that the system had given participants playing the role of the helper a feeling of being in front of the remote workspace and copresent with the remote objects and their remote partner. Comments from helpers include "I feel I was right there in front of the Lego toys and really wanted to grab them.". And after removing the HMD, "I was expecting to have the Lego toys in front of me; it surprised me not to see them." "When I was wearing the HMD there were times when I could see the legs of my partner as the continuation of mine; it took me a while to realize these were not my legs. So in this sense, I felt we were in the same space." A few helpers also commented on the feeling of almost embodiment of their partner. We observed that helpers referred to remote objects using words like "this" and "that" during their interaction with workers, this is consistent with a sense of physical presence.

User comments provided further evidence that the 3D interface improved perception of spatial relation and participants appreciated that. For example, "You can see the difference between 2 objects with the same base but different heights in 3D." "3D helped to see how the final shape looked. With 2D, I had to ask the worker to show me the piece in another angle." "It gives the depth of the objects, so remote guiding could be easier in some cases."

Participants generally liked the idea of having shadows of hands and objects, commenting that it would be easier to point and gesture in the remote workspace as hand shadows could provide them with an indication of hand location in relation to the remote objects. However, there were mixed responses when participants were asked whether the shadow feature actually helped. For example, "Yes, it helps. It makes a good stimulation effect. So I can do better communication with my partner." "The shadow helps me feel that the view is in 3D. But I think I can still understand without the shadow." "No, there are some real shadows in gray color. The black shadow is artificial and a little bit annoying." "The shadow could sometime cover objects and I think this could potentially lead to something wrong (maybe a transparent shadow)." "Yes, (shadow helps) for pointing only, but not much on rotating etc."

With regard to head tracking, participants commented that it enables them to change their point of view of the remote workspace. This in turn provides helpers with a better spatial understanding of the remote workspace. In most 2D systems, the camera looking at the workspace of the worker is either fixed or controlled by the worker. If helpers want to change their point of view, they will need to ask their partner to move the camera accordingly. In our view, providing helpers with a control of their view of the remote workspace and independently from what their partner is looking at, is important. One comment we received from a helper who did move their head during the guiding process is: "I could see how the toy blocks are connected without having to ask my partner to show it to me." This seems to suggest that the control of his own view allows the helper to focus on what needs to be done, which in turn may reduce the time on task.

Further, in comparison with the 2D interface, participants commented that 2D is fine with simple tasks, but 3D offers much richer user experience and is more preferable and useful for complex procedures when a high level of user engagement is required. For example, "3D is more realistic as I can see all angles." "In 2D, it seems like playing a game. When changing my viewpoints into 3D, I got a feeling of going back to real world." "(3D) helps more when I need to give instruction related to space concept." "3D interface makes it easy to match the screen and the physical objects." "3D feels real. 2D interface is enough for simple tasks but 3D interface helps more when the task gets more complicated."

Although the main purpose was to test the usability and usefulness of our 3D concept for remote guidance on physical tasks, user comments also gave some hints for further improvements. These include (1) using a lighter and more easily adjustable helmet; (2) increasing image quality and resolutions; (3) differentiating worker and helper hands by color and making them transparent; (4) providing a more dynamic and more immersive environment for helpers to interact with, such as when the helper moves closer to the objects, they become bigger; and (5) making shadows gray and transparent.

10.5 A Comparison of User Ratings between HandsInAir and HandsIn3D

The comparison of user ratings for the 2D and 3D versions of our HandsIn3D system revealed 3D interfaces provided much richer experience for remote collaboration. To further understand the pros and cons of 3D interfaces, we conducted a cross-study comparison of 2D verse 3D based on two different remote guidance systems: HandsInAir and HandsIn3D. HandsInAir is also a system developed for remote collaboration on physical tasks but built on mobile devices with 2D interfaces to support mobility [19]. The data we used for comparison for HandsIn3D were those reported in Section 10.4 of this chapter, while data for the 2D system were from the experiment reported in [20]. The comparison was made on the same items that had been reported by the participants of the two experiments: ease of learning, ease of use, usefulness, task satisfaction, perception of interaction, perception of pointing gestures, perception of representational gestures and copresence. For task satisfaction in 3D, the value was obtained as the average of the ratings of group task satisfaction and individual task satisfaction. To give a background, the 2D experiment is briefly described below.

There were 20 participants who were randomly assigned the role of either helper or worker. They were required to work together to collect toy blocks located in difference places and assemble them into pre-defined toy models. During the process, it was necessary for both the worker and the helper to move around the rooms to complete the tasks. Upon completion of their assigned tasks, questionnaires and interviews were administered to gather data about their experiences.

The overall ratings of the usability measures for the two systems are shown in Figure 10.6. As can be seen from Figure 10.6, the 2D system were rated slightly higher than the 3D system for both ease of learning and usefulness, indicating that the participants were quicker in learning how to use HandsInAir and thought that the 2D system was more useful than our HandsIn3D in its current form. On the other hand, the 3D system was rated higher in all other items, indicating that overall the users were in favor of using the system in terms of the individual items

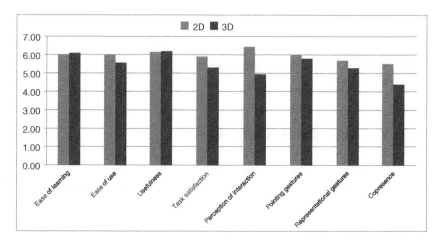

Figure 10.6 Overall usability ratings between the 3D and 2D systems.

Table 10.4 Statistical results of overall ratings between 2D HandsInAir and HandsIn3D.

	U value	*p* value
Ease of learning	253	0.54
Ease of use	196	0.06
Usefulness	279	0.98
Task satisfaction	180	0.03
Perception of interaction	58	0.00
Perception of pointing gestures	229	0.26
Perception of representational gestures	203	0.09
Copresence	110	0.00

rated. In particular, the users rated task satisfaction at 5.87, perception of interaction at 6.39, and co-presence at 5.46 for 3D, while users rated task satisfaction at 5.3, perception of interaction at 4.9, and copresence at 4.35 for 2D. To test the statistical significance of these differences, a nonparametric Mann-Whitney U tests for independent samples were conducted. The results of these tests are shown in Table 10.4.

As can be seen from Table 10.4, for most of the usability measures, the difference between the 2D and 3D systems were not statistically significant (their p values are greater than 0.05). However, the difference in task satisfaction was significant ($U = 180, p = 0.03$), indicating that participants were more satisfied with their experience and collaboration performance when the 3D system was used. Similarly, there was a significant difference between the two systems in perception of interaction ($U = 58, p < 0.01$) and in copresence ($U = 110, p < 0.01$). Again, these differences indicated that our 3D system provided a greater interactive experience to users and a higher sense of collaborators being together although they were in fact physically distributed.

10.6 Conclusion and Future Work

In this chapter, we presented our new remote guiding system which we developed with the aim to improve the remote helper's perception of the workspace. We presented the technical challenges we faced in designing and implementing this system. We also presented our findings from a user study that we conducted with our system. The results from the user study are encouraging and suggest that we are in the right direction; our proof of concept system demonstrates that the task of capturing, triangulating, and rendering a 3D world composed by the fusion of a worker space and helper hands is not only doable but also very intuitive. From a user's point of view, the unique feature of HansIn3D is the integration of the projection of the helper's hands into the 3D workspace of the worker. Not only does this integration give users flexibility in performing more natural hand gestures and ability in perceiving spatial relationship of objects more accurately, but it also offers greater sense of copresence and interaction. To the best of our knowledge, this is the first time that a 3D Mixed Reality system uses the hand gesture paradigm for remote collaboration on physical tasks, and our laboratory-based user study indicates this to be a great improvement over those previous 2D-based gesture systems.

However, it is also important to be aware of the limitations of our user study and the system. For example, the position of the helper's hands on physical objects (a broken machine, for example) means that the calibration/accuracy of machine, the hands, fingers and the types of possible gestures, such as pointing, turning, rotating, etc., becomes very important when we consider the broken machine is a big and complex piece of work with many small parts. However, the current system does not specifically provide a practical application since in both locations users have to sit and as such, calibration of two locations' systems on a real-world workspace could be a problem. Further, in our main user study, the results found that the ratings between workers and helpers were significantly different only in ease of learning ($t = 2.349$, $p = 0.035$) and in sense of copresence ($t = -2.386$, $p = 0.033$). This indicates that workers considered the 2D interfaces more intuitive to use, while the 3D interfaces gave helpers a higher sense of being colocated with their collaborators. This difference could be because the helper was immersed in the 3D situation while the worker was not. But this would be worthy of further investigation. In the experiments, the complexity of the task might have an impact on the responses of the participants. For example, a more complex task can require more intensive verbal and gesturing communications, thus likely reducing the quality of user experience, particularly on task satisfaction ratings. Finally, since the system was developed to provide a 3D environment to support the helper, we have asked only helpers to rate their experience in understanding spatial relations of objects in the remote workspace. However, workers might also have issues in understanding spatial relations when looking at the 3D virtual reality in a 2D screen and this should be examined in future studies.

Now that the fundamental benefits of having 3D gesturing are evident, our plan for future technical work is to improve the system on several fronts: on the rendering side, we are exploring how to speed up the capture, triangulate, and rendering loop, as to improve the general responsiveness of the system to the head tracking of the helper. We are, in particular, currently investigating the

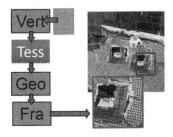

Figure 10.7   Introducing OpenGL tessellation shaders in our rendering pipeline; the amount of details used to render objects is adapted in real time to the amount of depth discontinuity present in the depth maps.

introduction of OpenGL tessellation shaders in our rendering pipeline, as this would allow for greater control on the way we use our polygonal budget, using smaller, more precise triangles for the edges of the captured objects, and a coarser triangle subdivision for those large areas where there is less geometric detail, as depicted in Figure 10.7. While we did not conduct yet an extensive measure of the performance advantage of this rendering technique, our very preliminary tests are showing a performance improvement of at least 50% over nontessellated depth meshes.

We are also aware of the critical importance to introduce in our system the ability compressing 3D data to achieve proper streaming in a real-world scenario where the worker and the helper are at different geographical locations. Even if we did not address this challenge yet, we are planning to add compression in the second phase of the research. There are various techniques that are suitable for our usage scenario, and we are in particular looking at two possible approaches: in the first one, standard image compressor is adapted to 3D meshes streaming, like in the work of Pece et al. [33]. An alternative method would be to apply more standard geometry compression schemes like in [34, 35]: this would probably lead to better compression rates, but the ability to compress/decompress at a 30 Hz rate needs to be verified.

With regard to evaluation, we plan to advance the prototype into a close-to-production system so that we can test it in a more realistic setting. For example, separate the two sides of the system and connect them through the internet, instead of hosting them by the same PC. We also plan to compare HandsIn3D with its 2D versions through rigorously controlled studies so that we can have more quantitative and objective information about the benefits of immersive virtual environments in supporting remote guidance.

Acknowledgment

This chapter is a reprint of Huang et al. [1] with minor edits. Permission is granted from Springer.

References

1 Huang, W., Alem, L., Tecchia, F., and Duh, H. (2018). Augmented 3D hands: a gesture-based mixed reality system for distributed collaboration. *J Multimodal User Interfaces* 12: 77–89. https://doi.org/10.1007/s12193-017-0250-2.

2 Clark, H.H. and Brennan, S.E. (1991). Grounding in communication. In: *Perspectives on Socially Shared Cognition* (ed. L.B. Resnick, J.M. Levine, and S.D. Teasley), 127–149. Washington, DC: American Psychological Association.

3 Liu, J. and Kavakli, M. (2010). A survey of speech-hand gesture recognition for the development of multimodal interfaces in computer games. *IEEE International Conference on Multimedia and Expo (ICME)*, Singapore (19–23 July 2010). IEEE. pp. 1564–1569.

4 Fussell, S.R., Setlock, L.D., Yang, J. et al. (2004). Gestures over video streams to support remote collaboration on physical tasks. *Human-Computer Interaction* 19: 273–309.

5 Kirk, D.S. and Stanton Fraser, D. (2006). Comparing remote gesture technologies for supporting collaborative physical tasks. *ACM Human Factors in Computing Systems*, Montréal Québec Canada (22–27 April 2006). ACM. pp. 1191–1200.

6 Ou, J., Fussell, S.R., Chen, X., et al. (2003). Gestural communication over video stream: supporting multimodal interaction for remote collaborative physical tasks. *Proceedings of the 5th Conference on Multimodal Interfaces*, Vancouver British Columbia Canada (05–07 November 2003). ACM. pp. 242–249.

7 Kuzuoka, H., Kosaka, J., Yamazaki, K., et al. (2004). Mediating dual ecologies. *ACM Conference on Computer Supported Cooperative Work*, Chicago Illinois USA (06–10 November 2004). ACM. pp. 477–486.

8 Kuzuoka, H. (1992). Spatial workspace collaboration: a SharedView video support system for remote collaboration capability. *ACM Human Factors in Computing Systems*, Monterey California USA (03–07 May 1992). ACM. pp. 533–540.

9 Yamashita, N., Kaji, K., Kuzuoka, H., and Hirata, K. (2011). Improving visibility of remote gestures in distributed tabletop collaboration. *ACM Conference on Computer Supported Cooperative Work*, Hangzhou China (19–23 March 2011). ACM. pp. 95–104.

10 Alem, L., Tecchia, F., and Huang, W. (2011). HandsOnVideo: towards a gesture based mobile AR system for remote collaboration. In: *Recent Trends of Mobile Collaborative Augmented Reality* (ed. L. Alem and W. Huang), 127–138. New York, NY: Springer.

11 Huang, W. and Alem, L. (2011). Supporting hand gestures in mobile remote collaboration: a usability evaluation. *Proceedings of the 25th BCS Conference on*

Human Computer Interaction, Newcastle-upon-Tyne United Kingdom (04–08 July 2011). BCS Learning & Development Ltd. pp. 211–216.

12 Kraut, R.E., Gergle, D., and Fussell, S.R. (2002). The use of visual information in shared visual spaces: informing the development of virtual co-presence. *ACM Conference on Computer Supported Cooperative Work*, New Orleans Louisiana USA (16–20 November 2002). ACM. pp. 31–40.

13 Mortensen, J., Vinayagamoorthy, V., Slater, M., et al. (2002). Collaboration in tele-immersive environments. *EGVE'02*. pp. 93–101.

14 Schuchardt, P. and Bowman, D.A. (2007). The benefits of immersion for spatial understanding of complex underground cave systems. *Proceedings of the 2007 ACM Symposium on Virtual Reality Software and* Technology, *(VRST'07)*, Newport Beach California (05–07 November 2007). ACM. pp. 121–124.

15 Mania, K., Badariah, S., Coxon, M., and Watten, P. (2010). Cognitive transfer of spatial awareness states from immersive virtual environments to reality. *ACM Transactions on Applied Perception* 7 (2): Article 9, 14 pages.

16 Huang, W., Alem, L., and Tecchia, F. (2013). HandsIn3D: supporting remote guidance with immersive virtual environments. *Human-Computer Interaction – INTERACT 2013. Lecture Notes in Computer Science*, Cape Town, South Africa (02–06 September 2013). Springer, vol. 8117. pp. 70–77.

17 Gurevich, P., Lanir, J., Cohen, B., and Stone, R. (2011). TeleAdvisor: a versatile augmented reality tool for remote assistance. *Proceedings of the SIGCHI Conference on Human Factors in Computing Systems (CHI'11)*, Vancouver BC Canada (07–12 May 2011). ACM. pp. 619–622.

18 Alem, L. and Li, J. (2011). A study of gestures in a video-mediated collaborative assembly task. *Advances in Human-Computer Interaction* 2011: Article ID 987830, 7 pages.

19 Huang, W. and Alem, L. (2013) HandsinAir: a wearable system for remote collaboration on physical tasks. *Proceedings of the 2013 Conference on Computer Supported Cooperative Work Companion (CSCW'13)*, San Antonio Texas USA (23–27 February 2013). ACM. pp. 153–156.

20 Huang, W. and Alem, L. (2013). Gesturing in the air: supporting full mobility in remote collaboration on physical tasks. *Journal of Universal Computer Science* 19 (8): 1158–1174.

21 Sakata, N., Kurata, T., Kato, T., et al. (2003). WACL: supporting telecommunications using – wearable active camera with laser pointer. *Proceedings of the Seventh International Symposium on Wearable Computers*, White Plains, NY, USA (21–23 October 2003). IEEE. pp. 53–56.

22 Kopácsi, S., Kovács, G., and Nacsa, J. (2013). Some aspects of dynamic 3D representation and control of industrial processes via the Internet. *Computers in Industry* https://doi.org/10.1016/j.compind.2013.06.007.

23 Brown, R.A., Recker, J.C., and West, S. (2011). Using virtual worlds for collaborative business process modeling. *Business Process Management Journal* 17 (3): 546–564.

24 Monahan, T., McArdle, G., and Bertolotto, M. (2008). Virtual reality for collaborative e-learning. *Computers & Education* 50: 1339–1353.

25 Higuch, K., Yonetani, R., and Sato, Y. (2016). Can eye help you? Effects of visualizing eye fixations on remote collaboration scenarios for physical tasks. *Proceedings of the 2016 CHI Conference on Human Factors in Computing Systems (CHI'16)*. pp. 5180–5190. San Jose California USA (07–12 May 2016). ACM.

26 Gupta, K., Lee, G., and Billinghurst, M. (2016). Do you see what I see? The effect of gaze tracking on task space remote collaboration. *IEEE Transactions on Visualization and Computer Graphics* 22: 2413–2422.

27 Otsuki, M., Kawano, T., Maruyama, K., Kuzuoka, H., and Suzuki, Y. (2017). ThirdEye: Simple Add-on Display to Represent Remote Participant's Gaze Direction in Video Communication. *Proceedings of the 2017 CHI Conference on Human Factors in Computing Systems (CHI'17)*, Denver Colorado USA (06–11 May 2017). ACM. pp. 5307–5312.

28 Huang, W., Alem, L., and Livingston, M. (2013). *Human Factors in Augmented Reality Environments*, 274. Springer Science & Business Media.

29 Gao, G., Yamashita, N., Hautasaari, A., and Fussell, S. (2015). Improving Multilingual Collaboration by Displaying How Non-native Speakers Use Automated Transcripts and Bilingual Dictionaries. *Proceedings of the 33rd Annual ACM Conference on Human Factors in Computing Systems (CHI'15)*, Seoul Republic of Korea (18–23 April 2015). ACM. 3463–3472.

30 Nguyen D. and Fussell, S. (2014). Verbal cues of involvement in dyadic same-culture and cross-culture instant messaging conversations. *Proceedings of the 5th ACM international conference on Collaboration across boundaries: culture, distance & technology (CABS'14)*, Kyoto Japan (20–22 August 2014). ACM. 41–50.

31 Maimone, A. and H. Fuchs (2011). Encumbrance-free telepresence system with real-time 3D capture and display using commodity depth cameras. *The IEEE International Symposium on Mixed and Augmented Reality (ISMAR)*, Basel, Switzerland (26–29 October 2011). http://dx.doi.org/10.1109/ISMAR. 2011.6162881

32 Tecchia, F., Alem, L., and Huang, W. (2012). 3D helping hands: a gesture based MR system for remote collaboration. *Proceedings of the 11th ACM SIGGRAPH International Conference on Virtual-Reality Continuum and its Applications in Industry (VRCAI'12)*, Singapore, Singapore (02–04 December 2012). ACM. pp. 323–328.

33 Pece, F., Kautz, J., and Weyrich, T. (2011). Adapting standard video codecs for depth streaming. *Proceedings of the 17th Eurographics Conference on Virtual Environments & Third Joint Virtual Reality (EGVE – JVRC'11)*, Nottingham UK (20–21 September 2011). ACM. pp. 59–66.

34 Bannò, F., Gasparello, P., Tecchia, F., and Bergamasco, M. (2012). Real-time compression of depth streams through meshification and valence-based encoding. *Proceedings of the 11th ACM SIGGRAPH International Conference on Virtual-Reality Continuum and Its Applications in Industry (VRCAI'12)*, Hong Kong Hong Kong (17–19 November 2013). ACM. pp. 263–270.

35 Marino, G., Gasparello, P., Vercelli, D., et al. (2010). Network streaming of dynamic 3D content with on-line compression of frame data. *2010 IEEE Virtual Reality Conference (VR)*, Waltham, Massachusetts, USA (20–24 March 2010). IEEE. pp. 285–286.

11

Supporting Tailorability to Meet Individual Task Needs

11.1 Introduction

In this chapter, we will discuss our research on tailorable remote assistance in the manufacturing industry. In general, systems that support collaboration must be tailorable due to individual differences of users with different experiences and preferences and due to the requirements of different collaborative tasks [1, 2]. In this regard, remote assistance is no different and must also be tailorable – "no one size fits all." For example, it is desirable to make the location of guidance tailorable to the worker, thus enabling him (in this chapter and Chapter 12, we refer worker as he or his, while we refer helper as she or her for clarity) to select from augmented reality (AR) and non-AR options depending on the collaborative task. Yet, recent related work [3–7] focuses on a key property of AR, merging communication space (the location of guidance) and task space (the physical task objects). On the contrary, we will demonstrate that separating the location of guidance from the task space can sometimes be advantageous.

We aim to answer the following interesting research questions:

- Which aspects of remote assistance need to be tailorable?
- How do users from the manufacturing industry tailor remote assistance?
- What are requirements and challenges for remote assistance in the manufacturing industry?

To answer these questions, we conducted an interview study intended to investigate current remote assistance practices in the manufacturing industry. In this interview, empirical data about remote assistance in the manufacturing industry was collected. User requirements for and tailorable aspects of remote assistance in the manufacturing industry were identified. Following that, a design response in the form of a component-based tailorable remote assistance system, RemoteAssistKit (RAK), was developed. RAK was also evaluated in a user study in the manufacturing industry and findings on how users tailor remote assistance

Computer-Supported Collaboration: Theory and Practice, First Edition.
Weidong Huang, Mark Billinghurst, Leila Alem, Chun Xiao, and Troels Rasmussen.

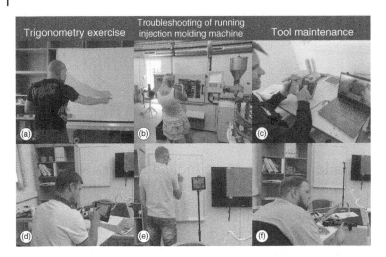

Figure 11.1 Three different example configurations of RAK used by participants during the experimental simulation. (d) A remote helper draws on live video of a whiteboard captured by a scene camera (not in view) in the worker's space, and (a) the drawings are projected onto the worker's whiteboard. (b) A worker captures the internals of a machine with a tablet camera, (e) a remote helper's drawings on a whiteboard are merged with live video of the machine, and (b) the worker sees the merged result on the tablet. (f) A helper uses hand gestures in a shared view of the worker's task space, and (c) the worker sees the helper's gestures in the shared view.

were analyzed, the results of which could in turn be fed into the next requirements phase for future developments.

The remainder of the chapter is structured as follows. First, in Section 11.2, we describe the component-based design of RAK from a technical perspective. Figure 11.1 shows RAK configurations in different realistic scenarios from the plastic manufacturing industry.

Next, Section 11.3 summarizes the results of interviewing employees from the manufacturing industry about their current remote assistance practices and requirements for remote assistance. The interviews revealed heterogeneous user needs and informed the tailorable aspects of RAK. To our knowledge, most research on remote assistance is driven by the latest advances in camera and extended reality (XR) technology [4–6, 8]. Instead, by conducting the interview we wanted to understand and empathize with users of remote assistance technology from a specific domain and base the design of a remote assistance system on their needs.

In Section 11.4, we discuss how domain experts from the manufacturing industry tailor remote assistance. To explore how remote assistance is tailored, an experimental simulation with RAK was conducted at a technical college for plastic manufacturing. This research differs from the sparse related work on remote assistance in manufacturing by focusing on how domain experts tailor remote assistance to different manufacturing scenarios with various requirements and physical scales, while related work on the other hand has focused on evaluating rigid (non-tailorable) remote assistance systems designed for specific industrial machines [9, 10]. In regards to tailorable remote assistance, Speicher et al. [11] also recently studied the use of a configurable remote assistance system, but the configurable aspects of their system are different from those of RAK. More specifically, RAK supports multiple guidance formats, drawings or hands, whereas their system only supports drawings, and RAK supports the use of smartphone/tablet cameras or multiple webcams for task space capturing, whereas theirs support a 360° camera. Similarly, we both support using a smartphone or desktop PC on the helper side and guidance in shared video or as a projection. Besides, we carried out an experimental simulation with domain experts to obtain qualitative and quantitative data on how users tailor remote assistance that were specific to the manufacturing industry, whereas they conducted an experimental simulation with university students in an office setting, so the studies differ in realism and industrial relevance.

As a result of the uncontrolled nature of our empirical studies, interesting new findings, which were not part of our original research questions on tailorability, were uncovered about user requirements and challenges for remote assistance in manufacturing. These findings concern user requirements for nonverbal guidance (Section 11.5), sharing of machine sounds (Section 11.6), and high-resolution views for remote quality control (Section 11.7), the challenges of 3D reconstructing a task space composed of metallic components (Section 11.8), and establishing workspace awareness in large industrial task spaces by using multiple cameras (Section 11.9).

11.2 Component-Based Design of RemoteAssistKit

RAK is a component-based system that enables users – one helper and one worker – to tailor important aspects of remote assistance. A helper can tailor interface mobility and guidance format, and a worker can tailor task space capturing and guidance location. More specifically, the software components of RAK are designed around four concepts: (1) interface mobility, (2) guidance format, which concerns the helper, (3) task space capturing, and (4) guidance location, which concerns the worker. RAK consists of two software components for each of the concepts, thus totaling eight components. A RAK configuration consists of one component from each concept, which equals 16 possible combinations $(2 \times 2 \times 2 \times 2 = 16)$. The details of these tailorable aspects of remote assistance

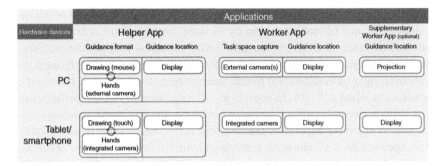

Figure 11.2 Relationship between hardware devices, apps, and software components of RAK. A user's selection of hardware device and app automatically configures the underlying function-oriented software components of which the apps are composed.

are described in Section 11.3. Users tailor their desired remote assistance solution by combining function-oriented software components through a selection of hardware devices and apps. RAK consists of three apps, the helper's Helper App, the worker's Worker App, and Supplementary Worker App, that run on a PC or tablet/smartphone device. The software components, of which the apps are composed, depend on the choice of hardware device, as shown in Figure 11.2. From the users' perspective, they need to be primarily concerned with the choice of hardware devices and mounts for devices without knowing about the underlying software components. Figure 11.3 shows the tailorable aspects of RAK from the users' perspective.

11.2.1 Helper App

To provide guidance to a worker, a helper uses the Helper App on a PC or tablet/smartphone device. The Helper App will detect the hardware device and select the correct software components to run. If the hardware device is a PC, the Helper App will be composed of a drawing (mouse cursor) component and hand gestures (external camera) component, which are guidance format components, and a display component, which is a guidance location component. If the hardware device is a tablet/smartphone, the Helper App will be composed of a drawing (touch) component, hand gestures (integrated camera) component and display component.

A guidance format component enables a helper to provide nonverbal guidance. The difference between the two guidance format components, drawings or hand gestures, is the strategy for capturing input from the helper. With the drawing component, a helper can use mouse input on a PC and touch input

	Helper			Worker		
No.	Interface mobility		Guidance format	Taskspace capturing		Additional guidance location (optional)
	Device	Mount		Device	Mount	Device
1	Tablet/smartphone	Tripod	Drawings	Tablet/smartphone camera	Tripod	Tablet/smartphone display
2	PC	Gooseneck	Hand gestures	Single webcam (PC)	Gooseneck	Projection (PC + projector)
3		Gorillapod	Verbal only	Multiple webcams (PC)	Gorillapod	
4		Handheld			Handheld	

Example configuration "11223"

Tablet/smartphone	Tripod	Hand gestures	Single webcam (PC)	Gorillapod

Figure 11.3 Tailorable aspects of RAK from the users' perspective. A RAK configuration is composed of hardware components (devices and mounts) and software components (guidance formats) for each tailorable aspect, i.e. interface mobility, guidance format, task space capturing, and guidance location. The numbers of the components are concatenated and used to uniquely describe a configuration as demonstrated with the example configuration "11223."

on a tablet/smartphone to make freehand drawings on video of a worker's task space, as in [12, 13]. With the hand gestures component, a helper can use an external camera on a PC or an internal back-facing camera on a tablet to make hand gestures on video of the task space, as in [14]. Drawings or hand gestures are then sent to the Worker App and *optionally* the Supplementary Worker App and displayed in a shared video of the worker's task space or projected into the task space, depending on choice of guidance location component. It is possible to select between the two guidance format components during runtime, which is indicated by the rotating arrow between these components in Figure 11.2.

A guidance location component is responsible for displaying a helper's guidance in a representation of a worker's task space. Both the Helper App and the worker's apps contain guidance location components. There are two guidance location components: a display component and projector component. A display component is necessary for the Helper App, as it merges live video from the Worker App's task space capture component and guidance from the internal guidance format component and displays the result on a tablet/smartphone or PC display. Similarly, a display component is necessary in the Worker App, as it merges live video from the internal task space capture component and guidance from the Helper App's guidance format component and displays the result on a tablet/smartphone or PC display. Thus, the display components in the Helper App and Worker App are responsible for creating a shared view between worker and helper and follow the "What You See Is What I See" (WYSIWIS) interface paradigm, as illustrated with two examples in Figure 11.4.

A projection component can only be used in the Supplementary Worker App, which is described in further detail later.

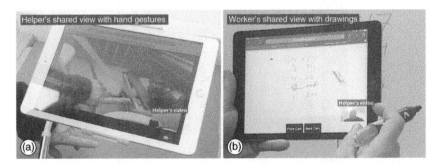

Figure 11.4 Two examples of WYSIWIS interface. (a) Helper's WYSIWIS interface on tablet: shared view of worker's task space with helper's hand gestures on top. (b) Worker's WYSIWIS interface on tablet: shared view of worker's task space with helper's drawings on top.

11.2.2 Worker App

To receive a helper's guidance, a worker runs the Worker App on a PC or tablet/smartphone. The Worker App will detect the hardware device and select the correct software components to run. If the hardware device is a PC, the Worker App will be composed of an external camera(s) component, which is a task space capture component, and a display component. If the hardware device is a tablet/smartphone, the Worker App will be composed of an integrated camera component, which is a task space capture component, and a display component.

The purpose of a task space capture component is to live-stream video of a worker's task space to the display components of the Helper App, Worker App, and Supplementary Worker App. The difference between the external camera(s) component on PC and integrated camera component on tablet/smartphone is the strategy for capturing the worker's task space. With the external camera(s) component, a worker can setup multiple webcams, which are connected to the same PC, in the environment to cover different areas of a task space or multiple perspectives of an area. Live video from the webcams and audio from the PC microphone are streamed to the display components of the different apps, where the video feeds can be selected from a list. The webcams provide the helper with a degree of view independence, because they give her visual access to areas and perspectives of a task space independently of the worker's whereabouts. Previous research has shown that view independence is preferred by the remote helper as opposed to only seeing the task space from the worker's point of view [15–18].

With the integrated camera component, a worker uses the back-facing camera of a tablet/smartphone to live capture and stream video from his task space to the different display components. In comparison to the setup with the webcams, capturing the task space with a tablet/smartphone gives untethered mobility.

11.2.3 Supplementary Worker App

Optionally, a worker runs the Supplementary Worker App on a PC or tablet/smartphone, whereby the Supplementary Worker App will detect the hardware device and select the correct software component to run. If the hardware device is a PC, the Supplementary Worker App will be composed of a projection component. If the hardware device is a tablet/smartphone, the Supplementary Worker App will be composed of a display component. The difference between the projection and display components is the location of a helper's additional guidance. The display component in the Supplementary Worker App works in the same way as the display components in the other apps, i.e. it merges video from the Worker App's task space capture component and guidance from the Helper App's guidance format component and displays the result on an additional tablet/smartphone display.

The projection component in the Supplementary Worker App projects guidance from the Helper App's guidance format component directly into the task space. Therefore, the projection component requires a PC with a projector. The projections are experienced as spatial AR similar to the system presented in related work by Adcock et al. [19]. The benefits of showing guidance directly in the task space using projections as opposed to showing instructions on an external display physically separated from the task space has been demonstrated in related research on AR instructions [20–23] and is due to reduced attention switching, reduced head movements, and a reduced amount of information to be retained in working memory.

11.2.4 Tailoring Mounting Equipment

Users of RAK can use a multitude of mounts like gorillapods, gooseneck mounts, and tripods to mount hardware devices in the environment. See Figure 11.3 for an overview of mounting options. The choice of mount is important because it influences the helper's interface mobility, i.e. the helper must decide whether the tablet or external camera of the PC must be handheld or mounted in the workspace, thereby affecting her mobility and ability to give guidance with one or two hands. In general, the tablet interface offers more mobility than the PC + external camera interface. However, by mounting the tablet in the environment, thereby reducing mobility, she can guide with both hands in front of the back-facing camera.

Similarly, the choice of mount influences the worker's ability to share the task space, i.e. he must decide whether the tablet/smartphone or webcam(s) connected to a PC must be handheld, worn, or mounted in the workspace. There are trade-offs. With a handheld camera, the worker can only use one hand for manipulating objects in the task space, the helper's view of the task space is constantly moving and unstable, and many possible close-up perspectives on task space objects are possible. In comparison, with one or more cameras mounted in the environment, the worker can use two hands for manipulating objects in the task space, the helper's view is stable, however, the possible perspectives on task space objects are limited. The worker also has to consider mounting possibilities in the environment and make sure that the view of cameras covers the relevant task space areas effectively, which requires additional and potentially continuous camera work throughout a remote assistance session.

11.3 Identifying Tailorable Aspects of Remote Assistance

The tailorable aspects of RAK were identified through empirical studies in the manufacturing industry (see [24] for more details about the studies). Six employees (E1–E6) at three different manufacturing companies (TRESU, LEGO, and Vestas) were interviewed using a semi-structured interview strategy. As part of their job, the employees supported remote problem solving on various machines: mold washing machine (E1), computer numerical control (CNC) machines for milling molds (E2), inline printing machines (E3 and E4), industrial robot arm for lubing bolts in a wind turbine production (E5), and turbines (E6). See Figure 11.5 for an overview of the relationship between companies, employees, and machines. The interviews lasted between 21 and 49 minutes, with an average of 27 minutes. Interview questions concerned the employees' daily work routine, current remote assistance practices, video communication technologies currently used for remote assistance, problems at work that required remote assistance, potential needs for being mobile, and requirements for a future remote assistance solution. All six employees had prior experience in the role of the remote helper, for example, they had experience helping field service technicians or customers on site. At least four of the employees (E2–E4, E6) also had prior experience receiving remote help as field service technicians, while E5 would sometimes receive guidance over the phone from the robot manufacturer. Since arranging meetings with employees at the companies was a time-consuming process, the interviews took place over the course of a year. During the visit to each company and before the interviews took place, the interviewer received a tour of the production facilities, and problems that might benefit from remote assistance support were discussed with the employees.

From the interview study, it was clear that the employees' requirements to interface mobility and task space capturing aspects of remote assistance were heterogeneous. For instance, E1 wanted to give guidance in nomadic use contexts due to the perceived urgency of a problem, *"I prefer helping them (the machine operators) right away, because it is very important that the machine is running [...] Often I am in the supermarket or picking up the kids (from school), when they (the machine operators) call me."* In comparison, E6 preferred a PC with a large monitor for giving guidance, *"On the PC I have the (technical) drawings and I can show them (his remote colleagues) the drawings [...] it will be too difficult to do it on the phone, I think."* Furthermore, E5 wanted an overview of a robot arm and thus installed a fixed webcam, which he could access remotely, while E6 needed to be able to see close-up views of the wiring on a turbine and thus were piloting a solution, where live video of the turbine was streamed from a handheld/head-worn camera. E1 suggested to use fixed cameras to make video recordings that could be used for troubleshooting, *"A bunch of cameras in the production that one could log into and use to go back in time because many of the errors are related to persons making mistakes."*

To further illustrate that users have various needs for interface mobility and task space capturing, we compare the needs of two participants from the interview

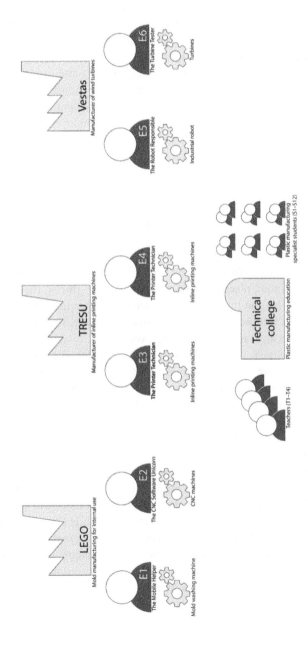

Figure 11.5 Overview of people involved in the interview studies in the manufacturing industry. Manufacturing employees (E1–E6) from company A (LEGO), B (TRESU) and C (Vestas) that participated in interviews, teachers (T1–T4) from the technical college that participated in a workshop, and manufacturing experts (S1–S12) that participated in an experimental simulation with RAK.

study, Bob and Charlie, who work at different companies. Bob, who works as a turbine tester and remotely assists colleagues in Russia, requires a close-up look of the remote turbine to compare the actual wiring inside the turbine to technical drawings. Therefore, his remote colleagues has to use either a handheld or head-mounted camera to capture the insides of the turbine. Bob expects to be in his own workspace while giving remote assistance. He prefers a desktop PC interface with a large monitor to a tablet/smartphone interface, because the PC is perceived as superior for making side-by-side comparisons of technical drawings and live video of the turbine. We compare this to the needs of Charlie, who is responsible for the purchase, implementation and maintenance of CNC machines for internal use. When particularly important machines stop working while he is not on site, he wants to use a mobile interface for giving remote assistance due to the perceived urgency of problem solving. He reported getting calls for help on his phone in everyday situations, e.g. while at the supermarket or picking up his children from school, which further emphasizes the need for a mobile interface and interactions that do not attract other people's attention, thus excluding mobile virtual reality (VR) interfaces. Additionally, he suggested to mount multiple cameras in the workspace to capture the most crucial areas of an important washing machine (used for cleaning CNC-milled metal parts) with the purpose of supporting both synchronous remote assistance and recordings for later retrieval and troubleshooting. Based on the accounts of these two employees, it is clear that they have different needs for interface mobility and task space capturing, and RAK can be configured to support both of them. In terms of exemplary RAK configuration, Bob, in the role of helper, uses a PC with mouse input to make drawings on top of shared live video captured by a worker with a handheld smartphone (configuration "20114," where the "0" describes that there is no mounting option for the PC). Charlie, also in the role of helper, holds up his smartphone and uses hand gestures on shared live video captured by multiple webcams mounted with gorillapods in the worker's space (configuration "14233").

Providing the users with the means to innovate and tailor a product to their needs have also been shown to improve user satisfaction with products in prior research, for instance, security software [25]. Due to these heterogeneous user needs, it was decided to support the helper's interface mobility on PC and tablet/smartphone and the worker's task space capturing with a tablet/smartphone camera or one or more webcams.

However, it was less clear from the interviews, how making the aspects of guidance format and guidance location tailorable would benefit users. The interviewees did not exhibit a good understanding of the pros and cons of different guidance locations, and they were mostly concerned with the presence of nonverbal guidance, not the specific format of it (see Section 11.5 on nonverbal guidance). This was likely because they did not have prior experience using specialized remote assistance apps with support for nonverbal guidance techniques. Yet, it was decided to make guidance format and guidance location tailorable in the final implementation of RAK. This was inspired by related work that compared guidance formats [12, 26–28] and guidance locations in non-AR and AR [13, 28, 29]. The comparisons of guidance formats have demonstrated some strengths and weaknesses of the drawing and hand gesture formats. Drawings on shared video are good for showing the end-state of an assembly step or process, which can be used for reference by the helper and worker [13], whereas this is difficult to achieve with hand gestures. In comparison, hands are good for making short-lived ephemeral gestures, such as pointing gestures, while pointing with drawings by making arrows or circling objects is not ideal because they clutter the worker's interface, unless automatically deleted after a short period of time [12]. Furthermore, the combination of hand gestures and drawings have been shown to improve remote assistance over just using one of the formats [30]. Thus, it makes sense to be able to alternate between the two formats during runtime, which is possible in RAK.

The benefits of using a specific guidance location, namely AR guidance visualized directly on the task space or non-AR guidance separated from the task space, we argue depends on the nature of the collaborative task, which will be demonstrated with two possible RAK configurations in two scenarios. Therefore, it was decided that RAK should support tailoring of the guidance location. In the first scenario, the RAK configuration shown in Figure 11.6 is used. A worker, who is doing maintenance of water pipes under a sink, has placed a smartphone camera under the sink in an area that is difficult to reach and outside his field of view. The smartphone live streams video of the area to a tablet held by the worker and to a helper on a PC. The helper uses the PC and mouse as input to provide guidance with freehand drawings on the video feed. These drawings are displayed on top of the video feed on the worker's handheld tablet. Hence, from the worker's and helper's perspective, this RAK configuration follows the "What You See Is What I See" (WYSIWIS) paradigm. The RAK configuration is convenient to the worker because it provides him with extra eyes on the water pipes and prevents him from having to do continuous manual camera work in a physically straining body pose. The scenario is a good example of how separating the task space (the water pipes

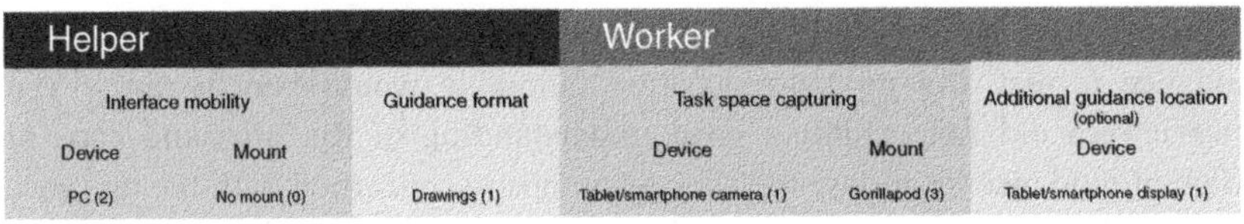

Helper			Worker		
Interface mobility		Guidance format	Task space capturing		Additional guidance location (optional)
Device	Mount		Device	Mount	Device
PC (2)	No mount (0)	Drawings (1)	Tablet/smartphone camera (1)	Gorillapod (3)	Tablet/smartphone display (1)

Figure 11.6 RAK configuration in scenario 1 from the users' perspective.

Helper			Worker		
Interface mobility		Guidance format	Taskspace capturing		Additional guidance location (optional)
Device	Mount		Device	Mount	Device
Tablet/smartphone (1)	Gooseneck (2)	Hand gestures (2)	PC + single webcam (2)	Gorillapod (3)	Projection (PC + projector) (2)

Figure 11.7   RAK configuration in scenario 2 from the users' perspective.

under the sink) and communication space (tablet display in the worker's hands) can actually benefit the worker.

In the second scenario (see RAK configuration in Figure 11.7), a worker needs help with injection mold maintenance at a workbench. He uses a semi-fixed setup with an overhanging web camera and projector mounted on a gorillapod. The webcam and projector are connected to the same laptop PC placed on the workbench. The webcam's top-down view of the workbench is live streamed to a helper. The helper uses a tablet mounted in a gooseneck to make gestures with both hands on the video of the workbench. The hand gestures are projected directly onto the workbench using the overhanging projector. Since the helper and worker see the guidance in the task space from different perspectives, this configuration does *not* follow the WYSIWIS paradigm, which can cause problems (see Section 11.4.2). On the other hand, the projected guidance combines task space and communication space on the workbench, which has been shown to be advantageous for quick consumption of instructions [20–22] because it reduces the worker's head movements and the amount of information to be retained in working memory in comparison to viewing instructions on a separate display. The benefits of a combined task space and communication space are one of the major reasons that AR technology is often used in recent remote assistance research [3–7, 31].

In regards to the separation or combination of task space and communication space, the configurations in the two scenarios represent polar opposites with each their strengths and weaknesses. Importantly, we believe that the scenarios illustrate why "one size does not fit all," and why designers should not blindly use AR technology for remote assistance.

In summary, the tailorable aspects of interface mobility and task space capturing were identified through the interviews with manufacturing employees, whereas tailorability of guidance format and guidance location was based on related work and the envisioned user scenarios.

11.4   How Users Tailor Remote Assistance

To study how users tailor remote assistance an experimental simulation was conducted with RAK at a technical college with experts in plastic manufacturing. Four

Figure 11.8 Three realistic remote assistance scenarios. (a) Trigonometry exercise, (b) troubleshooting of running injection molding machine, and (c) tool maintenance.

teachers (T1–T4) from the technical collage participated in a workshop to design and prepare scenarios for the experimental simulation. Later, 12 plastic manufacturing technicians (S1–S12) took part in the experimental simulation, which consisted of three realistic remote assistance scenarios: (1) trigonometry exercises ("trigo"), (2) troubleshooting running injection mold ("injection"), (3) maintenance of mold ("tool"). See images of the scenarios in Figure 11.8. The 12 technician participants were grouped into six worker–helper pairs and had to collaborate through each scenario by configuring RAK before and/or during a scenario. These participants were students at the technical college but had an average of 11 years of professional experience in plastic manufacturing industry at the time of the experimental simulation.

For scenario 1, the experts had to collaboratively solve two trigonometry exercises by calculating the values for the missing sides and angles of a triangle. They were allowed to use pen and paper and/or a whiteboard for the calculations, and only the helper had the trigonometry formulas (law of sines and cosines). This scenario is relevant, because knowledge of trigonometry is needed when designing injection molds. Scenario 2 had experts collaborate on troubleshooting a running injection mold by fixing an issue with the quality of the manufactured plastic parts. In scenario 3, they had to collaborate on disassembling and reassembling and injection mold.

The process of the experiment was as follows. Upon arrival, participants filled out a prestudy questionnaire about their demographic information and work experience. Then, a small training session was held: the experimenter introduced and explained the RAK and the three experimental scenarios. Some of the configurations of RAK were also demonstrated to the participants. The participants were also given a chance to explore and get familiar with the system and experimental settings. They could shortly explore the configurations to their liking in an open remote assistance scenario that required the assembly of LEGO bricks on a table. Afterward, the pairs assigned themselves roles as helper and worker based on their prior experience, and then the worker was taken to a room where the workspace was located for task performance of the three scenarios. After each scenario, a post-scenario was filled. And at the end of all scenarios, a post-study questionnaire was filled, in which we also asked participants to think of other scenarios in which they thought RAK was also useful and how RAK could be figured to support those scenarios.

Experimental sessions on both worker and helper sides were video recorded. We coded the video recordings to identify common and outlier usage patterns. The questionnaires administered during each session were to collect user feedback on the perceived usefulness of the RAK configurations and tailorability. In the following, the most important results of the experimental simulation are discussed.

11.4.1 Users Tailored Guidance Format and Task Space Capturing to the Requirements of Scenarios

The experimental simulation was conducted to explore the research question: *How do users tailor remote assistance to different problems with varying requirements?* For each scenario, we expected some patterns in the use of RAK configurations, which from the users' perspective are composed of selection of hardware devices and mounts. See Figure 11.3 for a list of components for each tailorable aspect of RAK. The numbers of the components are concatenated and used to uniquely describe a configuration.

Figure 11.9 illustrates the preferred use patterns, while Figure 11.10 shows all configurations used by the worker–helper pairs in each scenario. For the combined-configurations it is clear that there was no preference for a particular configuration in any of the scenarios. We speculate that this has to do with

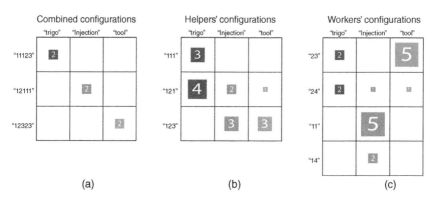

Figure 11.9 Configurations that were used two or more times in any of the scenarios. (a) Worker–helper pairs' combined configurations. (b) Helpers' configurations. (c) Workers' configuration.

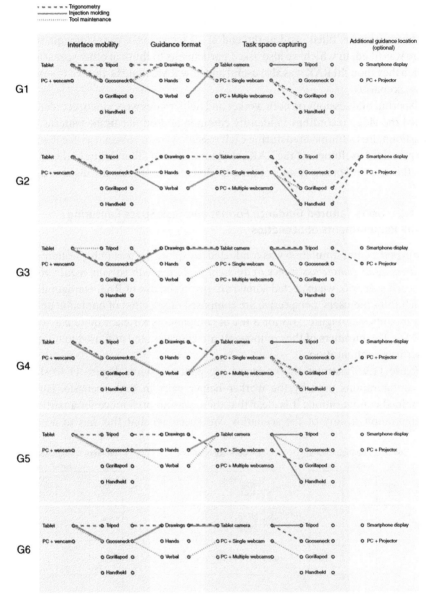

Figure 11.10 All configurations used by the worker–helper pairs (G1–G6) in each scenario.

worker–helper pairs not negotiating the use of a configuration, and thus, one user's choice of components did not affect the other user's choice of components.

Some helper configurations, "111" and "121," were preferred in the "trigo"-scenario, and the configuration "123" was slightly preferred in the "injection" and "tool"-scenarios. The defining difference between helper-configurations was the use of the guidance format drawings in the "trigo"-scenario and verbal guidance only in the other two scenarios, which likely had to do with the participants' high experience rating of the "injection" and "tool"-scenarios and shared terminology for task objects, thus reducing the need for pointing, in comparison to their low experience rating of the "trigo"-scenario and need to draw formulas and annotate triangles to establish common ground.

There was not a clear preference for a specific worker configuration in the "trigo"-scenario, but there were clear preferences for "11" (5/6 workers) in the "injection"-scenario and "23" (5/6 workers) in the "tool"-scenario, which likely had to do with the difference in size between the workspaces in the two scenarios. The workspace in the "tool"-scenario was table-sized, thus warranting the use of a webcam (connected to a PC) on a small gorillapod, which was occasionally picked up in one hand to show close-up details of the mold but otherwise used as an easy-to-move scene camera. In comparison, the workspace in the "injection"-scenario was larger, as the remote assistance revolved around a large injection molding machine, thus warranting a mobile solution with a tablet on a tripod. In both the "tool" and "injection"-scenarios, a common denominator was that configurations supported hands-free use. This need was further corroborated by qualitative statements from the post-scenario questionnaires were multiple participants stated that hands-free use was appreciated.

A common pattern in all scenarios was that workers used a scene camera most of the time (i.e. it was mounted in the environment), since it offered hands-free use, but every now and then, picked up the camera to move it to another location or provide a close-up view of an object. We will refer to this pattern as *movable scene cameras*.

In summary, we see that users tailored guidance format and task space capturing aspects of remote assistance to the distinct requirements of the problem scenarios, and some agreement was observed regarding suitable configurations for the problem scenarios. This is an important finding, because it points to the usefulness of tailorable remote assistance, confirms that tailorability is needed due to the varying requirements of collaborative tasks (e.g. different sizes of task spaces), and implies that remote assistance configurations that have been shown to be particularly suitable for a given collaborative task can be recommended to users in advance to speed up the configuration process and ensure a suitable configuration for a task. Furthermore, the pattern of using movable scene cameras is interesting, because it points to new research, where AR guidance is given through the perspective of one or more movable scene cameras, as in the work by Piumsomboon et al. [6].

We were not able to gain any insights on how users tailor guidance location to the requirements of different tasks and scenarios, because spatial AR guidance, i.e. projected annotations, was only used by two pairs in the "trigo"-scenario. However, their use of spatial AR revealed an interface challenge, mainly because the helpers found it difficult to annotate the shared view of the worker's whiteboard in adequate detail, which leads to the projected mathematical formulas taking up too much space and polluting the worker's whiteboard. A simple solution would be to add zoom functionality to the helper's video feed, allowing for annotations in zoomed-in areas of the task space.

11.4.2 Awareness of the Collaborating Partner's Composition of Components Is Important

During the "trigo"-scenario, two worker–helper pairs used a configuration that included a projector, meaning the helper's guidance was projected onto the worker's space, a whiteboard. One helper was observed to draw outside of the whiteboard and onto the surfaces in the background, which led to unintelligible drawings from the worker's perspective, as shown in see Figure 11.11. To the helper, however, the drawings looked perfectly understandable on the shared video on his tablet device.

This is a good example of how important it is that the composition of a component-based tailorable remote assistance system is made clear to the collaborating partners to avoid misunderstandings. For example, making the helper

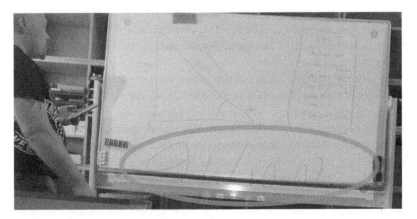

Figure 11.11 Helper draws outside of whiteboard and consequently the projected drawings appear unintelligible to the worker. The helper is not aware of this.

aware of the worker's choice of guidance location component should help avoid the above situation, where the helper applies the wrong mental model of how their guidance is visualized and perceived in the workspace. Furthermore, making the worker aware of the helper's choice of interface mobility component should help adjust their expectations. For instance, the worker may infer that certain resources, be it personnel or documents, is not easily available to a helper using a tablet/smartphone interface, because this interface is indicative of a nomadic mobile use context.

Generally, an important design consideration for tailorable Computer Supported Cooperative Work (CSCW) systems is the ability to tailor systems during run-time, to avoid shutting down a system for all users because one user is tailoring parts of the system [2]. However, it is our observation that run-time tailorability of a CSCW system, such as RAK, can pose a challenge, if a tailoring step by one user affects the collaborating partners. To remedy this problem, Herrmann et al. [32] implemented a feature in their tailorable CSCW system that allowed users to reject, accept, or modify a tailoring step made by a collaborator. Likewise, features must be implemented in tailorable component-based remote assistance systems that make users aware of their collaborating partner's composition either through a similar negotiation feature or persistent awareness cues in the interface, for example in RAK's case by displaying icons for the hardware devices used by the collaborating partner.

11.5 The Importance of Nonverbal Guidance Depends on the Knowledge Relationship

The interviewed employees from the manufacturing industry had little to no experience with specialized remote assistance solutions, and current practices involved sharing images over email or chat apps and video calls using FaceTime or Skype. Still, when they were asked about the ability to use nonverbal guidance (e.g. a pointer or drawings) in shared video, it was considered useful in situations, where the helper and worker did not share the same technical terminology, which was sometimes the case during communication with workers in other countries. E3–E4: *"At the same time you (the worker) show me something with the camera, you are able to see something on your screen, where I've added a layer of information. Often the OEM technician (Original Equipment Manufacturer) does not know the technical name of the mechanic components [...] their English is primitive [...] I don't always understand what he says."*

Additionally, in the experimental simulation the nonverbal guidance of RAK (drawings or hands) was used sparsely in the "tool"-scenario and "injection"-scenario. In those scenarios, it was clear that due to the participants' expert knowledge, pointing to task objects, and making deictic references were rarely necessary for establishing common ground because they shared the same technical jargon. For instance, they shared the technical terms for the different parts of the injection mold in the "tool"-scenario and were able to use the terms to effectively refer to the parts.

In conclusion, it seems that nonverbal guidance is most useful for situations when *no* community comembership within a field [33] can be assumed, and when there is an asymmetric knowledge relationship between the worker and helper. A good example is the asymmetric knowledge relationship between an expert technician and machine operator with different primary languages and educational backgrounds. Still, we agree with Fussell et al. [34] that a systematic comparison of the remote collaboration unfolding between expert workers and expert helpers to the remote collaboration between novice workers and expert helpers is needed to better understand the benefits of nonverbal guidance for different knowledge relationships.

11.6 Sharing of Machine Sounds Is Important for Remote Troubleshooting

It was revealed during the workshop with the teachers that sharing of machine sounds is perceived as important for the remote helper during troubleshooting of a running manufacturing machine because an expert helper is trained in listening to the manufacturing process and can potentially hear when the process is faulty. T1: *"You typically listen for uniform process sounds and react on sounds that sound off [...] It will be difficult for a remote helper to give guidance unless he can hear the process sounds and see the process at the same time, because you use all senses in troubleshooting mode."*

Furthermore, workshop participant T1 also considered other senses useful for troubleshooting a manufacturing process. For example, he considered it useful to be able to distinguish between the necessary smell of heated plastic from the production process and the erroneous smell of burned plastic from a control cabinet. So, the accounts of T1 provided interesting insights into the importance of using both the auditory and olfactory senses for troubleshooting machines in manufacturing and understanding the "full picture" of a problem. This is an interesting finding because research on task space sharing has predominantly focused on methods for sharing visual aspects of a task space [8, 15–17, 19, 35, 36] and dampening sounds from the environment perceived as noise to support clear verbal communication.

The concept of interactive machine sound enhancement, where sounds from machines are perceived as useful auditory information rather than the noise that must be dampened, represents an interesting future research direction for remote assistance in manufacturing. It is especially interesting to explore this concept with XR interfaces because they offer a natural way to select, enhance, and diminish sounds in a 3D environment.

11.7 High-Resolution Views Are Important for Remote Product Quality Optimization

A prevalent pattern in the "injection"-scenario was that all workers would show a close-up view of the imperfections on the small manufactured plastic pieces to the helpers by slowly rotating the pieces near the tablet camera, which is shown in Figure 11.12a. This observation implies that high-resolution close-up views are needed for remote product quality control during on-site manufacturing, which was further corroborated by a worker participant's feedback in the post-study questionnaire, S4: *"if the personnel on site does not have the required know-how to do quality control, you could do it remotely, but it must be possible (for the helper) to see small details."* This findings raise the questions: *What amount of detail must the helper be able to see to do remote quality control in a real-world setting and what camera setup is best suited for remote quality control?* For instance, the use of depth cameras for creating 3D geometric reconstructions has become popular in remote assistance systems from recent research [4, 5, 37, 38], but it remains to be seen whether they are useful in a manufacturing setting for capturing intricate details of small objects and small irregular motions of machine components – the devil is sometimes in the detail.

Figure 11.12 Tailoring patterns. (a) Workers show helpers a close-up view of imperfections in the manufactured plastic pieces by rotating the pieces close to the tablet camera. (b) Workers show close-up view of their interactions on the HMI. (c) Workers show close-up view of machine internals with the mold part. Workers alternated between (b) and (c), and the helper's workspace awareness suffered, whenever the worker forgot to bring the camera with him.

11.8   The Manufacturing Context Poses a Challenge for Creating 3D Reconstructions with Depth Cameras

As mentioned above, creating a static 3D reconstruction of a task space with one depth camera by panning over it [4, 5, 38, 39] or creating a live 3D reconstruction with multiple fixed-depth cameras [37, 40] has become popular in recent remote assistance research because it allows the helper to navigate the worker's space and give guidance in 3D. Yet, to the best of our knowledge, no remote assistance research has explored the challenges caused by the metallic material properties of industrial task objects, which is an obstacle to unleashing the full potential of depth sensors and reconstructions for remote assistance in manufacturing.

The interviews with manufacturing employees showed that remote assistance is needed on various machines, such as CNC machines and injection molding machines at LEGO, industrial robot arms and wind turbine engines at Vestas, and inline printing units at TRESU. Observations at the companies confirm that these machines consist of metallic components, some of which have specular, mirror-like surfaces, for example a mold inside an injection molding machine or the drill bits inside a CNC machine. However, it is challenging to use an optical depth camera to accurately measure the depth of such metal components with specular surfaces. This observation is supported by related work [41], where industrial objects with metallic specular surfaces were inaccurately reconstructed with a time-of-flight depth sensor, because the light emitted by the sensor was multiply reflected in the surfaces. Polished mirror-like metallic surfaces are even more problematic to measure accurately with an optical depth sensor without additional equipment [42]. Inaccurate depth measurements in turn lead to an inaccurate 3D reconstruction, which is problematic for multiple reasons. First, it makes it difficult for the helper to visually inspect imperfections in manufactured metal objects or misalignment of assembled components. Secondly, it makes it difficult for the helper to accurately annotate the reconstruction. This challenge illustrates why it is worthwhile to conduct research on alternative approaches of achieving view independency that do not make use of depth cameras, for example further research on multi-camera remote assistance systems that use only RGB video streaming or new techniques for creating 3D reconstructions from 2D images [43].

11.9   Multiple Cameras Support Workspace Awareness in Large Industrial Task Spaces

The two most important areas on the injection molding machine during troubleshooting were the human machine interface (HMI) and the machine internals

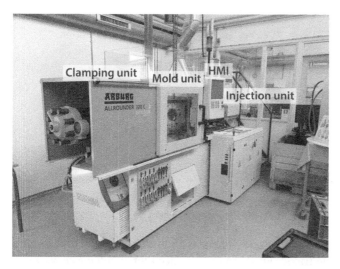

Figure 11.13 Areas on injection molding machine (without additional equipment). The HMI and mold unit are of interest during remote product quality optimization, whereas the additional clamping and injection units are of interest during remote repair.

including the mold, which became clear from the observation that 5/6 workers continuously moved a scene camera, a tablet camera on a tripod, to provide the helper with a view of one of the two areas. See Figure 11.12b,c for pictures of this use pattern and Figure 11.13 for an overview of the important parts of an injection molding machine.

The view of the HMI displayed information about adjustable production parameters, while the view of the machine internals showed the movements of the clamping unit and mold, which were controlled by these parameters. Thus, workers would move the camera back and forth between areas to show cause and effect of HMI user interactions. This use pattern of moving the scene camera from one area to another rarely occurred in the other scenarios, likely due to the smaller workspaces that could be covered in adequate detail from one view. Interestingly, it was observed that helpers lost awareness, when workers forgot to bring the scene camera as they moved to another area on the machine, of which an example is shown in Figure 11.14. A worker changed his attention from the HMI to the mold inside the machine but did not bring the camera with him, i.e. did not adjust the helper's viewpoint, and therefore, the helper expressed doubt about the worker's location. This observation shows that the helper may benefit from independent access to views of multiple areas, in particular views of the HMI and machine internals, for instance, through the use of multiple cameras each providing a close-up view of an area. Another possible RAK multi-camera configuration that could have prevented the helper from loosing awareness of the

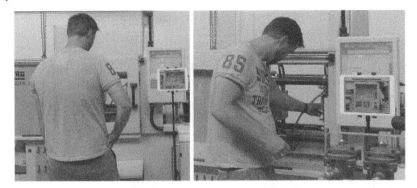

Figure 11.14 Worker forgets camera as he moves his attention from one area on the machine to another and leaves the helper wondering, where he is.

worker's location was to mount one camera as an overview camera and another camera on a movable tripod. Furthermore, the need for *simultaneous* views of the HMI and mold was brought up by the teachers at the workshop and by the helper, who was paired up with the worker in Figure 11.14. This helper stated the following when asked about missing features in RAK, S9: *"In the injection scenario, it would have been nice with a view of both the mold (inside the machine) and the HMI at the same time."* Obviously, simultaneous views of different areas on an injection molding machine can only be achieved with multiple cameras. A general pattern was also observed that large machines, be it injection molding machines, inline printing units, or specialized washing machines, grant the worker access to an HMI with which a worker can control and affect other areas on the machines, including movements of machine parts and the appearance of manufactured products. Thus, it could be generalized that multiple cameras are considered useful for remote assistance on injection molding machines to include other large industrial machines as well.

Additionally, during the interview with manufacturing employees the idea of using recordings from multiple scene cameras to aid with remote assistance was brought up by E1: *"It would be interesting with constant surveillance of the machines [...] A bunch of cameras in the production that one could log into and use to go back in time, because many of the errors are related to persons making mistakes [...] I know who used the machine last (because of digital logging of id) [...] but they used the machine in a different way (than intended) and we don't know why a certain error occurred and how the situation was (at the time the error occurred) [...] or when the error occurred, because they left the machine to itself."* The idea of the helper using video recordings of the worker's space for remote assistance is curiously absent from related research, although in the system by Speicher et al. [11] users could rewind a live stream by 10 seconds. This is interesting, because typically the research focus has been on supporting the helper in navigating the worker's space dimension by means of view independency [4, 16, 17], but not the time dimension. Thus, a potentially interesting direction for future research is the topic of time independence, i.e. a helper remotely accessing a workspace at a point in time independently of the worker's presence, which enables functionality such as annotating video recordings during synchronous remote assistance and asynchronous remote assistance with annotated video messages. However, concerns about surveillance of employees need to be taken seriously.

In summary, multiple cameras for remote assistance can offer the helper view and time independence on large industrial machines, compared to using one movable scene camera or head-mounted camera. However, instilling workspace awareness between the worker and helper during multi-camera remote assistance in large task spaces is challenging for several reasons. It is important that these challenges are addressed by using techniques for improving workspace awareness, which we will discuss in Chapter 12.

11.10   Concluding Remarks

In this chapter, we introduced a component-based tailorable remote assistance system called RAK. The design and development of RAK were informed by the results and findings of an interview study with employees of a manufacturing industry. Then, an experimental simulation with RAK that was conducted at a technical college for plastic manufacturing was briefly described. A large part of the chapter was devoted to our discussion and reflection on the results and observations of the user studies. It is encouraging that we are able to derive some meaningful and unexpected new insights, which could guide the directions of future work. These include the tailoring behaviors of both workers and helpers, sharing machine sound from the workspace to the helper and supporting workspace awareness with multi-camera setups. It is also important to note the limitations of our experimental simulation, which we outline below.

Helpers did not face the same level of changing requirements to remote assistance as the workers because they did not change location between scenarios. Helpers were in the same office space throughout the user study, which means they did not face nomadic use contexts where mobility of the interface was truly required. All of the hardware, devices, and mounts, were available to them at all times, which is not the case in nomadic use contexts. This is a limitation of the study design, as it did not enable helpers to reflect as much on the requirements to remote assistance in the different scenarios and the pros and cons of various configurations as the workers.

The observed pattern of frequently moving a scene camera back and forth between multiple areas in the "injection"-scenario required conscious effort, camera work and time by the workers that could have been avoided had they instead used a head-mounted/body-worn camera or two scene cameras, where one camera captured the HMI and the other camera captured the machine internals. Both configurations were technically supported by RAK but would have required more tailoring work prior to beginning the remote assistance session. Specifically, a shoulder-worn camera configuration required workers to mount a webcam on their own shoulder with a gorillapod and connect it with an extension cable to a PC to ensure mobility, and the configuration with two cameras required workers to connect two webcams to a PC with extension cables and mount them in the environment while ensuring a correct view of the areas.

There seems to be a trade-off between investing time on tailoring work before the beginning of a remote assistance session and coping with the deficiencies of the chosen configuration during remote assistance, which begs the question: Should a worker invest time on setting up a suitable configuration before a remote assistance session or go with a less ideal configuration and incur the cost of its deficiencies throughout the session? A third and probably better option is to change configuration during remote assistance in case the current configuration is unsatisfactory.

The workers could have changed configuration during remote assistance in case the current configuration was unsatisfactory – something that contrary to expectations was observed only once during the experimental simulation. This highlights a limitation of our method and the experimental simulation: participants were so focused on solving the problem scenarios because it was important to them and they had domain knowledge about the problems, that they chose one configuration before a scenario and neglected to explore the configurations of RAK during remote assistance, even though they were encouraged to do so. We could instead have opted for a method, where participants were forced to rethink their configuration halfway into a timed scenario, thus enabling them to compare and reflect on how well the chosen configurations supported the scenario contrary to their expectations. It is possible that this method could have led to a better learning experience for the participants in regards to understanding what constitutes a good solution in a given scenario and in turn could have generated richer data for the experiment.

References

1 Kaplan, S.M. and Carroll, A.M. (1992). Supporting collaborative processes with Conversation Builder. *Computer Communications* 15 (8): 489–501.
2 Stiemerling, O. and Cremers, A.B. (1998). Tailorable component architectures for CSCW-systems. In: *Proceedings of the 6th Euromicro Workshop on Parallel and Distributed Processing - PDP '98*, 302–308.

3 Lee, G.A., Teo, T., Kim, S., and Billinghurst, M. (2017). Mixed reality collaboration through sharing a live panorama. In: *SIGGRAPH Asia 2017 Mobile Graphics & Interactive Applications*, 1–4.

4 Teo, T., Lawrence, L., Lee, G.A. et al. (2019). Mixed reality remote collaboration combining 360 video and 3D reconstruction. In: *Proceedings of the 2019 CHI Conference on Human Factors in Computing Systems*, CHI '19, 1–14. Glasgow, Scotland, UK: Association for Computing Machinery.

5 Piumsomboon, T., Dey, A., Ens, B. et al. (2019). The effects of sharing awareness cues in collaborative mixed reality. *Frontiers in Robotics and AI* 6: 5

6 Piumsomboon, T., Lee, G.A., Irlitti, A. et al. (2019). On the shoulder of the giant: a multi-scale mixed reality collaboration with 360 video sharing and tangible interaction. In: *Proceedings of the 2019 CHI Conference on Human Factors in Computing Systems*, CHI '19, 1–17. Glasgow, Scotland, UK: Association for Computing Machinery.

7 Yang, J., Sasikumar, P., Bai, H. et al. (2020). The effects of spatial auditory and visual cues on mixed reality remote collaboration. *Journal on Multimodal User Interfaces* 14: 337–352.

8 Lee, G.A., Teo, T., Kim, S., and Billinghurst, M. (2018). A user study on MR remote collaboration using live 360 video. In: *2018 IEEE International Symposium on Mixed and Augmented Reality (ISMAR)*, 153–164. ISSN: 1554-7868.

9 Aschenbrenner, D., Sittner, F., Fritscher, M. et al. (2016). Cooperative remote repair task in an active production line for industrial internet telemaintenance. *IFAC-PapersOnLine* 49 (30): 18–23.

10 Fritscher, M., Sittner, F., Aschenbrenner, D. et al. (2016). The adaptive management and security system for maintenance and teleoperation of industrial robots. *IFAC-PapersOnLine* 49 (30): 6–11.

11 Speicher, M., Cao, J., Yu, A. et al. (2018). 360anywhere: Mobile Ad-hoc collaboration in any environment using 360 video and augmented reality. *Proceedings of the ACM on Human-Computer Interaction* 2 (EICS): 1–20.

12 Fussell, S.R., Setlock, L.D., Yang, J. et al. (2004). Gestures over video streams to support remote collaboration on physical tasks. *Human–Computer Interaction* 19 (3): 273–309.

13 Fakourfar, O., Ta, K., Tang, R. et al. (2016). Stabilized annotations for mobile remote assistance. In: *Proceedings of the 2016 CHI Conference on Human Factors in Computing Systems*, 1548–1560. ACM.

14 Huang, W. and Alem, L. (2013). HandsinAir: A wearable system for remote collaboration on physical tasks. In: *Proceedings of the 2013 Conference on Computer Supported Cooperative Work Companion*, CSCW '13, 153–156. San Antonio, Texas, USA: Association for Computing Machinery.

15 Fussell, S.R., Setlock, L.D., and Kraut, R.E. (2003). Effects of head-mounted and scene-oriented video systems on remote collaboration on physical tasks. In: *Proceedings of the SIGCHI Conference on Human Factors in Computing Systems*, 513–520. ACM.

16 Lanir, J., Stone, R., Cohen, B., and Gurevich, P. (2013). Ownership and control of point of view in remote assistance. In: *Proceedings of the SIGCHI Conference on Human Factors in Computing Systems*, 2243–2252. ACM.

17 Tait, M. and Billinghurst, M. (2015). The effect of view independence in a collaborative AR system. *Computer Supported Cooperative Work (CSCW)* 24: 563–589.

18 Kim, S., Billinghurst, M., and Lee, G. (2018). The effect of collaboration styles and view independence on video-mediated remote collaboration. *Computer Supported Cooperative Work (CSCW)* 27 (3–6): 569–607.

19 Adcock, M., Anderson, S., and Thomas, B. (2013). RemoteFusion: Real time depth camera fusion for remote collaboration on physical tasks. In: *Proceedings of the 12th ACM SIGGRAPH International Conference on Virtual-Reality Continuum and Its Applications in Industry - VRCAI '13*, 235–242. Hong Kong, Hong Kong: ACM Press.

20 Tang, A., Owen, C., Biocca, F., and Mou, W. (2003). Comparative effectiveness of augmented reality in object assembly. In: *Proceedings of the SIGCHI Conference on Human Factors in Computing Systems*, CHI '03, 73–80. New York, NY, USA: ACM.

21 Robertson, C.M., MacIntyre, B., and Walker, B.N. (2008). An evaluation of graphical context when the graphics are outside of the task area. In: *2008 7th IEEE/ACM International Symposium on Mixed and Augmented Reality*, 73–76.

22 Henderson, S.J. and Feiner, S. (2009). Evaluating the benefits of augmented reality for task localization in maintenance of an armored personnel carrier turret. In: *2009 8th IEEE International Symposium on Mixed and Augmented Reality*, 135–144.

23 Aschenbrenner, D., Rojkov, M., Leutert, F. et al. (2018). Comparing different augmented reality support applications for cooperative repair of an industrial robot. In: *2018 IEEE International Symposium on Mixed and Augmented Reality Adjunct (ISMAR-Adjunct)*, 69–74.

24 Rasmussen, T.A. and Gronbak, K. (2019). Tailorable remote assistance with RemoteAssistKit: a study of and design response to remote assistance in the manufacturing industry. In: *International Conference on Collaboration and Technology*, 80–95. Springer.

25 Franke, N. and von Hippel, E. (2003). Satisfying heterogeneous user needs via innovation toolkits: the case of Apache security software. *Research Policy* 32 (7): 1199–1215.

26 Kim, S., Lee, G.A., and Sakata, N. (2013). Comparing pointing and drawing for remote collaboration. In: *2013 IEEE International Symposium on Mixed and Augmented Reality (ISMAR)*, 1–6.

27 Li, J., Wessels, A., Alem, L., and Stitzlein, C. (2007). Exploring interface with representation of gesture for remote collaboration. In: *Proceedings of the 19th Australasian Conference on Computer-Human Interaction: Entertaining User Interfaces*, 179–182. ACM.

28 Kirk, D. and Fraser, D.S. (2006). Comparing remote gesture technologies for supporting collaborative physical tasks. In: *Proceedings of the SIGCHI Conference on Human Factors in Computing Systems*, 1191–1200. ACM.

29 Gauglitz, S., Nuernberger, B., Turk, M., and Höllerer, T. (2014). World-stabilized annotations and virtual scene navigation for remote collaboration. In: *Proceedings of the 27th Annual ACM Symposium on User Interface Software and Technology - UIST '14*, 449–459. Honolulu, Hawaii, USA: ACM Press.

30 Huang, W., Kim, S., Billinghurst, M., and Alem, L. (2018). Sharing hand gesture and sketch cues in remote collaboration. *Journal of Visual Communication and Image Representation* 58: 428–438.

31 Teo, T., Lee, G.A., Billinghurst, M., and Adcock, M. (2018). Hand gestures and visual annotation in live 360 panorama-based mixed reality remote collaboration. In: *Proceedings of the 30th Australian Conference on Computer-Human Interaction*, OzCHI '18, 406–410. Melbourne, Australia: Association for Computing Machinery.

32 Herrmann, T. (1995). Workflow management systems: ensuring organizational flexibility by possibilities of adaptation and negotiation. In: *Proceedings of Conference on Organizational Computing Systems*, COCS '95, 83–94. New York, NY, USA: Association for Computing Machinery.

33 Clark, H.H. and Brennan, S.E. (1991). Grounding in communication. In: *Perspectives on Socially Shared Cognition* (ed. L.B. Resnick, J.M. Levine, and S.D. Teasle), 127–149. Washington, DC: American Psychological Association.

34 Fussell, S.R., Kraut, R.E., and Siegel, J. (2000). Coordination of communication: effects of shared visual context on collaborative work. In: *Proceedings of the 2000 ACM Conference on Computer Supported Cooperative Work*, 21–30. ACM.

35 Kuzuoka, H., Oyama, S., Yamazaki, K. et al. (2000). GestureMan: A mobile robot that embodies a remote instructor's actions. In: *Proceedings of the 2000 ACM Conference on Computer Supported Cooperative Work - CSCW '00*, 155–162. Philadelphia, PA, US: ACM Press.

36 Ranjan, A., Birnholtz, J.P., and Balakrishnan, R. (2006). An exploratory analysis of partner action and camera control in a video-mediated collaborative task.

In: *Proceedings of the 2006 20th Anniversary Conference on Computer Supported Cooperative Work*, 403–412. ACM.

37 Joachimczak, M., Liu, J., and Ando, H. (2017). Real-time mixed-reality telepresence via 3D reconstruction with HoloLens and commodity depth sensors. In: *Proceedings of the 19th ACM International Conference on Multimodal Interaction*, 514–515.

38 Piumsomboon, T., Day, A., Ens, B. et al. (2017). Exploring enhancements for remote mixed reality collaboration. In: *SA '17: SIGGRAPH Asia 2017 Mobile Graphics & Interactive Applications*, 1–5.

39 Gao, L., Bai, H., Lindeman, R., and Billinghurst, M. (2017). Static local environment capturing and sharing for MR remote collaboration. In: *SIGGRAPH Asia 2017 Mobile Graphics & Interactive Applications*, 17:1–17:6. ACM.

40 Orts-Escolano, S., Rhemann, C., Fanello, S. et al. (2016). Holoportation: virtual 3D teleportation in real-time. In: *Proceedings of the 29th Annual Symposium on User Interface Software and Technology*, 741–754.

41 Kahn, S., Wuest, H., and Fellner, W.-D. (2010). Time-of-flight based scene reconstruction with a mesh processing tool for model based camera tracking. In: *VISIGRAPP 2010. Proceedings; International Conference on Computer Graphics Theory and Applications (GRAPP)*, 302–309. INSTICC Press.

42 Whelan, T., Goesele, M., Lovegrove, S.J. et al. (2018). Reconstructing scenes with mirror and glass surfaces. *ACM Transactions on Graphics* 37 (4): 1–11.

43 Gao, Y.-K., Gao, Y., He, H. et al. (2022). NERF: Neural radiance field in 3D vision, a comprehensive review.

12

Supporting Workspace Awareness with Augmented Reality-Based Multi-camera Visualization and Tracking

12.1   Introduction

As we discussed in Chapter 11, it is important to support ways in which helpers and workers can maintain awareness of each other in large workspaces for remote guidance. That is because both helpers and workers need to keep track of where their collaborator is and what activities he/she is carrying out to collaborate well. One way to support this awareness is to use multiple cameras mounted in the environment, since such a camera setup gives the helper view and time independence compared to using a head-mounted camera.

Multi-camera remote assistance is here defined as the use of two or more cameras that are mounted in the environment to capture different areas and/or perspectives of a workspace. A multi-camera system can consist of traditional RGB cameras or 3D cameras with depth-capturing capabilities (RGB-D). In this chapter, we discuss the advantages of multi-camera remote assistance systems over more traditional systems that use one camera from the worker's perspective. Further, we discuss some challenges for multi-camera remote assistance and demonstrate Augmented Reality research systems that address these challenges.

Imagine a basic multi-camera system, where the helper can access multiple live video feeds from cameras in the worker's environment, similar to the one described in [1]. Basic audio communication between the pair is supported. Even a simple system like this has some benefits over using just one camera from the worker's perspective:

1.   First, in a large workspace, multiple cameras enable the remote helper to navigate the space independently of the worker, i.e. they enable the helper to retrieve information about the state of a work area independently of the worker's location and focus of attention. In previous research, this view independence has been shown to be beneficial for remote assistance [2–5].

Computer-Supported Collaboration: Theory and Practice, First Edition.
Weidong Huang, Mark Billinghurst, Leila Alem, Chun Xiao, and Troels Rasmussen.
© 2024 The Institute of Electrical and Electronics Engineers, Inc. Published 2024 by John Wiley & Sons, Inc.

The benefits of view independence are particularly prevalent for large workspaces that cannot be covered using one camera from the worker's perspective.

2. Second, a view from a scene camera is stable while a view from a head-mounted camera is not. Research has shown that it is challenging for the helper to give guidance in a moving camera feed from a head-mounted camera [6].

3. By using a multi-camera configuration, the helper is able to achieve both detailed views of the worker's object manipulations in a work area and an overview of the relationship between multiple work areas. These overview+ detail views have been shown to be beneficial for workspace awareness [7].

4. Furthermore, multiple cameras also allow the helper to immediately view the consequences of the worker's actions in one area on another area. For example, it allows the helper to view the consequences of the worker's interactions with a human–machine interface (HMI) on the movements of the machine internals. On large machines, this can be potentially time saving because it reduces the need for the worker to walk between the HMI and machine internals.

5. Additionally, permanently installed scene cameras can make recordings, allowing workers and helpers to rewind and identify human errors that caused a particular problem. Thus, multiple cameras not only grant the helper view independence, but they also have the potential to grant *time independence*.

6. Finally, one or more permanently mounted scene cameras can be used for asynchronous communication and information sharing [8]. Thus, the helper can access a scene camera, even when the worker is not present possibly due to different time zones, and record a session where she shares information and makes annotations in the workspace, which can later be replayed by the worker.

Despite the above benefits, it is challenging to establish workspace awareness when using a basic multi-camera system, especially in large workspaces where the cameras are distributed across multiple areas. These awareness challenges are illustrated in Figure 12.1 and described below:

1. The helper lacks an understanding of the spatial relationship between camera views and the worker [1], since the views are presented to her in a disjointed manner. These spatial discontinuities can make it difficult for the helper to determine where the worker is in relation to other work areas and therefore complicates the use of natural spatial deictic references (e.g. "Look behind you"). It can also be difficult to predict where the worker is about to go.

2. With too many cameras to choose from, the helper may spend too much time looking at the "wrong" views, i.e. those that are not helpful for collaboration [9].

3. Due to spatial discontinuities, it is difficult for the helper to infer, which task object or location the worker points at, when the worker's distal pointing gesture is visible in one scene camera view, but the task object/location is visible

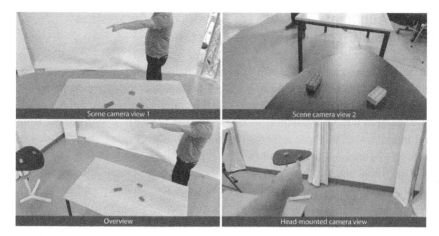

Figure 12.1 Possible views of the workspace. Scene camera view 1 and 2 illustrates the problem with spatial discontinuities in a multi-camera system when performing a distal pointing gesture. In scene camera view 1, the helper points to a distant task object in a work area, but the task object is only visible in scene camera view 2. An overview is provided to show the actual spatial relationship between work areas. A head-mounted camera view is provided to demonstrate one of its strengths compared to the other views: it is clear to which object the worker points.

in another scene camera view, as illustrated in Figure 12.1. Thus, referencing objects in a multi-camera setup lacks the ease with which two colocated persons can reference objects by simply pointing or gazing in the direction of an object, or the ease with which a worker can point to a task object in a head-mounted camera view [10].

4. Additionally, it is difficult for the worker to know which camera the helper views, and where her focus of attention is. This in return makes it difficult for the worker to know whether the helper can see his gestures and actions without verbal confirmation. Even if the worker knows which camera the helper views, he may not know the field of view of this camera, which can cause him to reference objects outside the helper's field of view. This is also a problem when using only one camera [11], but using multiple cameras exacerbates the problem [1].

5. Using multiple scene cameras requires conscious time-consuming camera work of the worker, because he has to place the cameras and make sure they capture areas of interest for collaboration, unless the cameras are permanently mounted in the environment.

These are the challenges that stand in the way of high workspace awareness during remote assistance with multiple cameras in large spaces. However these can be addressed using Augmented Reality (AR) for multi-camera remote assistance.

12.2 Augmented Reality for Supporting Awareness During Multi-camera Remote Assistance

AR technologies can be used to improve the helper's awareness of the worker and vice versa in large workspaces by alleviating some of the above challenges of the basic multi-camera system. Figure 12.2 illustrates a typical setup of AR-based multi-camera remote assistance. In this setup, the helper is able to choose from and interact with multiple cameras mounted in the workspace of the worker to perform gestures and draw sketches.

In the following, we give examples of this through the presentation of two AR research prototype systems, SceneCam and CueCam, which embodies the concepts of interactive focus-in-context views, automatic view selection, and awareness cues.

First, we describe the core AR functionality shared by SceneCam and CueCam. Secondly, we cover the design and implementation of SceneCam, a multi-camera AR system with techniques for making the helper aware of the worker in a large workspace. To this end, interactive focus-in-context views and automatic view selection techniques are introduced as means of making the helper aware of the worker's location and actions. Then, we discuss the design, implementation and user evaluation of CueCam, a multi-camera AR system that uses three AR awareness cues to make the worker aware of the helper's location and actions in a large workspace.

Figure 12.3 shows how workspace awareness information is communicated between the worker and helper. SceneCam is designed to communicate workspace awareness information about the worker to the helper, while CueCam is designed to communicate workspace awareness information about the helper to the worker.

Figure 12.2 Illustration of AR multi-camera remote assistance. (a) Helper's drawing on how to manipulate task object as seen from the worker's perspective in AR. (b) Third-person perspective of worker manipulating task object. (c) Helper's screen-based interface on PC with which he has access to multiple cameras and can draw in the selected camera view. The drawings are interpreted in 3D and shown to the worker in (a).

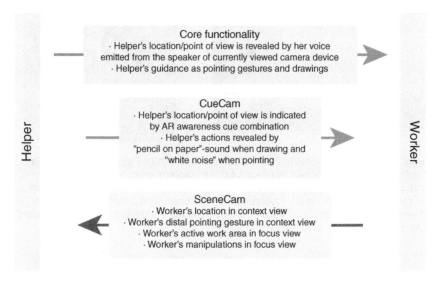

Figure 12.3 How SceneCam and CueCam are responsible for the workspace awareness information communicated between the helper and worker.

12.2.1   Core Functionality of SceneCam and CueCam

At their core, SceneCam and CueCam both support remote assistance through a Web App for the helper and an AR App for the worker. In the following, each application is described in more detail.

The AR App is developed in Unity3D[1] and runs on a head-mounted display, a Microsoft Hololens (version 1).[2] Using the Hololens, the worker can see the projection of the helper's 2D cursor and drawings into the worker's environment. These annotations are projected from the point of view of multiple scene cameras, as shown in Figure 12.4. Furthermore, the AR App supports standard voice communication with the helper.

Since SceneCam and CueCam are multi-camera systems, the worker must place and calibrate multiple camera devices (scene cameras) throughout the workspace to capture relevant parts of the environment, as shown in Figure 12.5. The calibration step involves aligning virtual cameras to the physical counterparts by tracking markers on them, which is a prerequisite for the projection of the helper's annotations. The camera devices include smartphones, tablets, or webcams connected to a PC. Each of the camera devices runs a Camera App, which is a web application developed in JavaScript. This Camera App has two purposes: (1) it sends live video from the worker to the helper and audio both ways; (2) it displays an AR marker

1 https://unity.com.
2 https://docs.microsoft.com/en-us/hololens/hololens1-hardware.

Figure 12.4 Core functionality of SceneCam and CueCam: (a) Helper's 2D web interface on a PC, and (b–e) worker's AR interface from the Hololens' point of view. (b–d) The wireframe models of the virtual cameras are aligned to the physical cameras, which in this case are a smartphone (b), a tablet (c), and a webcam with a marker (d). The helper's 2D sketches on live video from a scene camera in (a) are interpreted in 3D and visualized in AR on the work area, a whiteboard, in (e).

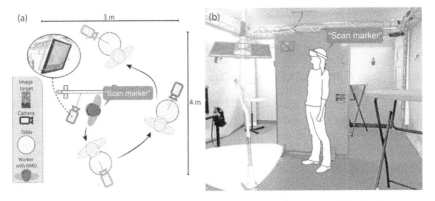

Figure 12.5 (a) Top-down view of camera calibration process, where a worker walks from scene camera to scene camera and scans a marker on each camera by issuing the voice command "Scan marker." (b) Worker scans marker on a scene camera to align virtual model and physical camera. Source: Troels Rasmussen et al. 2022/Elsevier/CC BY 4.0.

on camera devices that include displays (e.g. smartphones and tablets), which are used for calibration.

On the helper's side, the Web App which is developed in JavaScript, runs in the internet browser on a PC or tablet. It receives live video and audio from the worker's camera devices and presents these in the helper's 2D interface. Thereby, the Web App offers the helper access to different views of the worker's space through each of the camera video feeds, which are presented in thumbnail previews on the right side of the interface (see Figure 12.4a). When the helper clicks on a thumbnail, the selected video is shown in the large central video window, the primary view, on which the helper can make 2D pointing gestures and drawings that are interpreted in 3D and shown to the worker in AR. Additionally, the helper's audio is transmitted to the selected camera device in the worker's space.

12.2.2 Supporting the Helper's Awareness with SceneCam

SceneCam supports the helper's awareness of the worker by implementing focus-in-context views and automatic view selection. These awareness techniques addresses awareness **challenge 1**, i.e. serves to improve the helper's spatial relationship between the multiple camera views and the worker, and **challenge 2**, i.e. reduced the time spent on the wrong views.

12.2.2.1 Ad Hoc Creation of (Scene Camera, Work Area)-Pairs

A prerequisite for enabling focus-in-context views and automatic view selection is the ad hoc creation of (scene camera, work area)-pairs.

The worker tracks the pose and id of a marker on each camera using the Hololens. This links a camera to the physical work area it is pointing at via a ray cast from the camera onto the area, hence creating a (scene camera, work area)-pair as shown in Figure 12.6. A work area is defined geometrically as a tuple

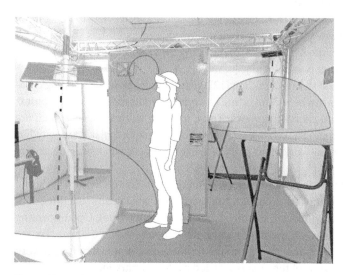

Figure 12.6 Camera-work area pairs, which are defined geometrically as the tuple (3D point, surface normal, and semi-sphere) at the intersection between a ray emitted from the focal point of the camera along its z-axis and the 3D reconstructed mesh of the work area captured by the camera. Upon the worker entering a semi-sphere, the camera of said semi-sphere is selected as the primary view in the helper's 2D interface. Source: Troels Rasmussen et al. 2022/Elsevier/CC BY 4.0.

(3D point, surface normal, semi-sphere) at the intersection between the ray and the mesh of the area. The radius of the semi-sphere is directly proportional to the distance between scene camera and area. Thus, when a camera and work area are close, the semi-sphere is small, and when a camera and area are further apart, the semi-sphere is larger, which accommodates that the camera covers a larger part of an area the further it is from it. This semi-sphere representing the extend of a work area is used later for detecting the presence of a worker inside an area during automatic view selection.

12.2.2.2 Focus-in-Context Views

Augmented Reality technologies, such as the inside-out tracking of a Hololens, can be used to map the environment and the wearer's pose. This can be utilized in a multi-camera setting to create interactive focus-in-context views presented to the remote helper. An interactive focus-in-context view is an overview (the context) of a workspace augmented with detailed camera views (focus views) of multiple work areas at their locations in the overview. The context view serves to make the spatial relationship between cameras, work areas and the worker explicit, thus improving the helper's ability to orient herself and use spatial deictic references and in turn addressing awareness **challenge 1**. The focus views are needed to see details of the worker's object manipulations. The focus-in-context view is interactive, because the helper can interact with the focus views, i.e. select a focus view to make it the primary view, from inside the context view. Previous research has shown that a combination of a detailed view and overview improves remote assistance over just having access to either one of them [7].

The multi-camera system, SceneCam [12], demonstrates how AR tracking capabilities of a Hololens can be used to create two types of interactive focus-in-context views: egocentric and exocentric views. These focus-in-context views, which are presented to the helper in a web interface on a PC screen or a tablet, are created by visualizing the (camera-work area)-pairs in the helper's context view as interactive focus views.

The exocentric focus-in-context view (see Figure 12.7) shows a top-down view of the worker (the dark gray head icon) and (camera, work area)-pairs (the three dark gray rectangles each connected to a light gray circle with a line). Upon hovering over a work area icon with the mouse cursor, the focus view of the area, i.e. the view from the paired camera, is shown in a thumbnail. Upon clicking on a work area icon, the focus view of the area is selected as the primary view.

Similarly, the egocentric focus-in-context view (see Figure 12.8) shows a live video feed of the entire workspace from an overview camera. The video feed is augmented with circles at the location of work areas. Upon hovering over a circle with the mouse cursor, the focus view of the area is shown in a thumbnail. Upon clicking on a circle, the focus view of the area is selected as the primary view.

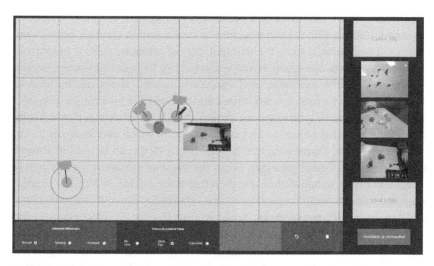

Figure 12.7 Screenshot of helper's screen-based interface. (Left) Exocentric focus-in-context view with a top-down view of virtual icons showing the live movements of the worker (the dark gray head icon) and (camera, work area)-pairs (the three dark gray rectangles connected to the light gray circles with a line). Upon hovering over a work area icon with the mouse cursor, the focus view of the area is shown in a thumbnail. Upon clicking on a work area icon, the focus view of the area is selected as the primary view.

Figure 12.8 Screenshot of helper's screen-based interface. Egocentric focus-in-context view shows live video feed from an overview camera. The video feed is augmented with circles at the location of work areas. Upon hovering over a circle with the mouse cursor, the focus view of the area is shown in a thumbnail. Upon clicking on a circle, the focus view of the area is selected as the primary view.

We discuss how the focus-in-context views contribute to three sources of awareness information: (1) consequential communication through body movements, (2) feedthrough from the manipulation of artifacts, and (3) intentional communication via conversation and gestures [13]. In regards to consequential communication, the exocentric and egocentric context views can be used to observe the worker's path in a workspace and thus predict his next location, and the focus views can be used to see the worker's hands, which may indicate that he is searching for a task object, or his facial expression, which may show confusion or comprehension. Concerning feedthrough, focus views can be used to observe the worker's object manipulations. Since the exocentric context view is purely virtual and object manipulations are therefore not modeled, it does not support communication of feedthrough information. Finally, intentional communication is supported by all types of views. In the exocentric view, the helper can see the worker point with his head to distant work areas combined with deictic pronouns (e.g. "this area over here"). Similarly, in the egocentric view, the helper can observe the worker point with his head or his outstretched hand and index finger to distant areas; however, the helper may have difficulties interpreting distal pointing gestures correctly [10]. In the focus views, the helper can observe the worker's proximal pointing to task objects and iconic hand gestures representing shapes or distances.

Existing research on focus-in-context views for remote collaboration in face-to-face office meetings include [14–16]. In comparison, the focus-in-context views of SceneCam [12] are for remote assistance on physical tasks in large room-sized workspaces, and thus, the focus views are oriented toward work areas with physical task objects instead of towards meeting participants. Furthermore, in comparison to this related work, where focus views are captured from the same angle as the context view due to the way cameras are configured, the implementation of focus views in SceneCam supports capturing multiple work areas that may be fully or partially occluded from each other by making one scene camera responsible for each focus view. In principle, this makes it possible to create an exocentric focus-in-context view of multiple rooms separated by wall dividers or shelves, which is not possible with the related systems.

In a recent AR system by Teo et al. [17], focus-in-context views are used for remote assistance in large workspaces. A static 3D reconstruction of a room-sized workspace is combined with a 360° live view from the worker's perspective. Thus, the helper can transition between the 3D reconstruction, an egocentric context view, and the 360° live view, a focus view. Their research supports that users prefer to have access to both views during collaboration instead of just one of the two view types because they complement each other. The 3D reconstruction offers view independence whereby the remote helper can focus on her task at her own pace, while the 360° live view better supports her awareness of the worker's focus.

Besides being based on 360° and 3D camera technology, the main differences between the work by Teo et al. and SceneCam is that while they give the helper access to only one focus view from the worker's perspective, SceneCam gives the helper access to multiple focus views independent of the worker's location and perspective, which can be helpful for establishing situation awareness.

12.2.2.3 Automatic View Selection

In addition to the focus-in-context views, SceneCam utilizes the (camera, work area)-pairs to enable automatic view selection based on the proximity of the worker to a work area. The idea behind automatic camera selection is to always ensure that the helper has the optimal focus view of the worker's actions and does not miss important object manipulations or gestures. Thus, it is designed to address awareness **challenge 2** by making sure that the helper does not spend time on the wrong camera views. Previous research has shown that automatic view selection improves remote assistance by reducing the time spent by the helper acquiring the right views of a work space [18].

Specifically, automatic view selection works by setting the primary view in the helper's web interface to the view of the area currently visited by the worker. The worker is inside an area, and thus visiting said area, whenever he is within a specific radius determined by the semi-sphere of a (camera, work area)-pair. Figure 12.6 illustrates the semi-spheres of the camera-work area pairs.

There is only a handful of research papers on automatic, event-controlled view selection for remote assistance, where the helper's view of a workspace is selected based on events in the worker's space [7, 14, 18–20]. Some of the research uses computer vision to make a pan-zoom-tilt camera follow the hands of the worker from one area to another in a table-sized workspace [7, 18], while other work uses machine learning to predict the helper's optimal view based on task properties and actions of the worker [20]. In comparison, SceneCam contributes an automatic view selection technique, where the information used for inferring the optimal camera view is the proximity of the worker (or rather his AR-HMD) to a work area in a room-sized workspace. Thus, our work is most closely related to that of [7, 18], the main difference being that they use one camera to follow a worker's moving hands between areas on a table, whereas we use multiple cameras to follow the worker's head in a large space that requires the worker to walk between work areas. This is an important distinction because we thereby extend their research to work in a large multi-camera setting.

12.2.3 Supporting the Workers Awareness with CueCam

CueCam contributes three augmented reality awareness cues, Virtual Hand, Spatial Sound, and Color Cue, for making the worker aware of the helper's

Figure 12.9 AR awareness cue combinations. (a) *Virtual Hand Only*, (b) *Virtual Hand +
Spatial Sound*, and (c) *Virtual Hand + Color Cue*. The illustration shows the virtual
wireframes of cameras and the *Color Cue* in AR from a third person perspective.

location, focus of attention, and actions in a large workspace accessed through
multiple cameras, and a controlled lab study comparing three awareness cue
combinations. See Figure 12.9 for an illustration of awareness cue combinations.
The table in Table 12.1 shows an overview of how the different awareness cues
contribute to the sources of workspace awareness information. These awareness
cues addresses awareness **challenge 4** by using AR to inform the worker about
the location of the camera currently viewed by the helper.

Virtual Hand is a three-dimensional virtual hand that appears in the worker's
physical environment. It appears in the location where the helper points with her
mouse in the currently selected camera video feed.

Spatial Sound is emitted as a white noise from the camera device set as the currently selected view whenever the helper moves her mouse over a work area in
this view. When the helper presses and holds the left mouse button to draw on a
work area, Spatial Sound changes to the sound of a pencil on paper.

Color Cue visualizes a persistent color in the worker's FoV – in our implementation we use the Hololens' AR cursor, a colored disc located at the intersection
between a ray cast from the head of the worker in gaze direction and the
mesh-reconstructed environment. In Figure 12.10, it is shown how the virtual
wireframe of each physical camera is assigned a unique color. Color Cue reveals
which camera the helper is viewing, as shown in Figure 12.9c. Thus, when the
helper changes camera view, Color Cue changes to the color of said camera,
which can be used to locate the helper's viewpoint.

A within-subjects lab study with 12 participants was conducted comparing
the efficiency of three awareness cue combinations (Virtual Hand, Virtual
Hand+Spatial Sound, Virtual Hand+Color Cue) for quickly locating the helper.
Figure 12.11 shows the worker's performance (a) and perceived ease of use
(b) for the different awareness cue combinations, while Figure 12.12 shows the
preferences for the combinations. As can be seen from the figures, the study
showed that Virtual Hand+Color Cue was superior, i.e. the fastest and preferred
choice, possibly due to the worker's ability to memorize the mappings between
colors and cameras, thus eliminating the need to do an auditory or visual search

Table 12.1 How the different cues contribute to the sources of workspace awareness information.

Sources of workspace awareness information	AR awareness cues		
	Virtual hand	Spatial sound	Color cue
Consequential communication	The helper adjusts her focus of attention in a view of the workspace and unconsciously moves the cursor. Thus, the virtual hand moves and reveals the helper's focus of attention to the worker.	The worker can listen to the white noise of helper's cursor or verbal shadowing to identify her location/viewpoint	The worker can map color of cursor to color of camera that serves as the current viewpoint of the helper
Feedthrough	NA	NA	NA
Intentional communication	The helper can use cursor to intentionally point to task objects and locations in the active area and draw in the active area, which is visualized to the worker in 3D AR.	The helper can call for the worker's attention with a verbal utterance, which will be emitted from the currently viewed camera device, thus giving away her location/viewpoint.	The helper can intentionally switch between two cameras quickly to create the effect of a flickering color-changing cursor in the worker's field of view which may catch the attention of the worker and inform him about a location change.

Figure 12.10 Colors of virtual wireframe in Color Cue condition.

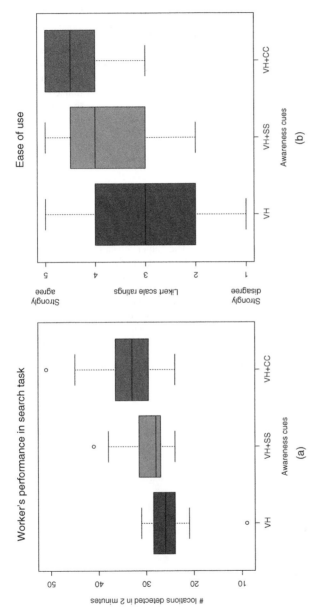

Figure 12.11 (a) The worker's performance was significantly affected by awareness cue combination. Performance was lowest in *Virtual Hand Only* condition (VH) and highest for *Virtual Hand + Color Cue* (VH+CC). Y-axis shows number of areas visited within a two-minute period. (b) Participants rated their agreement with the statement *"It was easy to use [awareness cue condition] to locate the helper."* *Virtual Hand + Color Cue* (VH+CC) was the easiest to use, while *Virtual Hand Only* (VH) was the least easy to use.

Figure 12.12 Users' performance with and preferences for the AR awareness cue combinations.

	VH	VH+SS	VH+CC
1st choice	1	2	9
2nd choice	2	7	3
3rd choice	9	3	

for the helper. The reader is referred to the publication on CueCam for more details on the study itself [21].

Related work has covered various view awareness cues for MR remote collaboration, including head pose, eye-gaze ray, and view frustum of the collaborating partner, which have been found to support awareness between helper and worker [22–25]. Recently, Yang et al. [26] used spatialized voice and auditory beacons to make a worker aware of the helper's location during a search task in MR. The spatialized voice is similar to CueCam's functionality of emitting the helper's voice from the speaker of the currently viewed camera device, and the auditory beacons are similar to the white noise emitted from the speaker of the viewed camera device whenever the helper moves her cursor over the camera view. The difference is that their sounds were spatialized by using the 3D speakers of an AR-HMD worn by the worker, whereas the sounds in CueCam are emitted from actual speakers placed in the workspace. The novel contribution of CueCam is the awareness cue, Color Cue, which is based on an assumption – inspired by observations in the manufacturing industry – that a workspace can be separated into a fixed number of work areas. For instance, we observed that there is a fixed number of areas of interest during remote troubleshooting of an injection molding machine, i.e. the HMI and injection mold. In the lab study mentioned above [21], we showed that by mapping Color Cue, which is persistently available in the worker's field of view, to the color and location of the camera currently viewed by the helper, workers could efficiently find their helper's viewpoint and focus of attention. We showed that *Virtual Hand + Color Cue* outperformed *Virtual Hand + Spatial Sound* and was preferred by the workers as shown in Figure 12.12. This finding thus contributes to the body of research on awareness cues for remote assistance in large workspaces. The finding is interesting, since Color Cue is applicable in a manufacturing context, where it may be difficult for the worker to use spatial sound to localize the helper because of loud noises from the production or the need to carefully listen to the sounds from the machines as a part of inferring their state [27].

12.3 Future Research on Multi-camera Remote Assistance

We have presented benefits and challenges of implementing multi-camera systems to improve workspace awareness for remote assistance on physical tasks. We also introduced our two prototype systems that use augmented reality technologies to address the challenges while achieving improvement in workspace awareness. Now, we turn to the discussion of possible future research directions for multi-camera remote assistance.

12.3.1 Improving Focus-in-Context Views

In this subsection, proposals for improving the focus-in-context views of SceneCam are discussed with the aim of improving the workspace awareness, i.e. collaborators' awareness of each other.

12.3.1.1 Supporting Accurate Distal Pointing in Exocentric Focus-in-Context View

The focus-in-context views of SceneCam are interactive. The helper can hover over a work area to see a detailed (focus) view of the area in a thumbnail (see Figure 12.7) and click on the area to access a large version of the detailed view. In the future, additional interactivity can be added to the focus-in-context views to improve the worker's intentional communication with hand gestures, thereby addressing **challenge 3**.

SceneCam already supports that a worker can point to a distant work area with his head in a context view (supported by the head pose tracking of Hololens v1). The area to which he points is easy to infer from the exocentric context view because his head direction is clearly visualized. However, distal pointing gestures with an outstretched hand and index finger are a common and natural way of creating a deictic reference and are currently not supported in the exocentric context view, and even if they were supported in the egocentric view, it could lead to misinterpretations because the view is not the worker's perspective [10]. The location of a distal pointing gesture with a hand can easily be visualized in either context views using a dedicated hand tracker (e.g. Leap Motion) or an AR-HMD with built-in hand tracking (e.g. Hololens v2). This added ability of the worker to use one hand to clearly point to an area in a context view increases the opportunities for intentional communication, and the worker is therefore

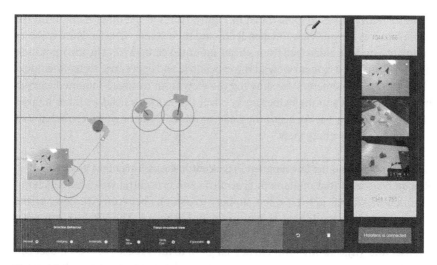

Figure 12.13 Helper's exocentric view of worker pointing to a specific task object, a tangram puzzle piece, in a focus area.

less constrained in the way he communicates. Moreover, the worker may want to make a pointing gesture with higher precision by pointing to a specific task object in a distant work area. His pointing accuracy can be improved by visualizing the focus view of the area in the AR-HMD at close proximity, so he can easily point to a specific task object in the focus view. At the same time the focus view appears as a thumbnail in the helper's context view (identical behavior to when hovering over an area with the mouse cursor) with a 2D circle on top of the task object being pointed to, as shown in Figure 12.13. This idea of emphasizing the object that the worker points to is similar to the work by Oda et al. [28], where a 3D reconstruction of a distant area is brought closer to the worker so he can more easily point to objects in the area. Their augmented reality pointing technique performed more accurately than existing AR referencing techniques at the time. Therefore, we believe that the proposed pointing technique, which takes advantage of the available focus views and is conceptually similar to their technique, will also improve accuracy of the worker's pointing gestures.

12.3.1.2 Combining Scene Cameras and Head-Mounted Camera

In SceneCam, the helper has access to views from multiple scene cameras in the focus-in-context views, but does not have access to the head-mounted camera on the worker's AR-HMD. The reason for combining both camera systems into one is to take advantage of each camera system's strengths and to cover their weaknesses. For instance, a worker's distal pointing gesture at a task object is easier for the helper to interpret from the point of view of a head-mounted camera than a scene camera [10], as demonstrated in Figure 12.1. On the other hand, multiple scene cameras provide view independency and their stable video feeds are easier to annotate a moving video feed from a head-mounted camera [6]. The incorporation of a head-mounted camera view in the exocentric and egocentric focus-in-context views can be implemented by showing the view from the head-mounted camera in a thumbnail, when the helper hovers over the worker's head with her mouse, and upon clicking on his head, the head-mounted camera view is selected and displayed as the primary view.

12.3.1.3 Modeling the Environment in Exocentric Focus-in-Context View

Research systems that implement egocentric and exocentric views for collaboration in VR/AR/MR [29–32] often support transitions between AR views of the physical 1:1 scale world and miniature/giant VR views. The VR user can scale himself down to become a miniature avatar in relation to the AR user, whereby the AR user and physical environment appear giant to the VR user, or the VR user can scale himself up to become a giant avatar in relation to the AR user, whereby the AR user and physical environment appear miniature to the VR user. Such systems need to model the physical environment to support collaboration tasks, like picking and placing of physical objects. However, a current limitation of the exocentric context view of SceneCam is that it is purely virtual, which means that only the 2D position and orientation of the worker, cameras and work areas are visualized (see Figure 12.7). Hence, the exocentric context view provides a clear overview of the spatial relationship between worker and (camera and work area)-pairs, but omits potentially important information about the physical environment, e.g. layout of industrial machines. Modeling the physical environment in the exocentric context view, possibly on the fly using the depth camera of the worker's AR-HMD, is important, because it will support collaboration tasks in areas not already covered by the focus views. For instance, it supports negotiation of where to move a scene camera to add or change a focus view. Furthermore, a model of the environment will presumably make it easier for the helper to predict the worker's path from one work area to another and will also avoid misinterpretations of pointing gestures, where the worker intends to point to a structure, e.g. a machine, in front of him, but to the helper, it mistakenly looks as if he points to a work area behind the structure.

12.3.2 Improved Automatic View Selection Algorithm

We discuss improving the automatic view selection algorithm of SceneCam with the aim of making better inferences of the optimal view in a multi-camera system that can be configured in different ways, as shown in Figure 12.14.

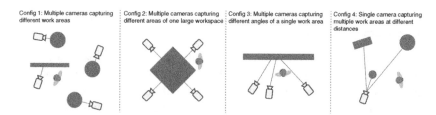

Figure 12.14   Examples of different large workspace setups and potential camera configurations to support multi-camera remote assistance.

It is possible to configure multiple cameras so they capture the same work area from different angles (see Figure 12.14, config 3). If the worker is close to the area, the automatic view selection algorithm must select one optimal view from a list of cameras. Since a worker may occlude one or more of the camera views in this camera configuration, an occlusion score must be factored into the algorithm for selecting the optimal view. In our work [12], an occlusion score was calculated per camera as the proportion of the camera's view taken up by the worker's head (i.e. pixel area of head in screen space divided by pixel area of camera video). It was a simple but reasonable metric intended to prevent the algorithm from selecting a camera that was almost or fully occluded by the worker's head.

The occlusion score can be improved by calculating the proportion of a work area taken up by the worker's head in screen space, i.e. the intersection of pixel area of head in screen space and pixel area of work area in screen space. The lower the score the smaller the intersection. According to research providing a helper with a steady over-the-shoulder view of a workspace is useful for remote collaboration [9], and collaboration is improved, when a worker and helper share the same perspective of a workspace compared to having opposing perspectives [33]. Thus, we want to avoid situations, where the automatic view selection algorithm selects the camera view with a perspective that is opposite from the worker's perspective, just because it happens to have the lowest occlusion score, *if* an over-the-shoulder view is also available and the important parts of the work area is not occluded from this view. Therefore, we introduce a perspective similarity score per camera. The perspective similarity score for a camera is obtained by calculating the dot product between the head-gaze vector (i.e. the direction of the worker's head) and the vector with direction along the z-axis of the camera's coordinate system. A score close to 1 indicates that the perspective of the worker and camera (helper) are similar (0° difference), whereas a score close to −1 indicates opposing perspectives (180° difference). The occlusion and perspective similarity scores are then combined by multiplying them to calculate the optimal camera view.

12.3.3 Automatic Camera Configuration Detection for Adaptive AR Multi-camera Remote Assistance

Automatic camera configuration detection is the idea that the spatial arrangement of multiple cameras can be automatically detected based on their geometric relationship to each other. Specifically, in an AR multi-camera remote assistance system like SceneCam or CueCam, an automatic camera configuration detection algorithm can be based on the tracked poses of the scene cameras and positions of work areas. We know of no related research that focuses on detecting the spatial arrangement of cameras for the purpose of remote assistance or for other application purposes. Automatic camera detection can be helpful for adapting the interactive behavior of an AR multi-camera system. For example, the logic of the automatic view selection algorithm of SceneCam can change based on camera configuration. If configuration 1 or 2 in Figure 12.14 is detected, then the current algorithm, which is based on proximity of the worker to a work area, will suffice for selecting an optimal view of the worker, whereas if configuration 3 is detected, then a more sophisticated algorithm is needed, which takes the similarity between camera perspectives and the perspective of the worker into account. Furthermore, automatic configuration detection can be used to detect use cases, where one scene camera acts as an overview capturing multiple work areas that are captured in detail by the other cameras, whereby the overview camera can be turned into an egocentric focus-in-context view automatically.

Automatic camera configuration detection can be implemented, due to the core functionality of the AR multi-camera systems, SceneCam and CueCam, collecting (camera, work area)-pairs, i.e. three-dimensional geometric information about the pose of each camera and position of the work area captured by the cameras. A vector starting in the position of the camera and ending in the center of the work area can be retrieved for each (camera, work area)-pair. These vectors and camera positions can be used to detect how the (camera, work area)-pairs are spatially arranged and thus detect the configurations in Figure 12.14. A simple example implementation that can distinguish between configuration 3 and all other configurations is one, where the distances between work area centers are calculated. Work areas, which are at a distance to each other below a certain threshold value, are captured by cameras that belong to configuration 3, whereas any other work areas are captured with cameras that belong to either configuration 1 or 2.

12.3.4 Improving Color Cue to Ease Mapping of Helper's Location

In CueCam, the awareness cue Color Cue works by mapping the color of a virtual cursor in the worker's field of view to the colors of the cameras' virtual models. A limitation of Color Cue is that it can be used to locate the helper's camera/

Single camera capturing multiple work areas at different distances

Virtual wireframe of camera has a color. Color mapping of Color Cue to camer color does not lead worker to the helper's focus of attention.

A colored virtual arrow hovers over each work area. Color mapping of Color Cue to color of area leads worker to helper's focus of attention.

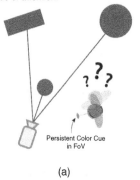

Persistent Color Cue in FoV

(a)

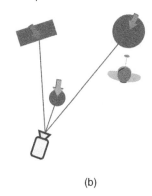

(b)

Figure 12.15 (a) Current way of mapping Color Cue to color of camera. (b) Proposed way of mapping Color Cue to color of work area to accurately identify the helper's focus of attention in this specific camera configuration.

viewpoint, but contains no information about the helper's focus of attention on its own. Therefore, in our lab, study Color Cue was combined with Virtual Hand to ease the search for the helper's focus of attention. Still, the usefulness of *Virtual Hand + Color Cue* presumably only holds for camera configurations, where the cameras are configured to capture close-up detailed views of work areas, similar to configuration 1 and 2 in Figure 12.14, thereby ensuring that the helper's Virtual Hand and focus of attention is close to her viewpoint. In comparison, for camera configurations, where a camera covers multiple areas of interest at a distance, similar to configuration 4 in Figure 12.14, *Virtual Hand + Color Cue* is a less attractive solution given the current design of Color Cue, because the location of the helper's viewpoint and Virtual Hand are too far apart, as illustrated in Figure 12.15. Consequently, the Color Cue may even have a misleading effect on the worker, when he searches for Color Cue instead of Virtual Hand. To combat this issue, Color Cue can be improved by color coding the 3D reconstructed mesh of work areas instead of the virtual camera models.

With this improvement of Color Cue, the worker's search task changes from mapping Color Cue to the location and color of cameras to mapping Color Cue to the location and color of work areas. A strategy for controlling the extent of coloring could be to only color the triangles of a surface mesh of a work area inside a radius proportional to the distance between camera and work area, similar to how the radius of the semi-spheres were calculated for the (camera, work area)-pairs. Alternatively, color coded arrows can be positioned to hover

over or next to the work areas, depending on whether they are comprised of a mostly vertical or horizontal surface mesh. See Figure 12.15 for an illustration of this concept.

12.3.5 Study of Multi-camera Remote Assistance in the Manufacturing Industry

For future work, two studies can be conducted to help us further understand the benefits of implementing awareness cues. The first study is the one where an AR multi-camera configuration, consisting of multiple scene cameras and the PoV camera from an AR-HMD, is compared to using only the PoV camera from an AR-HMD. The comparison will take place in a realistic manufacturing setting that requires a worker to troubleshoot in a large workspace. A realistic example is the troubleshooting and disassembly and reassembly of a large injection molding machine. The worker will be tasked to first identify a problem with the machine and then shut it down to solve the problem by disassembling and reassembling its machine internals. This comparison will improve our understanding of the benefits and challenges of providing the helper with simultaneous access to multiple dependent and independent views in a large and realistic collaborative setting.

The second study would be a comparison between an AR multi-camera configuration with no deliberately designed sources of awareness information and an AR multi-camera configuration with added AR awareness cues (e.g. *Virtual Hand + Color Cue*) and techniques (e.g. exocentric-focus-in context view and automatic camera selection) in a manufacturing setting. The goal of the study is to identify whether there are any benefits of using awareness cues and techniques in realistic collaborative scenarios, since our previous lab study suggests that there is [21]. More specifically, the study can be a within-subjects field experiment. Participants in the field experiment will go through the troubleshooting of different scenarios, including injection molding machine and mold repair, using the configurations with and without awareness cue and techniques. Qualitative data on the participants' satisfaction and opinion of awareness cues and techniques can be collected through post-condition interviews, while usage patterns can be video recorded for statistical analysis.

12.4 Discussion of 2D vs. 3D Workspace Information

As part of their core functionality, the AR multi-camera systems SceneCam and CueCam use RGB cameras and video with no depth information. However, we have to address the possibility of using multiple RGB-D cameras in a multi-camera setup, i.e. cameras that capture color and depth information and thereby enable the creation of a 3D reconstruction of the captured workspace. The helper can immerse herself in such a 3D reconstruction when she puts on a VR headset, move around the 3D reconstruction independently of the worker's movements, and use natural 3D hand gestures for pointing and expressing actions [25, 34, 35]. Hence, a helper immersing herself in a live 3D reconstruction of a space using a VR headset will experience some benefits that a helper consuming and interacting with multiple 2D videos on a tablet/PC screen does not. Still, we believe it is worthwhile to consider using RGB cameras in a multi-camera setup and equipping the helper with a PC/tablet screen for video consumption and interaction. First, RGB cameras are generally of higher resolution than the depth cameras, so from the same distance a RGB camera is able to capture details of a work area more clearly than a depth camera. Secondly, with SceneCam and CueCam, we sought to design a solution for the helper that is mobile and easy to use in different use contexts, where using an immersive headset is not desirable. VR immersion can be troublesome in situations where the remote assistance is competing with other activities for the helper's attention and in situations where it is not socially acceptable or awkward to dive into VR, such as in public spaces [27]. Thus, we have focused on improving the way the helper is provided with multiple 2D videos on a tablet/PC screen-based interface, rather than one 3D reconstruction in a VR headset. Thirdly, RGB cameras comes in all kinds of sizes and variants supporting different needs for focused and wide-shot angles, high-resolution capture, pan-zoom-tilt control, cordless use, etc., that the sparse selection of commercial RGB-D cameras cannot match at the current time. Finally, if a large workspace is fragmented into different areas, each covered by a camera, an exocentric context view of the spatial relationship between work areas, which is suitably presented in 2D on a tablet/PC screen, may support the helper in efficiently jumping from one area to another without losing spatial awareness compared to the more time consuming, fluent navigation, or "walking" from area to area in an egocentric VR view.

12.5 Concluding Remarks

Multi-camera remote assistance has some benefits over using one camera from the point of view of the worker, most notably view independence of the helper. However, in this chapter, we pointed out the challenges that stand in the way of obtaining good workspace awareness when using multiple cameras for remote assistance. In response to these challenges, we discussed two AR research prototypes, SceneCam and CueCam that demonstrate how AR visualization and tracking can be used to address these awareness challenges in various ways. SceneCam focuses on improving the helper's awareness of the worker and, vice versa, CueCam focuses on improving the worker's awareness of the helper. Further, possible improvements to SceneCam and CueCam and future research directions were discussed including conducting studies in a realistic manufacturing setting.

References

1 Gaver, W.W., Sellen, A., Heath, C., and Luff, P. (1993). One is not enough: multiple views in a media space. In: *Proceedings of the INTERACT '93 and CHI '93 Conference on Human Factors in Computing Systems*, CHI '93, 335–341. New York, NY, USA: ACM.

2 Lanir, J., Stone, R., Cohen, B., and Gurevich, P. (2013). Ownership and control of point of view in remote assistance. In: *Proceedings of the SIGCHI Conference on Human Factors in Computing Systems*, 2243–2252. ACM.

3 Tait, M. and Billinghurst, M. (2015). The effect of view independence in a collaborative AR system. *Computer Supported Cooperative Work (CSCW)* 24: 563–589.

4 Lee, G.A., Teo, T., Kim, S., and Billinghurst, M. (2018). A user study on MR remote collaboration using live 360 video. In: *2018 IEEE International Symposium on Mixed and Augmented Reality (ISMAR)*, 153–164. ISSN: 1554-7868.

5 Kim, S., Billinghurst, M., and Lee, G. (2018). The effect of collaboration styles and view independence on video-mediated remote collaboration. *Computer-Supported Cooperative Work* 27 (3–6): 569–607.

6 Fakourfar, O., Ta, K., Tang, R. et al. (2016). Stabilized annotations for mobile remote assistance. In: *Proceedings of the 2016 CHI Conference on Human Factors in Computing Systems*, 1548–1560. ACM.

7 Birnholtz, J.P., Ranjan, A., and Balakrishnan, R. (2010). Providing dynamic visual information for collaborative tasks: experiments with automatic camera control. *Human–Computer Interaction* 25 (3): 261–287.

8 Fender, A.R. and Holz, C. (2022). Causality-preserving asynchronous reality. In: *Proceedings of the 2022 CHI Conference on Human Factors in Computing Systems*, CHI '22, 1–15. New York, NY, USA: Association for Computing Machinery.

9 Fussell, S.R., Setlock, L.D., and Kraut, R.E. (2003). Effects of head-mounted and scene-oriented video systems on remote collaboration on physical tasks. In: *Proceedings of the SIGCHI Conference on Human Factors in Computing Systems*, 513–520. ACM.

10 Sousa, M., dos Anjos, R.K., Mendes, D. et al. (2019). Warping deixis: distorting gestures to enhance collaboration. In: *Proceedings of the 2019 CHI Conference* on *Human Factors in Computing Systems*, CHI '19, 1–12. Glasgow, Scotland, UK: Association for Computing Machinery.

11 Fussell, S.R., Kraut, R.E., and Siegel, J. (2000). Coordination of communication: effects of shared visual context on collaborative work. In: *Proceedings of the 2000 ACM Conference on Computer Supported Cooperative Work*, 21–30. ACM.

12 Rasmussen, T.A. and Huang, W. (2019). SceneCam: Improving multi-camera remote collaboration using augmented reality. In: *2019 IEEE International Symposium on Mixed and Augmented Reality Adjunct (ISMAR-Adjunct)*, 28–33.

13 Gutwin, C. and Greenberg, S. (2002). A descriptive framework of workspace awareness for real-time groupware. *Computer-Supported Cooperative Work* 11 (3): 411–446.

14 Yamaashi, K., Cooperstock, J.R., Narine, T., and Buxton, W. (1996). Beating the limitations of camera-monitor mediated telepresence with extra eyes. In: *Proceedings of the SIGCHI Conference on Human Factors in Computing Systems*, CHI '96, 50–57. New York, NY, USA: ACM.

15 Norris, J., Schnädelbach, H., and Qiu, G. (2012). CamBlend: An object focused collaboration tool. In: *Proceedings of the SIGCHI Conference on Human Factors in Computing Systems*, CHI '12, 627–636. New York, NY, USA: ACM.

16 Norris, J., Schnädelbach, H.M., and Luff, P.K. (2013). Putting things in focus: establishing co-orientation through video in context. In: *Proceedings of the SIGCHI Conference on Human Factors in Computing Systems*, CHI '13, 1329–1338. New York, NY, USA: ACM.

17 Teo, T., Lawrence, L., Lee, G.A. et al. (2019). Mixed reality remote collaboration combining 360 video and 3D reconstruction. In: *Proceedings of the 2019 CHI Conference on Human Factors in Computing Systems*, CHI '19, 1–14. Glasgow, Scotland, UK: Association for Computing Machinery.

18 Ranjan, A., Birnholtz, J.P., and Balakrishnan, R. (2007). Dynamic shared visual spaces: experimenting with automatic camera control in a remote repair task. In: *Proceedings of the SIGCHI Conference on Human Factors in Computing Systems*, 1177–1186. ACM.

19 Ou, J., Oh, L.M., Yang, J., and Fussell, S.R. (2005). Effects of task properties, partner actions, and message content on eye gaze patterns in a collaborative task. In: *Proceedings of the SIGCHI Conference on Human Factors in Computing Systems*, 231–240. ACM.

20 Ou, J., Oh, L.M., Fussell, S.R. et al. (2005). Analyzing and predicting focus of attention in remote collaborative tasks. In: *Proceedings of the 7th International Conference on Multimodal Interfaces*, 116–123. ACM.

21 Rasmussen, T., Feuchtner, T., Huang, W., and Grønbæk, K. (2022). Supporting workspace awareness in remote assistance through a flexible multi-camera system and augmented reality awareness cues. *Journal of Visual Communication and Image Representation* 89: 103655

22 Sodhi, R.S., Jones, B.R., Forsyth, D. et al. (2013). BeThere: 3D mobile collaboration with spatial input. In: *Proceedings of the SIGCHI Conference on Human Factors in Computing Systems - CHI '13*, 179. Paris, France: ACM Press.

23 Piumsomboon, T., Day, A., Ens, B. et al. (2017). Exploring enhancements for remote mixed reality collaboration. In: *https://dl.acm.org/doi/proceedings/10.1145/3132787*. *SA '17: SIGGRAPH Asia 2017 Mobile Graphics & Interactive Applications*, 1–5. ACM.

24 Gao, L., Bai, H., Lindeman, R., and Billinghurst, M. (2017). Static local environment capturing and sharing for MR remote collaboration. In: *SIGGRAPH Asia 2017 Mobile Graphics & Interactive Applications*, 17:1–17:6. ACM.

25 Piumsomboon, T., Dey, A., Ens, B. et al. (2019). The effects of sharing awareness cues in collaborative mixed reality. *Frontiers in Robotics and AI* 6: 5

26 Yang, J., Sasikumar, P., Bai, H. et al. (2020). The effects of spatial auditory and visual cues on mixed reality remote collaboration. *Journal on Multimodal User Interfaces* 14: 337–352.

27 Rasmussen, T.A. and Gronbak, K. (2019). Tailorable remote assistance with RemoteAssistKit: a study of and design response to remote assistance in the manufacturing industry. In: *International Conference on Collaboration and Technology*, 80–95. Springer.

28 Oda, O. and Feiner, S. (2012). 3D referencing techniques for physical objects in shared augmented reality. In: *2012 IEEE International Symposium on Mixed and Augmented Reality (ISMAR)*, 207–215. Atlanta, GA, USA: IEEE.

29 Billinghurst, M., Kato, H., and Poupyrev, I. (2001). The MagicBook: a transitional AR interface. *Computers & Graphics* 25 (5): 745–753.

30 Stafford, A., Piekarski, W., and Thomas, B. (2006). Implementation of god-like interaction techniques for supporting collaboration between outdoor AR and indoor tabletop users. In: *2006 IEEE/ACM International Symposium on Mixed and Augmented Reality*, 165–172. Santa Barbara, CA, USA: IEEE.

31 Piumsomboon, T., Lee, G.A., Ens, B. et al. (2018). Superman vs giant: a study on spatial perception for a multi-scale mixed reality flying telepresence interface. *IEEE Transactions on Visualization and Computer Graphics* 24 (11): 2974–2982.

32 Piumsomboon, T., Lee, G.A., Hart, J.D. et al. (2018). Mini-me: an adaptive avatar for mixed reality remote collaboration. In: *Proceedings of the 2018 CHI Conference on Human Factors in Computing Systems*, CHI '18, 1–13. Montreal, QC, Canada: Association for Computing Machinery.

33 Feick, M., Mok, T., Tang, A. et al. (2018). Perspective on and re-orientation of physical proxies in object-focused remote collaboration. In: *Proceedings of the*

2018 CHI Conference on Human Factors in Computing Systems - CHI '18, 1–13. Montreal, QC, Canada: ACM Press.

34 Tecchia, F., Alem, L., and Huang, W. (2012). 3D helping hands: a gesture based MR system for remote collaboration. In: *Proceedings of the 11th ACM SIGGRAPH International Conference on Virtual-Reality Continuum and its Applications in Industry*, VRCAI '12, 323–328. Singapore, Singapore: Association for Computing Machinery.

35 Adcock, M., Anderson, S., and Thomas, B. (2013). RemoteFusion: Real time depth camera fusion for remote collaboration on physical tasks. In: *Proceedings of the 12th ACM SIGGRAPH International Conference on Virtual-Reality Continuum and Its Applications in Industry - VRCAI '13*, 235–242. Hong Kong, Hong Kong: ACM Press.

13

Industrial Applications, Current Challenges, and Future Directions

13.1 Introduction

In the previous chapters of this book, we reviewed the current state-of-the-art research literature in the space of remote collaboration on physical tasks and presented major outcomes of our research in this space in the past decades. It is clear that, nowadays, existing technologies are consistently being advanced, and new ones are being invented and developed quickly to meet the demand of remote guidance scenarios in industrial sectors. Also, to validate the new ideas, advance our knowledge, and provide theoretical foundations to support the practice of remote guidance, it is encouraging that more research attention has been drawn into this space and even more effort has been devoted to the translation of those research findings into practical systems and tools that have been used in the real-world settings. In this chapter, we will introduce some of those commercial remote guidance systems that come to our attention on the Internet and through our research contacts, present current challenges, and propose some future research directions. Note that some of the challenges and future directions have been discussed specifically related to the chapter content. In this chapter, we will discuss them in more detail when we mention them again.

13.2 Remote Guidance Systems

In recent years, particularly during the pandemic, the need for remote collaboration has become widespread, and the demand for tools and systems that support remote collaboration has been increasing rapidly. In this chapter, we briefly describe some of these systems that appear in the market. These systems are mentioned because they are video-based, support the worker-helper scenario for physical tasks, and provide functionalities for hand gesturing and/or sketching on the video/static images. It should be noted we did not use these systems

Computer-Supported Collaboration: Theory and Practice, First Edition.
Weidong Huang, Mark Billinghurst, Leila Alem, Chun Xiao, and Troels Rasmussen.
© 2024 The Institute of Electrical and Electronics Engineers, Inc. Published 2024 by John Wiley & Sons, Inc.

personally, and our introduction is based on the information available on the company websites and the Internet. It is possible that our wording is not accurate. However, we provide the source of the information in the format of Internet links for each product.

The common functionalities and benefits that these products claim to have include:

(1) Reduce the need to travel, thus reducing the cost and time associated with travel arrangements and being able to access remote expertise quickly.

(2) Enhanced communication because these specifically designed interfaces will help collaborators to collaborate when normal phone calls or video calls are not enough.

(3) Some systems are designed to support their own customers, while others are more flexible and can be customized and integrated into the operational process of different organizations.

(4) Some systems use handheld devices on the worker side such as a tablet or smartphone, while others use rugged wearable devices so that hands of workers are freed for manipulation of physical objects safely.

(5) An augmented reality-based visual space is shared by both sides to achieve the common ground that combines the workspace of the worker and the digital guiding information of the helper. In addition to verbal communication, workers are able to see the guiding information visually such as pointing, drawing (sketching) and hand gestures over the video or static images provided by the helper.

13.2.1 Alfa Laval Remote Guidance

The Alfa Laval remote guidance tool [1] is designed to work with most of the Alfa Laval equipment. It allows physically distributed users to interact and collaborate in real time. The connection can be established quickly, and the interfaces are easy and intuitive, requiring no login details and little training. It provides an augmented reality-based visual space in which the collaborators can see visual guiding information such as hand gestures, digital sketching, still images, and real objects of the workspace while communicating with each other.

13.2.2 TEXO Remote Guidance Software

The product [2] is designed to support TEXO's service engineers and specialists in providing remote help to their customers remotely. The support is provided in the format of video chat. During this process, with the camera embedded in the tablet or smartphone, the customer is able to show the technician where help is needed. Then, with this software, the technician is able to take over the screen of the customer and use hand gestures or make digital marks on the customer's screen to demonstrate how the issue can be solved.

13.2.3 Kiber 3

Kiber [3] is designed to offer an all-in-one "hands-free" solution to companies when instant expertise is needed for field maintenance or manufacturing activities. It has a unique AR-based and fully software and hardware-integrated mobile industrial standard solution that allows both a local worker and a remote helper to collaborate with each other on physical tasks as if they were physically colocated. Kiber offers different versions that each are suitable for different real-world situations. These versions are Kiber 3 Kit's multi-feature wearable device, Kiber 3 Web's powerful web-based platform, and Kiber 3 Field's user-friendly mobile app. Each of these has unique features serving its own purpose. For example, Kiber 3 Web allows collaborators to connect with each other from their desktops, and no additional software installation is required. Kiber 3 Field allows an onsite worker to use his smartphone to see help from a remote helper for quick troubleshooting and technical assistance.

13.2.4 Shipznet RG 300 – Remote Guidance Service Connectivity

Shipznet RG 300 [4] provides service connectivity for remote guidance in vessels. The system offers high-speed mobile data communication with best-in-class network coverage and has a waterproof service case, power line communication adapters, and cables. It is designed to be easily integrated into the existing IT infrastructure onboard and enable service staff working on a ship in isolated areas to be connected to remote experts and engineers on shore for various remote guidance or inspection activities. It appears that RG 300 offers an extendable and customizable infrastructure. For example, shipznet extends its remote guidance product with the Kiber 3 Kit, which provides an out-of-box and hand-free solution, allowing it to be used in the deepest areas of the ship, where it is usually surrounded by thick steel and cut off from the Internet. Also, recently, the new Remote Guidance Toolkit has been added to the shipznet RG 300 plus. This makes it the first complete solution developed specifically for maritime use, offering the ideal conditions for remote use on-board.

13.2.5 HPE Visual Remote Guidance Service

The innovative HPE Visual Remote Guidance [5] is designed for a local user to solve issues quickly while collaborating virtually with a remote support engineer or expert. The collaborators can share information such as images, video, audio, and data in real time with each other in an intelligent and intuitive way. The support engineer or expert can see what the remote worker sees and does and can provide real-time remote visual guidance, all through the wearable computing display, such as M100 Smart Glasses, enabling convenient and hands-free interaction. The collaborators can also collaborate with text chat that is enhanced with real-time language translations, bringing further convenience to a support experience.

13.2.6 Dynamics 365 Remote Assist

Microsoft's Dynamics 365 Remote Assist [6] is designed to support people located in different locations to work together more effectively using HoloLens and other mobile devices. It supports a range of real-world remote guidance use cases including collaborative maintenance and repair, remote inspections, and knowledge sharing and training. Using heads-up video calling on Microsoft HoloLens and mobile devices, a local technician can be connected with a remote expert assigned to the session, receiving guidance in context from remote experts. With HoloLen's wide view angle, the local technician can see and talk to the expert, share the workspace, and see the visual guiding information such as free sketches, pointing arrows, documents, and images to perform the task at hand properly. It also allows onsite inspectors to work with remote inspectors to inspect, evaluate, and record the status and quality of assets.

13.2.7 ABB Remote Assistance

The ABB Remote Assistance for Electrical Systems [7] or RAISE is designed to allow remote ABB experts to guide local field workers in inspecting faulty devices or machines and pinpointing the defective parts and the source of the problem for maintenance quickly. This allows quick access to ABB experts who are at different locations and fast problem-solving, enabling significant improvements to field service. It provides a shared visual space in the format of a video application so that collaborators can have shared views for instruction and guidance. It also has a window of text chat, allowing augmented reality instructions to be placed in their field of view. At the same time, images, videos, and documents can be shared from their own mobile or wearable devices. Other key features include chatting with individuals or groups of experts, built-in zooming functionality of videos of the camera, and sharing documents such as files, photos, and videos.

All these features ensure quick remote maintenance, thus reducing downtime and increasing productivity. Despite the fact that smartphones and mobile tablets can be used in the field, in some cases, hand-free operations are more desirable during remote sessions. In this case, the ABB RAISE application is also made to support the usage of smart glasses.

13.2.8 CSIRO ReMoTe

CSIRO ReMoTe [8] is designed for onsite workers to have immediate access to expert help when needed. The system has two wearable units: the worker unit and the helper unit, which can be connected via communication networks such as 4G/5G, Internet, or WiFi and can operate in various environmental conditions. Each unit has a wearable minicomputer that connects to a helmet-mounted camera and a near-eye display. This setup allows for local worker to access the remote help anytime, anywhere. While both sides are able to talk to each other, the camera mounted on the helper side can capture the hand gestures of the helper, which will then be sent to the worker side as digital hands, combined with the workspace of the worker as a visual guidance for the worker to follow. The visual combination is also displayed to the helper as a shared visual space where the helper can see what the worker is doing and perform hand gestures when needed, and at the same time, the worker can see from his near-eye display the virtual hands of the helper and follow the instructions accordingly. The natural hand gestures such as pointing and rotating enabled by the system provide an easy and intuitive experience to the collaborators just like they were together onsite, thus performing tasks quickly and effectively.

13.2.9 Summary

There are also other systems such as XMReality's remote support solution [9], VMI Remote Guidance [10], and Argus Augmented Reality Remote Support Software [11]. Despite their differences in features and application areas, these systems use augmented reality techniques and mobile or wearable devices to enable the working model of worker-helper collaboration, allowing collaborators to communicate with each other verbally and at the same time to be able to share documents, perform gestures and insert sketches in the shared space provided. It is equally important in industrial application environments, in addition to the safety of their workers, when the collaboration session and data exchanged over the network are recorded, security and privacy concerns also need to be considered, and users are informed of the security and privacy statements and corresponding measures that have been taken.

13.3 Technical, Ethical, and Research Challenges and Future Directions

13.3.1 Ergonomically Tested Devices and Privacy and Ethical Aspects of Remote Guidance

Remote guidance has a wide range of use cases in industrial domains. In these use cases, mobility and hand-free skills are often required for the work to be done properly. Therefore, it is important to have mobile or wearable devices that meet the industry standards for the specific application domain and use scenario. Obviously, these devices are to be used by users at work. Therefore, they should be ergonomically designed and tested so that they are comfortable and safe to use and have no harm or negative implications to the end users in terms of workspace safety, human factors, health, and well-being.

In addition to the possible physical impact of technologies and tools, privacy and ethical aspects of the collaboration cannot be ignored. On the one hand, although many remote guidance technologies are developed as a proof-of-concept for testing and research purposes, a lot more factors on top of technical aspects should be considered when they are used in practice. For example, if their eye movement is going to be tracked and recorded, and their facial expression is going to be analyzed, then privacy and ethics issues will need to be addressed first. Some systems are based on a third-party cloud service platform or record the remote guidance sessions and store the data on the cloud. Again, in this case, data safety and privacy will be a concern, and precautionary measures need to be taken and put in place. The end users will also need to be well informed, and their possible concerns need to be addressed and communicated properly before the system can be used in the field. Further, when it comes to data safety and privacy, technical expertise is also needed. For example, the network connections and systems should have mechanisms to protect the data against possible external interference and manipulation such as virtual private networks (VPNs) and self-learning firewalls which have been proven to work effectively to ensure the confidentiality and authenticity of the data.

Possible research questions on these issues could be:

(1) What are the possible impacts, and how different could the impacts be when different types of devices are used on users physically and psychologically?

(2) Whether will users behave differently if they are told their personal data, for example, eye gaze or facial expressions, will be recorded, analyzed, and shared with their collaboration partners? And if yes, how?

13.3.2 Network Connection and Information Delay

Network connection is one of the basic requirements for any remote guidance systems and tools to work properly. This becomes particularly more important when collaboration is to take place in remote and isolated areas where network connection is usually poor. Poor network connection can have many implications that negatively affect the collaborating experiences of users and also make task performance difficult because poor connection means that no data can be transmitted fully between the two sides [12, 13]. If a connection cannot be established, then communication will not take place, and collaboration is no longer possible using the system. Therefore, the basic requirement for remote guidance work is to have stable WiFi/Internet or communication network connections (3G/4G/5G). However, even if the connection is established, if the connection quality is low, the voice of the verbal communication can be broken, and the video can be blurred or frozen. Donovan et al. [14] conducted a study to investigate the effects of network performance on user experience and task performance. In their study, they specified metrics for network quality of service (QoS), including delay, jitter, bandwidth, and packet loss, and then implemented five network conditions: Ethernet, Satellite, 3G, WiFi, and Fiber. Participants were asked to work in pairs to perform remote guidance tasks in each experimental condition. During and after the task performance, the quality of experience (QoE) was collected and recorded. The QoE metrics were both objective such as time taken and number of instructions repeated for task and subjective, but subjective ones can be measured quantitatively such as ratings of audio, video, and overall quality of experience. The results of the study indicated that, indeed, network performance had an impact on user experience, showing the importance of stale network connections for remote guidance, and more specifically, the packet loss rate in a network QoS metrics contributed the most to the loss in QoE.

Low-quality network connection not only causes blurred images but also leads to delays for users to receive audio and visual feedback. As mentioned by Gergle et al. [13], although immediately available shared visual information can be helpful for effective communication, the actual benefits will likely depend on the accuracy of it that is received by the user. If the visual information received is delayed, then the strategies and mechanisms with which users collaborate and communicate with each other will likely be changed and be different. To find out whether these assumptions held, the authors conducted two studies investigating the impact of delayed visual feedback on collaborative performance. Their first study investigated how task performance was affected by a wide range of visual delays, manipulated in the amount of visual delay for the helper's view of the workspace, along with linguistic complexity, manipulated in the amount of conversational grounding. How their conversational and communication behaviors were changed during the process was also investigated. In their second study, they investigated how the dynamics of the task such as color of the objects affected task performance together with visual delays. It was found that in their first study, when visual delay reached a specific point such as 939 ms, the task performance would be affected. When visual information was further delayed to a point, for example, 1798 ms, then the conversational grounding processes started to be affected. Further, in the second study, it was found that when the dynamics of the task objects increased, much shorter time intervals were needed for visual delay to start affecting task performance and user behaviors. The significance of this study is that the tolerable ranges of visual delay were identified before collaborative performance was affected and how the features of the task can interfere and interact with the visual delay for the effect. Therefore, as the authors suggested, it is not a simple task to pick up a threshold number to decide whether a given visual delay is tolerable or not. Instead, a detailed task analysis is needed to have a full understanding of the impact of visual delay on collaborative task performance. A similar study was also conducted by Vaghi et al. [15] in the context of a collaborative virtual ball game.

Instead of visual delay, Krauss and Bricker [16] conducted a study examining the effects of audio delay on verbal communication, in which the audio delay was manipulated into three delay conditions, and 14 pairs of male participants participated in each condition, performing a two-person communication task. It was found that the efficiency of communication was deleteriously affected by 1.8 seconds of audio delay, while participants performed in the condition of 0.6-seconds delay as well as they did in the no-delay condition. Tang and Isaacs [17] conducted a study as part of a larger research project investigating the possible impact of audio delay in a video conferencing setting. In that study, data were collected from a group of four members located in two different sites for over two months. The group conducted weekly video meetings, supplemented by occasional phone meetings. They also had a week of face-to-face meetings. The analysis of the data indicated that audio delay made it difficult to direct the attention of remote participants and negotiate turn-taking for speaking and actions, thus reducing the amount of interaction. The authors suggested that while video provided cues for grounding, audio also played a crucial and perhaps relatively more important role in supporting human interaction and that when network bandwidth is limited and requires trade-offs, it may be more desirable and usable to degrade video quality before degrading audio quality.

In practice, due to the changing environments where remote guidance takes place, stable network connections cannot be guaranteed. In designing and developing remote guidance systems, although a system can be designed in such a way that when the network quality is low, the system will limit the amount of the data transferred over the network, for example the lower resolution video will be used [18]. However, this is done at the cost of communication quality and user experience. Therefore, it is important for us to understand the implications of it and how these implications affect remote guidance task performance, user experience, and collaborating behaviors, decide on trade-off factors, and take action accordingly. The research we reviewed in this section is a good start in the direction. These studies laid a foundation for us to understand better how and when delay in information exchange can affect task performance and communication behaviors. More research is still needed, though, for more practical design guidelines so that we can use them to design and develop adaptable and resilient systems that are robust and effective enough to support remote guidance in changing network conditions. Examples of further research questions and technical development could be:

(1) How can information delay be detected in real time, and how should users be informed of the delay?

(2) What are the differences in terms of task performance and user communication strategies when information delay is informed by the system and when found by the users?

(3) How does information delay affect the collaboration user experience when their task performance is not affected?

(4) What other factors of task dynamics can interact with visual delay in affecting task performance and communication behaviors in addition to object color?

13.3.3 Reproducing the Environment of Face-to-Face Collaboration for Remote Collaboration

Relevant to the information delay we discussed in Section 13.3.2, currently, there seems a trend of making all communication cues that are available for colocated collaboration to be available for remote collaboration by providing colocated communication cues and reproducing colocated collaborative environments for remote guidance. It is argued that remote collaboration technology can enable people to work in distributed locations. However, when collaborators are physically distributed, connections between verbal and nonverbal cues, such as gaze or gesture, are interrupted, breaking natural colocated communication channels and making it difficult to build mutual understanding [19, 20]. To overcome these limitations, in addition to making communication cues that are available in colocated collaboration available to distributed collaboration as well (e.g. [21, 22]), various AR/VR/MR technologies have been used to build a visual space in which collaborators could interact and communicate like they were colocated. For example, a full 360 panoramic video view of the workspace can be made available so that collaborators can be immersed in the same task space to share and interact with the same virtual nonverbal cues [23]. Although this can be computationally expensive, causing a delay for systems to support remote collaboration in real time, with rapid advancements in networking and mobile and wearable technologies, providing rich colocated communication cues and reproducing the colocated environment for remote collaboration are now become feasible.

Much research has been done in this direction. This includes reproducing the 3D workplace of the worker to the helper and 3D gestures of the helper [24–26], providing implicit communication cues such as body movement, facial, and gaze information, which are missing in remote collaboration to the collaborators

[22, 27, 28], and encouraging natural and intuitive interaction such as hand gestures. For example, Wang et al. [24] developed a prototype system and conducted a study to explore the benefit of combining 3D gesture and CAD models of physical objects in a complementary manner. More specifically, the authors developed two interfaces: one having 3D CAD models only and the other having both 3D gesture and CAD models. Participants were asked to perform an assembly training task. Task performance time and accuracy, user experience and collaboration experience were measured and collected. It was found that participants performed significantly better with the combination condition of 3D gesture and CAD models and also had a better user experience. Extending previous research on eye gaze and 360 panoramic video systems, Jing et al. [19] introduced a new modality, speech, to trigger the visualization of shared eye gaze in order to improve task coordination. The authors conducted two studies to evaluate the design of the MR gaze-speech interfaces, and they found that participants preferred to have an explicit visual form to directly connect the collaborators' shared gaze to the contextual conversation. They also found that participants took a short time to attend to the object or location of interest when the gaze-speech modality was used, making more natural and more effective communication and collaboration.

Despite the progress being made in producing more complex workspace and nonverbal cues so that they can be shared with collaborators, more research questions, specifically in the context of remote collaboration in physical tasks, could be:

(1) How can different modalities of communication cues supplement each other for better communication and collaboration?
(2) What are the design principles and guidelines for producing more realistic shared space for collaborators?
(3) What are the effects of making more nonverbal communication cues and more realistic shared space available on task performance and user experience?
(4) How will those effects change or not change when different domain tasks, or different ways of presenting communication cues and the more realistic shared space, are used?

13.3.4 Replacing a Communication Cue with Another Cue of a Different Modality

As we have mentioned, reproducing an environment of face-to-face collaboration for remote collaboration can be computationally expensive, requiring more advanced computing devices and more network bandwidth. Although technical advancements have enabled us to make more realistic environments for colocated collaboration available for remote collaboration, there are always situations where computing devices and network bandwidth are limited or unstable, and this is particularly true in some industrial settings such as in mining sites. In these situations, natural questions to ask would be: do we have to reproduce the same environment for remote collaboration to be effective? Can a communication cue that is effective in a colocated situation be replaced by another cue in the remoted situation but is equally effective? Is it possible to use currently available AR/VR devices and technologies to produce a new collaborative environment or a new communication cue for remote guidance, which is more feasible to produce and set up in certain situations, and with proper training if necessary, can lead to a similar level of user experience and task performance as colocated?

Some attempts have been made in this direction. For example, although eye contact and eye gaze information are important for effective interaction and communication in collaboration, Wang et al. [29] conducted a study with a spatial AR remote collaboration platform in which possible effects of a cursor pointer, head pointer, and eye gaze were investigated with respect to task performance, workload assessment, and user experience. It was found that there was no significant difference in performance time and workload assessment between head pointer and eye gaze and that head pointer served as an effective referential pointer. It was therefore suggested that in some circumstances, head pointer could be used as a good proxy for and an alternative to eye graze in remote collaboration as eye-tracking hardware is more expensive than head pointing, which is relatively more accessible and cheaper to use. This research is promising as it shows that it is feasible to support remote guidance without having to reproduce everything that is available for colocated collaboration. Other examples include Remote Manipulator (ReMa) [30], BeHere [31], and GestureMan [20]. In ReMa, instead of a physical object, a proxy object was created at a remote site to reproduce orientation manipulations so that a shared understanding of what each person is talking about can be built between the collaborators. The authors also conducted evaluations of the technology, and it was found that the shared perspective is more effective and preferred and that the physical proxy can be used as a complement to video chat in a combined system: the proxy is for understanding objects while video chat is for the performance of gestures and confirmation of remote actions. In BeHere, a remote helper is enabled to work in a 3D virtual environment (VR) to guide a local worker through a procedural task in the real world. The 3D VR has both virtual replicas of objects and virtual avatars of the worker. The helper can perform gestures to manipulate replicas and see the avatar. The authors found in a user study that users had s better user experience with the combination of gestures, avatars, and virtual replicas than with just gestures and virtual replicas. GestureMan [20] is a system in which a mobile robot is situated at the worker site and has the ability to express the pointing actions of the helper, shifts in orientation, and reference to the space and objects. The mobile robot has a camera unit capturing the scene of the workspace for the helper and determining body and head orientations. It also has a laser pointer and a stick for pointing to space and objects.

13.4 Conclusion

In this book, we have presented reviews of a few important topics that have been researched extensively in the literature and are relevant to remote collaboration on physical tasks, including communication models, communication cues, and eye gazes. We have also introduced various systems with user studies that we developed to support remote collaboration in various ways using AR/VR and the latest mobile and wearable technologies. These include four systems, with each supporting different communication cues: allowing helpers to perform hand gestures on the video of the object so that natural hand gestures can be conveyed to the other side of the collaboration, allowing helpers to perform hand gestures in the air so that the helper is no longer limited within the office and expert's expertise can be accessed anytime anywhere, allowing helpers to both perform hand gestures and draw sketches over the object on a touch display so that natural interactions with devices can be performed and richer help information can be communicated for better task performance and user experience, allowing helpers to perform hand gestures in a virtual 3D immersive environment reconstructed from the workspace of the worker. Two additional systems were further introduced, with one focusing on making the system adjustable to accommodate individual differences of users with different experiences and preferences for different requirements of collaborative tasks and the other aiming to provide better situation awareness of the partner's working environment and task status by using multiple cameras mounted in the environment.

We conclude the book with the introduction of some industrial systems that support remote guidance on physical tasks. Each of these industrial systems was designed to meet specific design and/or business purposes. Currently, more and more systems have been made available in the market due to improved availability of advanced networking and computing technologies and devices, wide applications of remote guidance, and high demand for collaboration support in the business world and in our daily lives. These systems are diverse in nature. For example, some systems are developed to support remote collaboration of their own employees or between their employees and clients, while others are developed and sold for business users who are in need of the system. Some systems are flexible allowing system components and devices reconfigurable to meet different task needs, while others come with a fixed set of devices and components to meet specific task or application requirements. Finally, current challenges and possible future directions were discussed. Apart from what has been mentioned in the chapter, other possible directions include artificial intelligence and cloud-based remote guidance support, embedment and integration of cognitive, physiological, empathic, and multimodal communication cues, investigation of possible effects of human factors, language, social and cultural factors, and empirically validated evaluation frameworks, design principles, metrics, and methodologies for remote collaboration on physical tasks.

References

1 Alfa Laval. Alfa Laval Remote Guidance. https://www.alfalaval.co.nz/service-and-support/remote-guidance/ (accessed 2 July 2022).

2 TEXO. TEXO Remote Guidance Software. https://texo.se/support/support/ (accessed 3 July 2022).

3 Kiber. Kiber 3. https://kiber.tech/company/ (accessed 3 July 2022).

4 Shipznet. RG 300 – Remote Guidance Service Connectivity. http://shipz.net/solutions/rem-serv/ (accessed 3 July 2022).

5 HP. HPE Visual Remote Guidance Service. https://www.myroom.hpe.com/vrg (accessed 3 July 2022).

6 Microsoft. Microsoft's Dynamics 365 Remote Assist https://docs.microsoft.com/en-us/dynamics365/mixed-reality/remote-assist/ra-overview (accessed 3 July 2022).

7 ABB. ABB Remote Assistance for electrical systems https://new.abb.com/medium-voltage/service/technical-support-and-repairs/remote-assistance-for-electrical-systems-raise (accessed 3 July 2022).

8 CSIRO. CSIRO ReMoTe. https://research.csiro.au/robotics/remote/ (accessed 3 July 2022).

9 XMReality. XMReality's remote support solution https://www.xmreality.com (accessed 3 July 2022).

10 VMI. VMI Remote Guidance. https://www.vmi-group.com/specifications/vmi-remote-guidance/ (accessed 3 July 2022).

11 Argus. Argus Augmented Reality Remote Support Software. https://argus-remote.com/ (accessed 3 July 2022).

12 Kraut, R.E., Gergle, D., and Fussell, S.R. (2002). The use of visual information in shared visual spaces: informing the development of virtual co-presence. *Proceedings of the 2002 ACM conference on Computer supported cooperative work*, New Orleans Louisiana USA (16–20 November 2022). ACM, pp. 31–40.

13 Gergle, D., Kraut, R.E., and Fussell, S.R. (2006). The impact of delayed visual feedback on collaborative performance. *Proceedings of the SIGCHI Conference on Human Factors in Computing Systems* (22–27 April 2006). Montréal, Québec: ACM, pp. 1303–1312.

14 Donovan, A., Alem, L., Huang, W. et al. (2014). Understanding how network performance affects user experience of remote guidance. In: *Collaboration and Technology* (ed. N. Baloian, F. Burstein, H. Ogata, et al.), 1–12. Cham: Springer International Publishing.

15 Vaghi, I., Greenhalgh, C., and Benford, S. (1999). Coping with inconsistency due to network delays in collaborative virtual environments. *Proceedings of the ACM Symposium on Virtual Reality Software and Technology.* London, United Kingdom (20–22 December 1999). https://doi.org/10.1145/323663.323670.

16 Krauss, R.M. and Bricker, P.D. (1967). Effects of transmission delay and access delay on the efficiency of verbal communication. *The Journal of the Acoustical Society of America* 41 (2): 286–292. https://doi.org/10.1121/1.1910338.

17 Tang, J.C. and Isaacs, E. (1992). Why do users like video? *Computer Supported Cooperative Work (CSCW)* 1 (3): 163–196. https://doi.org/10.1007/BF00752437.

18 Alem, L., Tecchia, F., and Huang, W. (2011). HandsOnVideo: towards a gesture based mobile AR system for remote collaboration. In: *Recent Trends of Mobile Collaborative Augmented Reality Systems*, 135–148. New York: Springer.

19 Jing, A., Lee, G., and Billinghurst, M. (2022). Using speech to visualise shared gaze cues in MR remote collaboration. *2022 IEEE Conference on Virtual Reality and 3D User Interfaces (VR)*, Christchurch, New Zealand (12–16 March 2022), pp. 250–259, https://doi.org/10.1109/VR51125.2022.00044.

20 Kuzuoka, H., Kosaka, J.I., Yamazaki, K. et al. (2004). Mediating dual ecologies. *Proceedings of the 2004 ACM Conference on Computer Supported Cooperative Work.* New York, NY, USA: Association for Computing Machinery, Chicago, Illinois, USA (06–10 November 2004), pp. 477–486, https://doi.org/10.1145/1031607.1031686.

21 Huang, W., Kim, S., Billinghurst, M., and Alem, L. (2019). Sharing hand gesture and sketch cues in remote collaboration. *Journal of Visual Communication and Image Representation* 58: 428–438. https://doi.org/10.1016/j.jvcir.2018.12.010.

22 Li, J., Wessels, A., Alem, L., and Stitzlein, C. (2007). Exploring interface with representation of gesture for remote collaboration, *Proceedings of the 19th Australasian Conference on Computer-Human Interaction: Entertaining User Interfaces* (28–30 November 2007). Adelaide, Australia: ACM, pp. 179–182. https://doi-org.ezproxy.lib.uts.edu.au/10.1145/1324892.1324926.

23 Lee, G.A., Teo, T., Kim, S., and Billinghurst, M. (2018). A user study on MR remote collaboration using live 360 video. *2018 IEEE International Symposium on Mixed and Augmented Reality (ISMAR).* 16–20 October 2018, pp. 153–164. https://doi.org/10.1109/ISMAR.2018.00051.

24 Wang, P., Bai, X., Billinghurst, M. et al. (2021). 3DGAM: using 3D gesture and CAD models for training on mixedreality remote collaboration. *Multimedia*

Tools and Applications 80 (20): 31059–31084. https://doi.org/10.1007/s11042-020-09731-7.

25 Huang, W., Alem, L., Tecchia, F., and Duh, H.B.-L. (2018). Augmented 3D hands: a gesture-based mixed reality system for distributed collaboration. *Journal on Multimodal User Interfaces* 12 (2): 77–89.

26 Teo, T., Norman, M., Lee, G.A. et al. (2020). Exploring interaction techniques for 360 panoramas inside a 3D reconstructed scene for mixed reality remote collaboration. *Journal on Multimodal User Interfaces* 14 (4): 373–385. https://doi.org/10.1007/s12193-020-00343-x.

27 Kim, S., Billinghurst, M., Lee, G., and Huang, W. (2020). Gaze window: a new gaze interface showing relevant content close to the gaze point. *Journal of the Society for Information Display* 28 (12): 979–996. https://doi.org/10.1002/jsid.954.

28 Xiao, C., Huang, W., and Billinghurst, M. (2020). Usage and effect of eye tracking in remote guidance. *Presented at the 32nd Australian Conference on Human-Computer Interaction*. Sydney, NSW, Australia (02–04 December 2020). pp. 622–628. https://doi.org/10.1145/3441000.3441051.

29 Wang, P., Bai, X., Billinghurst, M. et al. (2020). Using a head pointer or eye gaze: the effect of gaze on spatial AR remote collaboration for physical tasks. *Interacting with Computers* 32 (2): 153–169. https://doi.org/10.1093/iwcomp/iwaa012.

30 Feick, M., Mok, T., Tang, A., et al. (2018). Perspective on and re-orientation of physical proxies in object-focused remote collaboration. *Proceedings of the 2018 CHI Conference on Human Factors in Computing Systems* (21–26 April 2018). Montreal QC, Canada: Association for Computing Machinery. p. 281.

31 Wang, P., Wang, Y., Billinghurst, M. et al. (2023). BeHere: a VR/SAR remote collaboration system based on virtual replicas sharing gesture and avatar in a procedural task. *Virtual Reality* 1–12. https://doi.org/10.1007/s10055-023-00748-5.

Index

Note: *Italicized* and **bold** page numbers refer to figures and tables, respectively.

Computer-Supported Collaboration: Theory and Practice, First Edition.
Weidong Huang, Mark Billinghurst, Leila Alem, Chun Xiao, and Troels Rasmussen.
© 2024 The Institute of Electrical and Electronics Engineers, Inc. Published 2024 by John Wiley & Sons, Inc.

IEEE Press Series on

Human-Machine Systems

The IEEE Press Series on Human-Machine Systems is to publish leading-edge books that mainly cover the following topics integrated human/machine systems at multiple scales, including areas such as human/machine interaction; cognitive ergonomics and engineering; assistive/companion technologies; human/machine system modeling, testing and evaluation; and fundamental issues of measurement and modeling of human-centered phenomena in engineered systems. Our target audience includes human-machine systems professionals from academia, industry and government who are interested in enhancing their knowledge and perspectives in their areas of interest.

Series Editor:
Giancarlo Fortino
University of Calabria, Italy

1. *Handbook of Human-Machine Systems*
   Edited by Giancarlo Fortino, David Kaber, Andreas Nürnberger, and David Mendonça
2. *Computer-Supported Collaboration: Theory and Practice*
   Weidong Huang, Mark Billinghurst, Leila Alem, Chun Xiao, and Troels Rasmussen

Printed and bound by CPI Group (UK) Ltd, Croydon, CR0 4YY

16/04/2025

14658595-0003